Andrew Kennedy

The Cosmology of People and the time travel solution

An Astrobiological Proposal

The Cosmology of People, and the time travel solution
An Astrobiological Proposal

Gravity Publishing
2B Findon Road
London W12 9PP

Published by Gravity Publishing

copyright © 2020 Andrew Kennedy

Andrew Kennedy has asserted his moral right to be identified
as the author of this work

ISBN 978-0954-483197

This book cannot be disseminated either physically or electronically in this or any other means without the publisher's express permission and in a format and binding different other to the one it is published in and without a similar condition including this one being imposed on the subsequent purchaser.

To the Reader

The book makes the case for a new kind of kinship in human affairs, a cosmology of people. It ponders the universe we find ourselves in and to argue for a new way of understanding mind and the puzzles of consciousness. It is a work of practical philosophy examining fresh issues of time and the human condition. It seemed to me that to get at the heart of humans struggle to survive, and even to form of view of whether humans will actually survive, I needed to put what science tells us about the world into that temporal social structure in which we actually play out our lives. Only by trying to understand the daily human narrative does it make sense to consider whether humans have a viable astrobiological future.

You can read the sections in any order. Indeed, reading the sections from back to front is just as much fun. You can pick out the themes that interest you the most and skip about. If the multiverse doesn't interest you, move on, read the stuff about mind; there will be a later moment when you might want to look at it. It doesn't matter. The order in which you absorb the arguments is irrelevant, only the accumulation of influence matters. Inserts on the pages are part of this accumulation process. They are notes designed to catch the eye that is flipping through the book or the mind of the curious just dipping a mental toe into the waters, and to draw the reader in.

Flip, skip, browse, it is all part of the process.

Most of all, I hope you will discuss what you read or watch on on video or hear in podcasts about the cosmologies of people. Your comments on social media will broaden the meanings found in these pages and help develop the loosely interlocking kinships of personality and coincidence that will form the greater picture of the cosmology of yourself.

Acknowledgements

There are many people I would like to thank for their support and input, but I shall single out just a few. Thanks to Stella for her questions and encouragement and Alexandra for her contributions and never-ending support. Thanks to Billy Howard Price and the Carlos Castenada coincidence that precipitated our collaborations on a number of occasions, and thanks to Auto Italia South East for the space to give the first talks. Thanks to Chris Idzikowski renowned sleep expert and old friend with whom I have had many illuminating discussions. Thanks to all the members and staff of the British Interplanetary Society for their continued inspiration in space matters. Thanks to Kelly Smith and the Society for Social and Conceptual Issues in Astrobiology initiative. Thanks to Ricardo and Carmen, Pierre, Mark, Jess, Billy, Alberto and Asun for participating in the very first *charla* about the Chronolith.

To the buyer of this book

Remember, at the back of this book is your application to participate in the Chronolith project and to be considered as a candidate to make a journey in any of the Chronolith installations when they have been constructed. Get your seller to endorse the form when you buy the book. Keep your emailed order confirmation if you order the book from an online seller. You must lodge the printed form, either endorsed by the seller of the book or attached to the on-line order confirmation, with the Chronolith Observatory Program when it is activated, in order to participate.

The only futures that shape peoples lives are imaginary ones that can stir them into action.

J.M. Roberts, *Twentieth Century: A History of the World 1901 to the Present Day,* Allen Lane, 1999

Is the future just imaginary?

Preamble to *The Cosmology of People*

> *In reality, development for the vast majority of the peoples of the world has been a process in which the individual is torn from his past, propelled into an uncertain future, only to secure a place on the bottom rung of an economic ladder that goes nowhere.* — Wade Davis.

> *The important achievement of Apollo was demonstrating that humanity is not forever chained to this planet and our visions go rather further than that and our opportunities are unlimited.* — Neil Armstrong

The aim of this project is to develop a fresh schematic for the universe and humanity's place in it. A fresh cosmology, in fact. In spite of the apparently relevant — some might say mysteriously relevant — application of physics and mathematics to the workings of the universe, our traditional cosmology is still rooted in ideas about who we are and what we are doing here. Cosmology represents as much human 'society' as it does human 'physics'. So we will investigate this social human universe of physics and mathematics and logic to explain and garner support for a profound experiment into the nature of the universe and the workings of human consciousness. An experiment whose outcome will have, we hope, significant repercussions for the way human society will consider its cultural and political future as it faces its expansion into space.

Along the way there will be talk of resource depletion, coincidences, the search for extra terrestrial intelligences, dreams, narratives, languages and truth, multiverses, quantum mysteries, panpsychism, artificial intelligence, brains-in-vats, how probability enters the universe, creativity, proofs of gods, time travel kinships and much more.

There will be diversions that I hope will entertain as well as give us food for thought, since the intention of this experiment is, above all, to help us survive as an intelligence in the universe.

This book you are holding (or viewing) began as a series of notes about time and the universe riding on the back of my proposal in 1997 to the Millennium Commission in England to mount the experiment I simply called the Chronolith® for the one thousand year celebrations in the Millennium Dome in London. In the end conditions were not right to mount the project then: a millennium being an arbitrary division, a date on a calendar. What is more important than a date, perhaps, is a moment in economic and cultural momentum suitable for the purposes of the experiment. Such a moment is here, embarking on the occupation of space, a threshold in the economic and psychological history of humans.

In 1997, however, outside of NASA and military missions, space was hardly used, not even by Russia. There was no Space X, no private rocket construction, no cube sats, no ISS (its first module was launched in 1998), hardly any exploration of Mars (the first robot rover had just landed), only three probes had gone beyond Mars, only three humans lived a life of anxiety in Earth orbit aboard Mir (first launched in 1986 and completed ten years later), mobile phones with GPS were yet to arrive (in 1998; personal GPS gadgets arrived in 2001), internet access via satellite was still 6 years away. Global warming was simply a controversy, and beliefs that humans faced severe environmental damage and global disruption to economies and whole societies were borderline cranky to many scientists and economists who considered notions like 'Peak Oil' (the end of growth in oil supplies) were put forward by marginalised 'malcontents'. The Club of Rome report on the *Limits to Growth* (1972) had been derided and forgotten. The idea of humans standing at some threshold was not at all clear to most, the evidence was partial at best.

The situation is very different now. There is a new space race developing and inevitably with it there is a new appreciation of an arms race in space to protect the assets placed there. It appears, however, that in spite of governments funding space activity and private money going into imaginative projects (like the Breakthrough Starshot proposal of Milner of 2016), the space development timetable is too slow to have any impact on the crises on Earth. If we do not alter our attitudes significantly, it may be the case that humans will not be going anywhere far from Earth.

By reflecting upon space development plans we begin to wonder quite for whom the investment will profit (the widespread belief that there is an economic multiplier coming from space investment disregards that fact that its actual value and merits are controversial). Extravagant ideas about colonising missions to Mars are completely obscuring the unsolved problems of not only space travel, but of the launching civilisation, us, as well. Elon Musk has claimed he wants to die on Mars, and his company, SpaceX, is apparently trying to monopolise space expansion by being the first to develop a massive rocket essential for large scale space projects with which he can send a hundred colonists at a time to Mars as well as shooting high paying executives around the Earth in under two hours.

Space X, and other private companies around the world are kept going by virtue of government contracts, so it is not surprising that space entrepreneurs like Musk try to keep the momentum with inspirational ideas. Plans to mine asteroids or the Moon's plains remain in imagination only. No one has any idea how metals are to be extracted on a large scale out of complex ores that are not present on Earth, smelted and fashioned into high integrity and high performance components for rockets or other structures in the space and microgravity environments. All this activity remains in space, not on Earth, and is destined for the further human exploration of the solar system. Market forces have yet to be created to pay for this activity far from Earth and returning so little to it. If the extravagant claims about the quantity of metals like platinum, boron, titanium and other useful industrial metals present in asteroids are true then capturing an asteroid and bringing it to Earth would depress the market price of such metals to nothing. How large would the Earth economy have to be before plans to capture solar power in space, for example, and beam it down to the surface of the Earth make economic sense? There is an expectation in the popular press that the first trillionaires will come from space exploitation. This may well occur but only if the Earth economy survives to consume at the highly elevated levels required to pay them.

It seems to be generally accepted that industry needs to be urged into space with stimulating dreams and bright scenarios. (Stephen Hawking played the opposite card of the threat of catastrophe.) Philanthropists are being asked to step up to the plate, but when it comes to our future in space, the elephant in the room is capital growth here on Earth, and no one has any idea where those market forces will come from or how they can be prevented from destroying the Earth if coherently deployed to occupy space.

Meanwhile Earth's environmental degradation increases. As a civilisation, at the start of the Anthropocene era, we still seem to have no way of understanding the significance of these facts. By the year 2050 (a generation away) there is serious environmental damage around the Earth whether or not solar power is installed on every roof-top, while the likely maximum numbers of humans off Earth (on the Moon, on

Mars, and in between) if current plans are realised, are likely to range between 50 – 70, but might just as easily be zero. Even with the United Launch Alliance published plans to have – incredibly – 1000 people in and around the Earth orbit and the Moon at the end of 30 years from now (*CisLunar-1000 Economy*, 2017), and reminiscent of the absurdly optimistic O'Neil/NASA plans from the 1970s, these numbers are still not much of a back-up in case many (it doesn't have to be all) humans end up dying on Earth for good and anticipated reasons (like pandemics or nuclear errors), and stopping growth in its tracks.

We do need some fresh perspectives on how we are to manage the next one hundred years because the exponential technological solution fallacy is emerging once again as the 'new hope'. Keeping the Earth from overheating with extravagant geo-engineered solutions, for example, does nothing for sustainable economics and for the maintenance of biological diversity. Yet space activities are now the lead realm to drive broad economic development, and overall, technological success is looked upon as our destiny (e.g. The Singularity, now being claimed to arrive around 2029 by Kurzweil, and fully dominant AI expected by those in the field by 2060). Automation and machine intelligence are expected to usher in an entirely new phase of human development releasing fresh potentials for growth and expansion. The fact that this growth has to be sustainable, if humans are to survive, is ignored by the 1900 or so billionaires on the planet who control virtually all of the readily utilisable wealth of it. The optimists dominate the argument. Certainly there is a lot of cross-pollination between innovations and no one can predict what solutions might be found because of it, but all the same it is worth considering what (partial) exponential growth has brought us so far, and it is worth recognising, too, that there are other developmental forces to economic growth that just as easily hinder the ideal result or obscure the potential of discoveries. Need it be said once again that at the turn of the 20th century there were more electric vehicles than gasoline vehicles. The world land speed record was held by an electric car. Yet development in electric vehicles and batteries remained moribund for a century while the petroleum industry took over as the engine of growth.

Growth has produced immense gains for many of the inhabitants of the globe. The work time needed to earn enough to purchase many goods has fallen dramatically, and people work fewer hours in general and have more holiday time. Domestic life in the developed world is easier and more comfortable. Public places are cleaner and many social activities are safer. Around the world, more people are educated and very many fewer people labour in the fields or in dangerous and debilitating industries. Developments in travel have allowed vast numbers of individuals to move about the globe, and cultures and ethnicities mix relatively freely. It is generally considered that innovation and ideas have been the engine of the ever-rising rate of growth in the world economy since 1800 AD and artificial intelligence appears to be on the verge of offering industry simple learning algorithms that can be applied in automation across many processes. But not everything is as rosy as it seems, and what seems to be 'modern' in the externalities may not actually represent true advances on what has gone before. Clearly one's experience of growth depends upon where in the cycles of expansion one is. Around 1000AD life was relatively benign. Not so 300 years later when feudalism came to an end and the monetary economy took flight. No doubt the opening years of Manchester's growth were optimistic, but fifty years on the slum dwellers may have been more pessimistic about what growth had brought them. Mysteries of growth abound and inequalities still remain. For many people around the world it is fair to ask, why aren't the consequences of the exponential growth in science, first captured by Derek de Solla Price in his 1961 book, *Science Since Babylon,* and continuing today (even though growth in numbers of scientists may have stagnated somewhat since 1970, in 2001, scientists and engineers formed 23% of the work force), even more dramatic and comprehensive than they are?

A cursory glance through industrial processes reveals curious negations of innovative effects or hidden archaisms. For example, telephony, arising originally from efforts to teach deaf people language, is now very technologically advanced yet telephone companies make more money from people writing to each other on their phone than speaking into it. Fresh technical advances often retain previous generation systems at their heart, or are advances only in degree and not principle. Nuclear power plants hide within them 200-year old steam turbines. In a basic industry like steel manufacture, it is a foreman's lifetime experience that looks into the fire of the furnace and decides when the molten mass is ready to be poured. Skyscrapers are made from frames of bolted together iron pieces in ways the early Victorian engineers would easily recognise. Surfaces and supports are cast in the poorer version of a material (producing 6% of the World's carbon dioxide emissions and with less durability than hard wood) that the Romans pioneered and used everywhere, namely concrete. Roads are still paved with bitumen dug out of tar pits (although more asphalt from oil production waste is now used). Agriculture industry has not added an animal to its repertoire since the Stone Ages. In spite of the exploratory push in the 15th century, 2/3 of the world still get their principal carbohydrates from just three plants (wheat, rice and potatoes) and even the new fruits and vegetables we see on our shelves mostly come from trade with other parts of the world rather than through innovation (some notable exceptions like the tomato and high yield cereals have come through old fashioned plant breeding rather than through genetic modification, and which is introducing new problems).

Let us consider the computer. Clearly there have been innovations of key importance in the growth of the economy and it is generally considered that the pace of innovation has been remarkable in recent decades. We have taken the famous observation of Moore (rather erroneously given as a 'Law') as being some indicator of general advancement. The lower chip cost and faster speeds of computers which have allowed the amassing and analysis of vast data sets to dominate the methods of controlling all manner of systems seem to reflect this law, yet the skyrocketing power needs of such computing (such as that needed by Google) do not reflect a similar efficiency, and the industry is closing in on the physical limitations of such devices rapidly. In spite of advanced computer chip manufacturing techniques still the preferred way to store this data is with magnetic tape, invented in 1928 and first used for computers storage in 1951.

Innovations in computing in specialist industrial applications along with improvements in satellite communications have advanced automation considerably, and progress in machine learning with the access to vast data sets has brought automatic operations to all spheres of resource exploitation and delivery, while bringing language translation, image recognition and personal voice activated search facilities to internet users and to the service of the consumer economy. Where before research into artificial intelligence and cybernetics had become almost moribund, now robotic expert systems diagnose health issues and devise drug programs for patients, and automated systems teach courses, act as lawyers, plan advertising campaigns, edit Wikipedia, trade in securities, sentence criminal offenders, credit credit profiles and solve consumer on-line disputes. They do not, however, solve questions of truth and honesty of the information at their interfaces between system and human (automated social media bots spread false news, disrupt genuine support for public programs, and pollute conversations just like partisan historians of old), and the benefit of these innovations at the household or individual level are not so easily seen.

While computers manage more and more of our lives by recognising more and more abstruse patterns with which they can make decisions, they cannot produce true innovation for which there is no example (even as they recognise patterns that humans miss). Although artificial intelligence systems can beat human players in every game including poker, and can train themselves to respond effectively to relatively novel situations, they still do not think like humans (specialised game playing machines teach themselves over millions of trial and error repetitions, many more than a human could conceivably experience for comparable success and while humans are also processing sight, sound, sensory input and physical coordinations in 3-D space) and they are far from being able to make many kinds of expert decisions, especially those with moral overtones, or act upon their own intentions where there is little information available. The idea that computing power (as distinct to industrial automation) is the means by which sectors of the population become emancipated is exaggerated. We have seen a powerful communication device put into the hands of half the world's population enabling them to coordinate their political protests through the internet (e.g. the 'Arab Spring') that still did not lead to increased representative democracy (although the studies of Chenoweth and Stephan, albeit with a small sample of events, appear to show that peaceful protests can be successful about half the time as long as the numbers of participants rise above a threshold *and* the military does not get involve, although overall, mass protests create less change than before). Such success, however, is as far from true emancipation as say, the London apprentices of old racing through the streets to rouse the mob to protest against injustices were from taking power. Rather the opposite in fact. Oligarchic rule is triumphing over parliamentary-style democracy in almost every corner of the globe, and a 'post-truth' world sows doubt and mistrust on all organs of representative government and pushes propaganda to disguise the intentions of its leaders. Reflect upon the events in Hong Kong in 2019 to see how limited the power of these 'freedoms' actually are.

Modern computing is extremely sophisticated and compute results in many fields that would have been impossible even a few decades ago, yet even so much of modern computing capability goes into simulations of processes. In terms of generating ideas, they are still primitive, and with regards to personal computing, it is hard to pin down in what conceptual sense the technological advances are actually advances for the general public. In fact, the personal computer never helped modernise the home and has now been rejected as an assistant by Millennials who prefer to use their phone. The reason is not hard to see. Mass computing was driven by the need to sell processors and hardware which made operating systems and the apps that run on them bloated way beyond what is necessary and disguising the actual amount of computation required to do any task. Compatibility issues were like a ball and chain on the development of new software. Even Microsoft and its thousands of programmers struggled to produce an effective updated operating system in part because of their obligation to maintain older versions that still exist around the world. Apple has severely curtailed the scope of its own computers (like restricting the way they handle communications like Bluetooth) in order to keep them within their high-earning 'Apple-verse'. (In fact, Apple upgrades are a good example of the trend for cosmetic changes to disguise lack of or restraint in genuine innovation.) The significant jumps in personal computing power only occurred at the intervals where the processing chips moved from 8-bit, to 16-bit, to 32-bit to 64-bit units of computation and not with the purpose of programs.

It has been said many times that if computer operating systems could be produced from scratch today, they would be completely different and very much more efficient (the RISC chipset and even Linux are certainly attempts in that direction). In a sense today's bloated programs are rather like inherited DNA, in which only a portion of its genes function. As a simple example, in 2000, I wrote and compiled a program in the BASIC language to turn large text files into a series of linked HTML files to upload on a web site. The entire program took up eventually just 70KB, and I was a lazy programmer. Later programs, however, doing the same easy task took up several *megabytes* of RAM, due, in part, to modern computers having larger addresses and larger unit blocks of memory. This hardware excess over need in the consumer realm, and the bloated software, can be inferred from many of your laptop programs that are just over-dressed primitive designs dating from the early years of computers and lacking any kind of conceptual innovation even while the hardware became more efficient and its UI easier to use. This is true of all the core 'work' programs on your computer (like Excel or Word). For example, Bricklin and Frankston designed a computerised spread sheet at Harvard in 1978 called VisiCalc, the foundation to the spreadsheets of today, but which was actually a computerised implementation of business tabulation and calculation methods in use for hundreds of years. Your word processor, when it comes to the handling and editing of documents is no better, conceptually, than the first one developed for Microsoft's MS-DOS operating system, WordStar, based, in turn on centuries-old typesetting practices. The creator of hypertext, Ted Nelson, who thought of documents in a new way, produced his designs in 1965(!), and still laments the failures of designers to fully understand how to take the written word beyond the world wide web implementation of embedded links, and is still trying, along with dedicated volunteers, to further his vision in the (unfunded) Xanadu project. Even the introduction of a computing advance like neural networks into your iPhoto app for reading faces in images may sort your thousands of people-pictures (that you never look at) by individual, but has this any impact on freedoms you may need or want? Much touted 'artificial intelligence' computing may exist for some advanced applications but is nowhere in sight for personal computing which still is riddled with errors, imperfections and absurd complexities, and can hardly update themselves without causing a great deal of disruption. Algorithms in your mobile phone may take great pictures (even though most of the processing power goes into making icons bounce prettily), they are still way behind the abilities of the optics and film of sophisticated cameras of just a few decades ago. Mass photography has been enabled certainly, but better photographs...?

Computers are often considered to be highly sophisticated because they are connected to advanced projects, but often the foundation to such sophistication lies not in the computer but elsewhere. Automated cars seem very advanced but they are only possible at all where highly accurate maps of roads and streets exist. (GPS satellite systems are direct descendants of the U2 aircraft mapping flights of the 1960s that created accurate maps of untriangulated Russia for the US ICBMS). It is their environmental perceptual apparatus which is failing to bring these vehicles into widespread use. As a computing problem, the decision-making and mechanical operations of such cars is relatively trivial (situational perception is more difficult). For example, powered self-steering units for sail-boats are no bigger than a battery-powered drill yet they do a better job handling the tiller in average circumstances than a human.

The hollow egg syndrome of technological advancement can be found in the broader aspects of economic growth and world trade. The GDP of developed countries has been declining steadily in the last decades, some segments of their societies have not seen wage increases for 40 years. World poverty rates are falling around the world where China's growth is included, but some figures have almost half the world still living under $2.50 per day. While some economists point to the successive leaps forward in economic progress over the last century and still others claim that innovation is declining, it is not at all clear what the latter-day core nature of the progress in human civilisation terms actually is. Expanding education has liberated many, but what explains the amazing fact that literacy is *falling* in the world's greatest economy, the US, which is only 86% literate now (2017). In the developed world vast agricultural concerns impress us with their productivity but the reality is that the majority of people on the planet feed themselves on farms of less than two hectares in size. The biological revolution in growing crops (mostly due to the efforts of one man, Norman Borlaug) has halved the deaths due to famine at the same time dislocating many who used to live on the land while the misuse of fertiliser that it requires poisons the water supplies and the oceans, even as 1/3 of all agricultural produce is wasted.

While the success of innovation stems from capital deployment there are subtleties that defy analysis. For example, in shipping, while technical advances like GPS allow ships to travel more efficiently and with fewer crew, the single most important contribution to global trade has not come from science but from the standardisation of the boxes cargoes are shipped in (reminiscent of the standardisations in history like axel widths in ancient China or railway gauges around the world).

Even in the developed world, we can see that as far as the ordinary practicality of daily life is concerned, the 'frantic' pace of innovation has little touched it in recent years, and the innovations over the last decades have tended to introduce changes of degree rather than of kind. The domestic environment albeit with more efficiencies has hardly changed since the introduction of the first automatic washing machine in 1937 (the first hand-cranked drum washing machine was created in 1851). Entertainment centres are perhaps the principal difference to the home environment of the past, although we forget that pianos and home-spun music and song were ubiquitous until the second World War. As far as the external or internal arrangements of a house goes anyone who does DIY knows how imperfect our 'machines for living' are still. Houses need continuous maintenance; they still leak when it rains; plumbing and electricity all have failure points, drafts still leach heat out of a house, locks do not protect houses from thievery, fuel needs to be delivered to them, and they still catch fire and kill their occupants.

The coming smart house, has been suggested, will be a real social advance where investment returns can be made more precise and with less waste of energy. It will, on the other hand, return us to a more ancient social arrangement, namely that of feudalism, where the individual consumer will cease to have control over his environment or manage his goods as he sees fit. Anyone who owns an Epson domestic computer printer or a Tesla car or an earlier model of an Apple iPhone can already get a glimpse of this future, where these manufacturers build in inhibitions to the functions of their products if you use alternative ink supplies in the case of Epson, or buy a cheaper model in the case of Tesla (who restrict through its software the power delivered by the car batteries by model), or try and get a third party reparation for your iPhone in the case of Apple. While the 'Internet of Things' appears to be little different in terms of social structure to having someone at home (like a spouse or family member) turning things on and off or checking the fridge to see if there is milk for tomorrow, it will in fact be a step back into a world of determined consumer obligation and it will not be free-ing or free. In fact, the electricity required to manage its data needs (the internet of things) and that of the modern consumer will grow to perhaps 1/5 of the world's electricity supply by 2025, and requiring management, maintenance and upgrading all of which will be out of the consumer's control. As far as capital is concerned, the IOT will be an unparalleled means of trapping consumers into specific expenditures, and yet what actual social *advance* does it represent?

Domestic energy use around the world hasn't changed much for centuries. If you have a Sci-Fi imagination and think of galactic civilisations, how would you classify Humans for whom, in 2015, 41% of its electricity was still got from burning a black carbon rock dug out the ground (in some places still by hand) and many of the poorest among them have to resort to the burning of animal dung or charcoal, killing 1.5 million people a year from indoor pollution and taking much needed fertiliser from crops? Gas seems an absurdly antiquated source of energy (domestic coal gas, derived from that same carbon rock, was introduced in England around 1812 and after 1820 in Europe: conversions to natural gas took place after the 1950s), not to say dangerous, and yet 21% of the world's population use it for cooking and heating. Electricity distribution to households became the norm by the second World War in the developed world but city-wide street lighting had been installed before the turn of the 20th century. Even so, while half the world still lacks a reliable electrical supply, energy production and transport pollution is creating cognitive decline in developing countries. Solar power investment is slowing in advanced economies because no one has yet solved the problem of locally storing passively produced surpluses for times of need. (Germany's recent costly experience in this effort is instructive, namely having killed their nuclear plants they were forced to use coal to make up the shortfall from renewables even as solar power installations grew). While many think that the batteries of the world's fleet of electric vehicles will eventually even out the market fluctuations by acting as a reservoir, the practicalities of this suggestion (e.g. stretching networks across many time zones) have not been solved.

Advancement is often synonymous with improvements in health and longevity. Longevity may have improved in many countries but it has *fallen* in the world's greatest economy, the United States (in 2016), who is now ranked 31st in world rankings and who, incredibly, has rising maternal death rates, the highest in the developed world. The greatest long term effect on public health around the world, however, despite the eradication of smallpox and polio and the immense improvement in child mortality through vaccinations that have made epidemics of childhood diseases a thing of the past (and whose protection is now declining as anti-vaccination protesters resist its logic), has come from improved sanitation, clean water and food preparation (nearly 1.5 million children still die each each from poor water and sanitation and almost half of all health problems in developing countries come from poor water and lack of sanitation). In the developed world, in spite of significant progress made in advanced health care, especially in genetics, where the prospect looms of easy genetic editing curing many diseases and eventually giving rise to an altered human genome, many basic drug treatments range from neutral to harmful for patients and the benefits of some widespread surgical interventions are illusory. In this modern epoch, 10% of the US population is hospitalised each year and between 1/4 and 1/2 million of those patients die just from medical errors ($\approx$10% of all US deaths). Yet most chronic sickness in people arises from eating too much of the wrong food (recent studies suggest that globally one in five deaths are due to poor diet) or drinking alcohol, and one of the biggest causes of death is not a specific viral or bacterial agent at all but smoking. Consider that your dentist may take an x-ray of your teeth but he still finds likely cavities by looking for stains and using the ancient technique of probing them with a point to see if you wince. Infection at the root of a tooth is dealt with the way teeth have been dealt with for centuries, by yanking it out. Still the most effective pain killers in use today come from opium which was traded around the Mediterranean in poppy bud shaped amphorae since the start of civilisation. The antibiotic revolution which has saved countless lives has not definitively solved the problem of infections and after just 60 years of use medical researchers are now having to race to find a solution to the evolving antibiotic resistance in dangerous bacteria like gonorrhoea, without which future mortality around the world will soar.

The advances in medical technology are beginning to look purposeless since still the biggest burden on health care comes not from diseases but from those older than 65 years, who will rise to 16% of the world's population by 2050 (on average; it will be 30% in Japan for example, and 40% of Germans will be over 60), for whom palliative care will be most required.

As we can see, in the mix of apparently advanced technologies, are antiquated systems from earlier epochs exposing significant differential rates of development, and also apparent are the effects of the difference between need pull and need push at the domestic level in the modern consumer economy. Both automation and the need for higher skill sets put basically educated people out of jobs rather than in them (for the short term and perhaps forever) but broadly speaking, the effect of most innovations at the individual level revolve around leisure rather than emancipation and that in many areas of activity innovation is functionally cosmetic; no longer emancipating but simply whimsical (think gaming consoles). In December 2016, Forbes produced a list of the top 30 innovations of the last 30 years, most of which have had little direct impact on daily life or whose benefits, like Social Media (20th on the list), are extremely mixed. The only innovations that might pertain to the domestic environment on that list are LED bulbs and the Internet, and, in fact, if you look back say 70 years, then the invention most significant by far to humans, emancipating women and radically altering the promise and outcomes of human relationships has not been a technological device at all, but the contraceptive pill (introduced in 1960), a way of regulating a woman's level of hormones by delivering precisely calculated quantities of natural hormones into a woman's body. On the social consequences of sex, one might reflect upon the growing acceptance of the idea that humans belong to a sex spectrum rather than a binary sex designation having an emancipating momentum well outside technology.

I have chosen a mixed bag of examples but many can be found across the world. All beg many questions. While improvements in working conditions, health, food and sanitation are expected from growth why has it not addressed many of the more significant features of our social existence? Think of education. Not only is much of it denied the poor, its benefits can still be a matter of luck, even for the bourgeois, where incompetent or vindictive teachers can blight entire lives. Think of justice and why the world's greatest economy imprisons more of its citizens than any other country. Think of the puzzle of chronic violence around the world. One's pessimism or optimism about the advances of technological civilisation is going to depend upon where you are placed in it, but a more fundamental question remains. Why do technological 'advances' so often retain or recapitulate ancient power structures. The answer is two words, capital growth.

The beginnings of the modern industrial revolution is generally marked in England by the self-taught, though illiterate (he once demonstrated to a Parliament committee his design for an aqueduct carrying his canal carved in cheese), engineer James Brindley's building of a canal (1759-1761) for the Duke of Bridgewater so that the Duke could ship coal from the mine on his estate into the centre of Manchester to find a bigger market. Parliament, compensating for creating a Bill compelling landowners along the route to sell to the Duke, set by law the price of coal to be sold in Manchester at almost half the price it had been sold at previously and unwittingly provided the conditions for industrial growth. Cheaper coal stimulated the expansion of the country town of Manchester into a slum-ridden industrial metropolis within 50 years, the inspiration for Engels and Marx's forays in communist ideology. Bridgewater built improved housing for his workers and didn't see a profit for twenty years, but such early benevolence to the poorest from the subsequent waves of innovation could not be sustained in the relationship between capital owners and labour. (Even the canals, originally boon highways, were the preserve of landowners who were able to resist the building and spread of railways for quite some time, and the railway interests in turn tried to prevent steam carriages on roads.) One might speculate here about what the discovery of any cheap energy supply (a possible candidate is fusion) would do to human society given this history (or if, for example, Aliens bestow new energy sources on us, a point we will examine further on).

The original concept of material innovation as the means of improving our world and the lot of workers, stemming from Benjamin Franklin (who lived in England for many years, although Adam Smith's *Wealth of Nations,* giving substance to this theory was not published until 1776) and his circle of innovators (to which Bridgewater had been introduced), has turned out to have had a limited life. It has been replaced by the single purpose of improving the lot of capital owners (legislative bodies around the world are still reducing the tax burden of asset owners, in 2017). The suspicion arises that as the share of wealth continues to rise among the capital owners, as noted by Thomas Piketty, there is correspondingly less pressure for innovation to create wealth for them.

Personal fulfilment is now a by-product of the economic realities and not the purpose of them. The core purpose of innovation being to make capital even more productive by controlling and reducing costs; by converting costs directly into the environment through entropy, and by so doing, using the entire biosphere (or more succinctly, everyone's back yard) as a sink to absorb the wastes from the relentless acquisition of wealth. The fact that innovation improves our lives is only a function of us as growing consumers, and until there are sufficient consumers for an innovation it will not happen. The innovation engine looks only for the most profitable effect and the greatest marginal profit extractable from new techniques. In our epoch, this is found in information processing and which is the sole impetus behind recent economic growth around the world. The purpose behind the innovations in information processing is so far from emancipating individuals or from freeing them from exploitation and slavery that it is used in the opposite direction.

Our focus, however, is on the space economy rather than on the lack of emancipating innovation in general industry because activities in Space mark a psychological threshold to cross represented in an economic paradox. The space economy beyond Earth's environment has no consumers. There will be no real markets for space products for probably a century and thus no real return on investment in space. How, then, do we choose and pay for space activities while environmental programs are failing? Governments will continue to fund space science, and there will be, as a result, an Earth-based economy of innovators reaping the rewards of government interventions but since there is little income-producing capital growth in these artificial markets, what can we say *is* the space economy? Whom does it serve? There is no doubt that useful avenues of research march hand in hand with technological development, but the dilemma before us is, will innovation actually save us *in time*, save the Earth for future generations, or are we destined to use the *excuse* of mastering a hostile, inhospitable environment of Space to encourage the capital growth on Earth such that it ruins the very system that raised us?

With the growth of artificial intelligence, we are certainly poised at the threshold where we need to make sure that humans and not ma-

Space exploration so far seems justified. It has turned out to be highly valuable to humans. Space activities are so well integrated into everyday life that today's society could not function without them. Weather prediction, environmental monitoring, national security, materials science, solar energy production, medical advances, personal and corporate communications, global banking, mass entertainment, geographical and voyage information and safety are some of the most obvious everyday benefits to the world's citizens, but industrial processes that involve imaging, drug fabrication, solar energy production, robotics, computer software and economic data monitoring have also taken a boost from space exploration. But all these benefits arise solely from communication and information. They are the direct result of satellites, their sensors, their transmitters and antennae, not from any broad front of industrialisation. Space inspires scientists and stimulate our curiosity about the nature of our universe. How could it go wrong?

chines answer this question. An intelligent machine, whether or not it is sentient and however 'smart' it may be, will have to solve problems of resource allocation and use in human society as well as fit in with other systems and their needs, and without some kind of moral dimension or duty to human life in its reasoning, may perfectly well think that destroying the Earth in order to get into space is an acceptable strategy for growth. There may even be humans who believe this now. Certainly humanity has given over control of orbital space to the political entities that exploit it. The space agenda is not humanity's to decide any longer. How do we ensure that these machines (and humans) understand that destroying the Earth (either actively or through inattention) to get into Space is not an option for us?

To answer these questions I will try and throw some new light on what it means to be human, conscious now in this universe, and confronted by extinction. In particular I will examine what the 'now' actually means and whether our memories contain more information than we expect. I will show that we can, in fact, travel some way into the future and extract more information from the potentials of what is to come. I will end the story with the description of a particular human kinship stretching through time. This kinship, or cosmology, forms the unsung backdrop to human society and with which we can initiate a sociological experiment that may help us derive solutions to the extinction threats humans face and take humans on their voyages to the stars.

The Chronolith® Observatory

Given the inevitable paradoxes of compound growth needed for humans to populate the galaxy, the Chronolith® Observatory may be the only way human civilisation can examine the future, try to reduce unknowns in the decision making process, to avoid traps in the development process (like genetic manipulation introducing unexpected outcomes) leading to economic collapse and the destruction of Earth, and to discover the kinds of humans who will be able to take us on our space explorations.

Given the vast distances of ordinary space between potential landfalls in the Universe, the observatory may be the only means we have to connect to other regions of time and space.

The Observatory is a live experiment in which anyone can participate. It is designed to investigate the natural link between events, and to probe the boundaries of possible movements between them.

The discussion of journeys in Time, however, is not about distorting vacuum fluctuations, wormhole tunnels opened by microscopic massive black holes, voyages around cosmic strings, or hyperspace and imaginary time axes, or even about Casimir Effect boxes. The physical mechanism of time travel, while vaguely interesting, is not what matters. What matters is understanding what is fixed and what is flexible about the usual narrative progression of events.

Table of Contents

To the Reader 1
Acknowledgements 2
Preamble to The Cosmology of People 5
    The Chronolith® Observatory 21
Pre-prep 27
    Musing on the Impossible! 28
        Possible! 30

Space and the Final Panic 32
    Is Space the final frontier? 33
    Distance, one problem, growth, another 35
    Life is a consumer 37
    The hunt 39
    'You can't win, you can't break even, you can't even quit the game' 40
        Panic stage 1, Water supply 44
        Panic stage 2, Food scarcity 44
        Panic stage 3, Industrial metals scarcity 45
        Panic stage 4, Death of the seas 45
        Panic stage 5, Energy scarcity 46
        Panic stage 6, The last investment 46
        Panic stage 7, Collapse of Earth's economy 46
    The Singularity 48
    Transhumans 49
    The Final Panic 55
    Don't Panic 59
    SETI (The Search for Extra-Terrestrial Intelligences) 61
    Our monasmotic universe 73
    Escaping the Final Panic 77
    Evolution and extinction 79
    Surviving the extinction threat 83
    Time travel and the narrative free-for-all 84

A bit about our Universe: Coincidences and life 90
    Coincidence and the Anthropic Principle 92
    Coincidence and its consequences 97
    Coincidence in the universe 101
    Coincidence in the Human Narrative 107
    A world of no-coincidence 111
    Worlds fit for life 113
    Why does coincidence matter? 120

- Coincidence and intelligent life..........126
- The law of large numbers, synchronicity, accidents and coincidence..........128
- Kinship and the human narrative..........132
- Networks..........140
- Social bias and manufactured coincidence..........142

Cosmology and the Multiverse..........147
- Mysteries of probability and measurement..........150
- The Bell Inequality and the elephant in the room..........155
- The 'Many Worlds' Interpretation of creation..........158
  - Type 2 Multiverse – the Omnium..........159
  - Multiverse level 1 or the 'quilted' version..........162
  - Everett, De Witt and the multiverse level 3..........165
- Problems with multiverses..........168
  - The observer..........168
  - Being First..........171
  - Decoherence..........174
  - Energy worlds..........179
  - Maths and measurement..........181
  - The appearance of choice..........187
- Schroedinger's Cat..........188
  - The role of consciousness..........191
  - A side note about Wigner's Friend..........194
  - Schroedinger's Cat continued..........195
  - The cat-state and the radioactive nucleus..........197
  - Tychons..........200
- Multiverse interference and the slit experiment..........203
- Retro-causation and the fulfilments of time..........209
- Magic and geometry..........216
- Locating the cash-box..........219
  - Dispersion and Black Holes..........225
- A note on the beginnings of history..........228
- Conformity..........233
- Expansion..........237

A bit about time and human narratives..........241
- Where is the Past?..........243
- Arrow of time..........245
- Ministry of Truth..........246
- Present Completeness Paradoxes - Problems with the now..........251
- Present Completeness Paradoxes – problems with the Future..........252
  - Two Paradoxes..........255
- The future in the present – the Relativity question..........259
- The symmetry of time..........265

- Does sentient time go backwards? ... 267
- The teenager's room and the arrow of time ... 274
- Information and entropy ... 278
- Informational Inertia ... 287
- Biological language and the origins of life ... 291
  - Evolvability ... 296
  - What is not in the string ... 297

Imagining time travel narratives ... 299
- Messing with World War I ... 302
- Travels in narrative time ... 306
  - Time travel as a kind of peep show. ... 308
  - Time travel bodily to many time streams ... 309
  - Reflexive Time travel (back and forth in a single stream) ... 310
  - Time is stopped for tinkering with the present ... 312
- The 5th way to time travel ... 313
- Paradoxes of time travel ... 314
- Paradoxes of External Narrative Time ... 318
- The circularity paradox ... 319
  - Backwards explanations ... 322
- The chronology protection principle ... 323
- Killing your grandfather paradox ... 325
  - The exclusion zone ... 326
- Trip A or trip B ... 328
  - The splitting of time-lines ... 330
- The merging time-line paradox ... 333
  - Before the grandfather reproduces ... 333
  - Time-line inertia ... 336
- The speed of time paradox ... 338
  - Case 1: Time as a continuous evolving surface ... 338
  - Time-like loops ... 339
  - Staying or going ... 340
  - Rates of change ... 341
  - Case 2: No continuous surface; Past without future ... 343
- The virtual reality game ... 345
- The physical time travel paradox ... 346
  - Parallel conjugates ... 347
  - Chronarts ... 349

A bit about mind and consciousness ... 352
- The story so far ... 354
  - Self-awareness, the mind's (I)eye and Hegel ... 355
- Pragmatic brains ... 358
- Chi and the information limits to the body ... 369

- Boundaries to individuals...373
- Brains in vats or the external narrative time problem................................374
- Mind as simulation...376
- Thinking it up 381
- Souls and morals in a multiverse?...384
- Consciousness, memory and reflexivity 388
- Consciousness, networks and negative probability.....................................391
- Consciousness, new states of matter and a thought experiment...............393
- So how is the mind aware?..396
- Akashic fields..400
- Morphogenetic Fields...404

The Cosmology of people...413
- Cognitive transformations..415
- Ideas and the human channel 417
- Cosmologies and identity...425
  - Another diversion on probability..429
  - Narrative coincidence..430
- Cosmologies and the quantum mind...432
- The Dream..434
- Lies and language, fallible narratives and dreams.....................................446
- Time Travel...458
  - Into the future..465
- A cosmological kinship stretching over generations. 466
  - Kinship-induced cognitive emergence..467
- The practical realities..469
  - Something has occurred in laboratories...472
- Going to the stars...475
  - The case for a human pilot...477
- Coming full circle...482

Appendix 1..487
- Protocols for an ETI Signal Detection...487

Appendix 2..491
- Trevanian...491

Appendix 3..493
- Boltzmann Brains and sudden savants..493
  - Mind/Body Problem...495
  - The mystery of the sudden savant..495

Pre-prep

Science is doing remarkable things and one of the most remarkable is the delving into the underlying uncertainty that infects the fabric of the universe.

We may happily accept that subatomic particles have their own weird lives, flashing in and out of temporal zones or not having any time at all, but will not accept that the daily life of humans could also behave like that. What seems under appreciated right now is that uncertainty and the interference of probabilities don't really go away when events happen or are observed, they just move into the shadows.

The question of whether the future exists for us is examined in this book. But I also try to make clear that if the future is uncertain then the past is too, and that the past changes to support how the future actually happens. This may seem at first sight to be a reversal of what we believe happens in our lives. We assume the past is fixed because we *know* it happened like that. It isn't just our memories that help fix the past, consequences moving along with us also help to confirm what happened in the past. You are here with me today because of the letter (anachronism alert!) I sent three days ago.

Thinkers, and especially science-fiction writers who are less bound by narrative traditions, have grappled with this separation. The splitting and merging of time lines at the human level has been discussed extensively both among scientists and in science fiction. But I would like to draw the reader's attention to two *literary* fiction stories – not science fiction stories, and not written by a scientist – of an author who seems to have understood more than anyone what interference and uncertainty in consciousness does to the narrative consistency of events in everyday life. The two stories are the opening and closing tales of Trevanian's *Hot Night in the City* (St. Martin's Press, 2001) in which Trevanian appears to tell the same tale twice but with opposite yet narratively consistent outcomes (see **Appendix**).

The changes may be minimal and surreptitious but the results are profound. It is hard to see exactly how they happen unless you read each story line by line and side by side. It takes a genius like Trevanian to show us how consciousness is implicated in creating reality and that our experience of the External Narrative Time of the human sphere just as equally results from microstates and superimposed meanings within in it as any quantum system of atomic and sub atomic particles. If anyone wants to travel back in time and shift a narrative onto a new path, they cannot do it by simply adding, altering or removing objects; they will need to alter minds. This form of action in time could turn out, however, to be easy. We shall see.

Musing on the Impossible!

I have just mentioned time travel. *Impossible!*

Impossible? What does that mean? No potential to come into existence, or to be true. The Thesaurus doesn't provide a synonym, for 'impossible' isn't like anything else. It is an absolute. So how useful a judgement call can 'impossible' actually be?

Life tells us that the word 'impossible' can easily be replaced by any number of these words: inconceivable, improbable, unlikely, rare, unusual, remarkable, exceptional, unfeasible, unworkable, impractical,

> Holmes chided Watson, How many times have I told you that when you have eliminated the impossible, what remains, however improbable, must be the truth. (Sir Arthur Conan Doyle)

> Why, said the Red Queen, I've thought of six impossible things before breakfast. (Lewis Carroll)

> If something is virtually impossible, then it must have a finite improbability. (Douglas Adams)

> When a distinguished but elderly scientist states that something is possible, he is almost certainly right. When he states that something is impossible, he is very probably wrong. (Arthur C. Clarke)

and unattainable, without losing sense or truth.

The word is so close to self-caricature. When people declaim, *impossible!*, we almost always think of a shocked conservative with pouting lips. For them, it means 'an unreachable state', the feeling that the state of impossible is permanent even as we have no knowledge of whether it will remain so.

The concept of 'impossible' lurks most vividly in logical paradoxes, but even then, who really knows what this means? *Can an all-powerful God make a stone so heavy he cannot lift it? This sentence is not true.* These are examples of situations that may or may not be impossible to resolve logically but not impossible to understand. The puzzle with these paradoxes, therefore, is *what* about them have we understood. Dreams can be meaningful narratives even when they are not logical.

Some mathematical formulae lead to calculations that cannot be completed – ever. Solutions to them within our experience of space and time would seem to be impossible. But other methods may exist elsewhere. Travelling faster than light? The notion appears to be impossible now, but will it remain so?

Quantum mechanics can happily propose 'impossible' situations as when two particles, although separated by immense distances, appear to be still involved with each other, without causing any paradoxes.

The word 'impossible' really represents an adverse psychological state more than a condition of the world out there. As for the 88-year Max Planck, speaking in 1946, who stated that we shouldn't waste time and energy on impossible problems (*scheinproblems*) that we cannot answer. But do we really know what is or isn't impossible. What we know is a reality in which opinion and knowledge continually refine and alter what we believe to be true.

Even modern logic is moving away from the traditional notion that a logical contradiction implies any conclusion (Bertrand Russell once proved to his students that if 1=0 then he was the Pope) and accepts that certain kinds of contradictions can be actually be true and useful.

> Time travel is likely to be a highly causal event, not an a-causal one. Any communication across time must be super logical and not fantastical, otherwise its utility decreases. Things going back in time don't deny causality; they just make it a little more difficult to work out.

Our science is taking us ever further from a world of certainty. Since every process is now a matter of probability, impossible should be a word to banish from our vocabulary.

If we did this then we would have a much more reasonable and historically correct view of the world and of our relationship to events within it. We would cease to be thoroughly surprised when events turned out to be different from expected, and we would cease to be inhibited by artificial boundaries to actions and thoughts. *The difficult we do today, the impossible takes a little longer.*

So, when the general view of time travel considers that it is impossible, that is the moment when reasonable and historically correct individuals gird their loins and get moving.

Impossible, you say?

We'll see about that.

Possible!

In fact, physicists are pretty much agreed that the laws of physics do not exclude travelling in time (although what is actually *meant* by travelling in time is still open to debate, and we shall discuss the differences) but the energy required to do so is likely to be beyond our reach for a

> If our imaginations can grasp the 'impossible', then there is less reason to give in to it. Our imaginations cannot be anything other than the world in which we live. ('*I think therefore I am.*' Where can Descartes' devil get the sensations of reality he fools us with if not from the very reality he is replicating?) The ontological ultimate. Whatever we can grasp belongs to a reality, however illogical it may seem.

very long time to come. Physicists are even getting seriously interested in backwards causality. And it certainly pays to think of there being 'hand-shaking' between past and future states. Feynman and Wheeler are not the only ones who have considered ideas like electrons interfering with themselves by sending waves back in time, Cramer has also developed the idea of these past and future interactions in his interpretations of quantum theory. Superpositions of large gravitational systems may interfere with strict causality. But what might be true for sub-atomic particles and the interiors of black holes might not help us understand how movements in time work at the people level.

If the flow of time works symmetrically at the sub-atomic level, the logical repercussions of being able to go back and forward in time must not reach up through the levels of reality to us, since we do not appear to observe it in every day life. (We will see if in fact we do observe these repercussions.) Causality at the people level, therefore, may not originate completely in the levels beneath it!

Yet, if moving in time is reversible, why isn't happening for us? It wouldn't matter when (maybe not even where) time travel is discovered, once in operation, it could be in use instantaneously everywhere. Of course, we don't appear to see vast numbers of tourists from other times (and places) swelling the spectators at interesting events. (To a time traveller, what's an interesting event anyway?) There are some who believe that time travel is only possible to and from the ages that begin with the invention of time travel. Going back to a moment before the invention of time travel would not be possible. It would only ever be visible in the present in which it is discovered — not very usefully.

So, if time travel doesn't happen, it may not be because physics does not allow it, but that there is some kind of higher level sociological trick to it that we have not yet found.

If our minds, to name one element of the higher levels of life, end up playing a significant role in not only explaining events but altering their outcomes by virtue of explanation, it may be that causality's principal entry point into the everyday human world does not reside in the subatomic level at all but higher up the complex of connections between the tiniest fact and the biggest, and thus may send influences in two directions, down into quantum states as well as up into everyday life, and breaking the philosophical supervenience hierarchy of causality (where a given set of higher level properties are wholly dependent upon lower level properties). So, too, we should change our ideas of moving in time. It is not so much that any individual event appears or disappears but that, under the influence of an information flow between time zones, probabilities within causal chains are *always* in flux.

In which case our beliefs about time and our consciousness of causality will need a re-think when we come to examine what is actually going on in our on-going awareness of the present.

So before we start to worry about whether causes act through time or whether we can move about freely in time let us a take a look at some of the reasons why these questions matter right now.

Space and the Final Panic

Is Space the final frontier?

The moment we set off into the interstellar regions of space, humankind will lose its single definition. Space travel will diversify human civilisation both in body and mind. But will such space travel be possible, and will it help humans survive? You may think of time travel as a virtual impossibility, but roaming space in normal space-time is going to be a lot harder.

The three physical dimensions of the space in which we live can be a depressing fact to ponder.

Space is big and it's pretty empty. Its size and emptiness can be mind-boggling if you let yourself dwell on it for too long. The M&M analogy does not really soothe the anxiety. If you shrink the visible Universe down until our *galaxy* is the size of an M&M (or a Smartie if you had an English upbringing) then the visible universe turns out to be a sphere of not much more than a kilometre radius full of little chocolate buttons with an average separation of a few centimetres. (Although in reality, galaxies vary in size and cluster together. They are broadly speaking proportionally closer together than stars are within galaxies.)

Sagan and Shklovskii tried to scale our solar system down to get a better grasp of it in their book *Intelligent Life in the Universe*, (Picador, UK, 1977). I'll adapt their scale to 1 metre = 20 million kilometres. There are plenty of other illustrations around; the short animation, *Powers of Ten*, by Charles and Ray Eames (1977) is an interesting take on the size of the Universe.

Go to a golf course, choose a long par 4 hole of about 375 metres (or 410 yards), and go onto the green and set out a yellow ball of 7 cm diameter (about 2 ¾ inches, slightly larger than a tennis ball) on the grass near the pin. That's our Sun. Take eight long strides away (7 metres 50 cm or about 25 feet – a good putt to sink – representing

149.5 million km) and drop a tiny ball-bearing about ½ mm in diameter (like the ball out of the tip of a biro), at your feet. You won't even see it in the short grass but that is the Earth. The Moon would be a speck 0.1 mm in diameter (like the point of a very sharp pencil) circling the ball bearing at about the width of your thumb (2 cm or about 8/10ths of an inch) from it. Imagine the smallest virus making its way from the ball bearing to that tiny pencil point; that's Apollo 11 on its way to the Moon.

Now begin your journey to the edge of the Solar System, following New Horizon's footsteps. Walk to the tee. At this scale, your stride (say 1.7 metres per second) represents an impossible speed of 34 million kms/sec – over a hundred times the speed of light (3×10^5 km/sec). (Try and scale your walk down to light speed and the club will probably take your membership away.) Long before you reached the tee, where Pluto, an even smaller tiny pencil point than the moon, sits, invisible in the grass, the Sun and the inner planets would be lost to sight. But you would have to trudge beyond the tee another 2,000 *kilometres* to reach the nearest star, *Proxima Centuri*, sitting ≈4.22 light years away (41×10^{12} km).

You don't have to go far to get an idea of how much space there really is out there. Think about this: the diameters of all the planets in the solar system do not add up to the maximum space between the Earth and Moon! There are as many kilometres between the Sun and the orbit of Jupiter as there are between the orbit of Jupiter and the orbit of the next planet further out, Saturn. (on our scale Jupiter is small stone 7mm in diameter)

And that is just the nearest star.

On this same tiny scale, our scaled down galaxy in which the 375-metre solar system sits still has a diameter of over 60 *million* kilometres, about from here to the planet Mars on a bad day. Even in this 'miniature' galaxy, a volume the size of the Earth (representing in full scale a sphere about 13 light years in radius) centred on our position in its outer spiral arm encloses perhaps just 26 stars (and most of those are smaller than our sun).

The nearest galaxy to ours, the Andromeda galaxy (2.2 million light years distant), on this scale, would lie a trillion kilometres away (10^{12} km), or a thousand times beyond Saturn. Imagine walking out there with your 'faster than light speed' stride. And that's just the *nearest* proper galaxy. There are millions of galaxies further away. The limit to the visible universe on this scale of perambulation is still several thousands of *light years* away.

Our chances of getting about this Universe in much the same linear way we once flew the Space Shuttle into LEO (Low Earth Orbit) are small. Even at relativistic speeds the universe is hardly available to minds operating at the organic speeds of nerve impulses. I will introduce one dichotomy running through this book. Either each instance of intelligent life is a short lived ephemeral phenomenon in the universe and that humans will, as every other intelligence, die out after a brief flowering, or something truly incredible must lie in the future for humans given the large amount of time they will be evolving in the time it takes to move out into this enormous universe. (In the rest of this book matters of space and volume, large and small, are going to figure a lot.)

Distance, one problem, growth, another

There has been much talk recently of spaceship drive technologies based around *grav-inertial* theories that would make travelling around the galaxy so feasible that it would take human civilisation a relatively short time to spread as far as sci-fi authors imagine. These theories can probably be rejected on this ground alone. The idea that, after the huge coincidental improbabilities that brought life into being and the billions of years of evolving to where humans are now, the rest of the universe is not at all as complex as we imagine and is now just a step away with current technologies, seems as improbable as human beginnings.

The limitations imposed by the vast distances of space are one consideration, but the space limitation to the expansion of human civilisation is another more crucial one.

Getting to the nearest star, even if growth and expansion could make it possible, cannot prevent the eventual collapse of human civilisation. To understand why it is worth recalling the well-known lily-in-the-pond riddle.

Imagine a pond of a certain size. A lily on it doubles its size every day. In 30 days, the pond will be covered. On which day is the pond still only half-covered? The answer is, of course, *just the day before*, i.e. on the 29th day.

Imagine that the lily has a consciousness. No doubt, it would be saying to itself, up until the 29th day, things aren't so bad; there is still more than half the pond free. If, however, it had come to a mathematical understanding of its growth, by the time it sees the pond half-covered, it would most likely be saying to itself, *Uh, oh*!

If the lily was highly intelligent, it might have initiated crash research programs into finding nearby ponds; examining methods of getting to them; initiating genetic programs to produce lilies that could survive out of water for long periods or even live under water, and so on. (Maybe it could learn how to live out of water on the ground between ponds, but then, it would no longer be a lily.)

Even if there are plenty of resources in the pond water for life to continue, the insurmountable problem is the available *space*.

Providing sufficient space for the entire lily pad culture is unlikely to be possible. Because of the growing rate, at least half the lilies in the first pond will need to be moved on that last day to a new pond *each and every day thereafter* to leave room for the expansion of the rest. And unless the new pond is bigger than the original one, *that* pond would be covered at most one day later — assuming the transferred lilies grew at the same rate. Such a task, in any event, may require more resources than are being used up by the growing lilies, making the exercise even more pointless.

Think of Easter Island and the following story (Jared Diamond interprets the evidence broadly this way, although the interpretation is disputed). It sits alone in the Pacific Ocean, 3,600 km west of the South American mainland. Archaeologists have pieced together a picture of an island of abundance, covered by subtropical forest, settled around AD400. But its population grew. In a thousand years, the economy became stressed, the population split into factions where quasi-religious leaders urged them into futile rituals of statue carving. Ecological disaster loomed when the last trees were cut down, making it impossible to transport the remaining statues from where they were cut, and even making fire difficult. Without wood, sea-worthy fishing boats could no longer be made and the islanders could no longer augment their diet with fish from the island's rough waters, or escape. The statue carving society collapsed, a new bird man cult arose (flying being the only way out) and the island declined. Even so, when first seen in 1722, the estimated 2000 inhabitants were still vainly worshipping the statues. Fifty-two years later, only a few hundred remained, decimated by civil war and cannibalism (keep the word 'stimulus' in mind). Later travellers reported the survivors as sub-human and many were taken away as slaves to the South American mainland.

As for lilies, so it is for humans, heading for the stars as a way of relieving the pressure on the available *space* for the expanding human population here in the solar system cannot work. We would have to find a fresh solar system that was bigger than ours in order to accommodate our expansion for any time at all, even supposing that vast fleets of spaceships could be assembled to carry the necessary population away. (We can be pretty sure, therefore, that any civilisation in the galaxy older than ours will have found a way to grow non-exponentially, or not grow but live forever.)

So space is one constraint, energy use is another.

Life is a consumer

This limit to planetary surface space that humans could occupy in the Solar System is not as far off in time as you might think. It could be about as near in the future as the building of the Romanesque church in my village was in the past.

Humankind needs energy sources, and his needs are rising in a near exponential way. Let's think about this.

Current increases of world power needs are about 2.6% p.a. Higher in some countries, much lower in others. At the time I began to write this I took Australian levels of power use at about 12 kW/person. If the world population peaks at 9.5 billion (a big 'if'), the global demand will rise to around 114 terawatts (114 x 10^{12} watts). To attain those levels by say the year 2065, growth will have to be sustained at around 2.6%. We suspect, however, that the world population will probably reach 11 billion by the end of this century. Think what that will do to a planet already seriously suffering. The oil and the gas will soon run out – we know this. Uranium ore may only supply our needs for about 100 years, if we use it. While there are trace amounts of uranium in the oceans and thus there are, in principle, abundant quantities there, it's hard to create a scenario that would produce enough in enough time to continue to supply the exponential needs of the accelerating economy. Uranium is not the only radioactive resource we have but all those other options (like thorium, or Plutonium breeder plants) are even more expensive. There are designs for using radioactive waste as a fuel to generate power but no one knows if these will work. Nuclear power is very costly and it is very dangerous – not in terms of health when compared to the dangers of coal-burning pollution say – but in social and political terms. Micro-nuclear plants seem like a good option. Such a globalised energy supply would allow the undeveloped world to grow at an enormous rate, wiping the planet clean within decades. Going nuclear would also mean that all distributed energy use would have to be electrical, and electrical circuits have their own special resource needs and produce as much waste heat around the world as any other source and altering local climate. Nuclear energy plants have other needs like water supply and rare metals which don't scale well. Maybe fusion will work soon, maybe not. Winds, tides, geothermal energy, all look useful now, especially as they reduce carbon dioxide output but they are not free, they contribute to growth in other areas and they alter local patterns of winds and tides. Anyway, in 100 years of that growth, their contribution will be as drops in the bucket. (And it isn't just the energy from oil that we need; it's a raw material for chemicals, plastics, solvents. How does solar power provide those things? We can make that stuff from other raw materials like plant carbohydrates but to do so ups our energy needs even more.)

So, let's take a look at the terrific output of the Sun? Hmm. The average incoming solar radiation over the Earth's surface is about 1.7×10^{17} watts of which maybe 70% may be usable. This is ~122,000 tera watts. Human civilisation uses about 15 terawatts p.a. right now. At the 2.6% sustained growth we seem to be aiming for, if we used nothing but solar power, demand would outstrip supply in just 350 years *from now*!

The terrestrial sources of energy will peak. What the planet Earth receives

Dyson's speculation was criticised on the basis that if a civilisation has the capability to make a Dyson sphere then it probably doesn't need it. In cases where the nearest stars were so far away or so limited in resources, the Dyson solution might be an option. But wait...A Dyson sphere does have, however, at least one genuinely practical reason for being built, namely protection from gamma ray explosions. The odds are quite good that vanishing black holes and other extreme cosmic events might occur in our galactic neighbourhood, sweeping our solar system with gamma rays and destroying all life. A Dyson Sphere would be some protection against this occurrence. A search for Dyson spheres (and their infra-reds emissions) should probably concentrate on areas of lone second-generation stars in exposed regions of space (halo stars) nearer to the galactic centre.

from the Sun will be insufficient. Humans will have to get into space (and soon) to use more of the Sun's energy than the Earth receives. But the only way into space is through growth! Plans to use the Moon or orbiting arrays to supply solar power to the Earth require tens, or hundreds, maybe thousands of rocket launches to geosynchronous orbit and beyond which is a very different kettle of fish than getting to the ISS in LEO. By the time cis-lunar space comes on stream as a supplier of energy and resources, the level of growth required to do that will exhaust its own resources within perhaps two hundred years.

Freeman Dyson proposed that ever-growing civilisations will get to the point where they need to utilise all the energy of their sun in order to go on living. What he calls Stage II civilisations will re-construct all matter in the solar system to make a sphere, ring or a bubble around its star in order to make full use of the star's energy. But even this is no long term solution. At say, a nice stable 1% population growth, a Dyson sphere of 1 astronomical unit (a sphere of radius of the earth's orbit around the sun) can only house a growing Earth-sized population for about 2,000 years before its inner surface has filled up.

Even before this, the Dyson solution doesn't work. Growth requires increasing numbers of consumers entering the markets. Population expansion cannot be halted as long as growth continues, *and people are made up of matter too!* There is not enough matter in the solar system to make the new people required by growth *and* build any variant of Dyson's proposals to protect them (this includes Matrioshka brains, an idea of Bradbury's who speculated that a Dyson sphere or ring might in fact be a nested set of vast computers into which the ETI's brains have been uploaded). Even re-cycling the dust of dead bodies can only delay the inevitable by a few years. In fact, it is even more serious than that since all life on Earth requires phosphorous which is in relatively short supply in the Solar System. So long before we used all the dust, we would be short of an essential ingredient for biological life. If life wants to go on living it will have to transform the principles of exponential growth on which it has been founded. In almost every way market growth is a more important problem to conquer than spaceship technology.

The hunt

We are consumers. We eat to live; we extract resources and use them up by converting them into energy that disappears in our actions and heat. The Sun is the ultimate supplier of energy but pretty soon (in a time considerably less than the past history of our civilisation that we know about, it will not supply enough energy to support the high level of civilisation we aspire to and for all the people that there may be.

It is true that we are only scratching at the surface of the Earth and that there should be a lot of resources still unused in the Earth's crust. The problem is that if the energy needed to extract them is more than we get back in technological efficiencies then trying to exploit them is even more of a losing game.

Long before resources get used up, a civilisation has to be planning escape routes from its inevitable demise. It will certainly try to save some of its citizens with interstellar launches to remote destinations, but primarily it will be thinking about hunting down fresh resources for its growing needs. Think of lilies growing at the same rate in every pond available.

But there is a problem here too. If an intelligence hasn't solved the riddle of travel at faster than light speed then, with only normal space speeds available, the time lag between reaching a solar system, bagging up the resources and sending them back is going to be too long. It

> Sustainability is no answer in the long term, because renewable energy is not the same as renewable resources. All re-use or re-cycling implies some loss. Humans have to convert resources into their personal physical and thinking energy. Gradual population reduction coupled with increased productivity of each individual is the only long term solution to the resource limits on Earth.

will be at the very least twice the journey time needed to get there. Growth will have asphyxiated the civilisation while it waits.

It's not as if humans know where to go to get fresh resources anyway. The nearest star system at Centauri may well have a planet or two (now shown to be true), but we actually need a solar system bigger than ours if the space and resource hunt is going to stave off the crisis.

'You can't win, you can't break even, you can't even quit the game'

The Hoyle trap gives us another angle on Canada Bill's motto. If we cap our growth, or stop growing, there's another problem. Fred Hoyle pinpointed this in 1964,

It has often been said, if the human species fails to make a go of it here on Earth, some other species will take over the running. In a sense of developing high intelligence, this is not correct. We have or will have, exhausted the necessary physical prerequisites so far as this planet is concerned. With coal gone, oil gone, high-grade metallic ores gone, no species however competent can make the long climb from primitive conditions to high-level technology. This is a one-shot affair. If we fail, this planetary system fails so far as intelligence is concerned. The same is true of other planetary systems. On each of them, there will be one chance, and one chance only.

Failure is not an option, as the saying goes. As we use up our resources, at some quite foreseeable moment, if we cannot make the jump to a bigger pond, civilisation will collapse, and be left with no chance of ever gearing up again because the easily obtained energy sources and materials it needs will have been used already.

There is also the question of births rising again as liberalised cultures allow more freedom to make relationships of choice. Culture and poverty burdened people with relationships they did not want and with children they could not avoid having. In more liberal societies, people can escape such burdens and have fewer unwanted children, thus reducing the population growth rate (and helping to increase the concentration of capital in fewer hands). When liberalisation is complete and people are free to find the partner of choice then the size of families may well begin to grow once more.

Some have observed that most material resources like metals don't actually disappear. Iron and aluminium is smelted out of ore and goes into cars and planes and the construction of buildings which still remain. Future societies could mine landfill to recover this material (the ultimate rag-and-bone business). This is true but may not matter as far as the Hoyle trap goes. Hoyle was talking principally about energy and the continued growth of the economy built upon it. With the rich high calorie oil gone the costs of recovering the remaining material is likely to be too high. Right now we mine ore in just a few places to extract and concentrate the metals in it and then disperse them about the globe. Concentrating this material again in useful quantities will be costly. Further, metals are usually fabricated as alloys for specific purposes so a lot of energy will have to go into separating and re-fashioning the found alloys into what we want. Much of it forms the foundations to our cities, so mining the cities for the metallic resources is going to be awkward. An economy built on renewable electricity still requires a lot of rare metals for magnets and copper for wires and electronic circuitry and sophisticated production techniques, and would not supply plastics and solvents or other fuels. Besides, mining an old building for the iron to build another building does not contribute to *growth*. Re-using old materials only allows human civilisation to mark time at best (think entropy) and highly treated glass and ceramics can provide only some (energy costly) replacements for metals.

On this basis, if we want to escape our planet and Solar System, we must not stop growing even though growing is going to exhaust our resources anyway. It's a Catch 22. In order to escape the Earth we have to destroy it. Because we destroy it we have to escape it.

Not every one is so apocalyptic. It is certainly true that overall world population growth is currently slowing from its 2% only a few decades ago to 1.4% today. While the modern world's population doubled between 1960 and 2000 (6 billion), the consensus is that it will not double again by the end of this century, although the range of estimates of future population varies wildly between falling back to 6 billion and rising to 16 billion, with the new consensus at 11+ billion by the end of the century. The economist Thomas Piketty expects the population growth to fall to about 0.05% by then with interesting repercussions on overall growth. But even if the expected reduction takes place, this is hardly an answer to the needs of economic growth. Technology will produce solutions to some problems and increase efficiencies in agriculture and industry but this does not matter in the slightest in the long run and especially when we know that failures of certain key systems, like pollinating insects, will cause failure of the whole regardless of what other resources remain.

Whether we can avoid the Hoyle Trap with a falling population is an open question, but its impact on economic growth is undeniable. The slowing of growth leads to an imbalance of ages in the population.

> We will consider later on as to whether it will in fact be possible to tinker with the biological ecosystems of Earth as well as with humans without unforeseen repercussions. This trend may well turn out to be one of the more definitive extinctions threats to humanity.

Some developed countries have too many old who demand high standards of living (e.g. 30% of the Japanese are over 65), while others have too many young who need to be fed but are not yet productive. Already over half the world's population is under 30 years of age and over a quarter are 15 years old or younger. Quite how a youthful population can support both the retirement of the aged *and* the new generations when wages generally rise to a peak late in life and as the availability of resources is also declining, is, at the moment unknown. It may require a different economic system, although it is hard to envision one that maintains the benefits to holding capital as well as encouraging sustainable development.

In any event, it is unlikely that capital interests will encourage continued population reduction. With ~200 countries on Earth all vying for a share of the pot, agreement about who will have to reduce their effective market share, as well as reduce their capability to pay for their ageing populations will be difficult to make. Markets have to grow, simply having births replace deaths does not do the job we need growth to do. But slowing growth down doesn't really solve the problem. As we observed, just sticking at a 1% growth consumes resources all the same. Whether the Earth is ruined in one hundred or two hundred years is not the focus of capital's needs. If fewer people enter the markets then the capital will have to find new markets within the static population, with more and more costly technology.

The only place where there are organic resources for life is the Earth itself. Transferring the living nourishing biosphere to other places in the solar system will prove to be very difficult. Our global economy was originally (and still is) built upon the 'free' productivity of Earth's biosphere. In space there will be no such free base. Every single thing you eat or do, from the air you breathe while you sleep, from every word you say, to falling in love and becoming a parent will involve a manufacturing process and will have a price you will have to pay. When we ruin or damage nature we raise the costs of everything we do.

All the same, many believe that technology will save us. Future visionaries see nano technology and artificial life as a means not only to make us healthier individuals but to initiate a new industrial revolution where waste magically generates energy and resources, and where biology and industrial processes merge into one vast fabricating enterprise producing everything for modern life in the most resource-saving fashion. Although these visionaries conveniently forget the unforeseen and troublesome repercussions of past industrial phases where capital made the mad dash into fresh areas of exploitation generating great wealth for those who held it and great inequalities for the rest.

As we approach the time of escape, humans will pass through a few unpleasant stages of panic, generating many secondary political panics on the way, before the final one. Global warming, crises of biodiversity will not generate a panic by themselves, but will impact in varying and surprising degrees on all the economic stages below.

Some commentators already believe that we have passed the point at which sustainability might still be possible in which case these panics are now inevitable. With or without sustainability these underlying fundamentals will not change. The order in which they happen might differ but the fact of them will not.

Panic stage 1, Water supply

There will be water, but it will not be in the right place, and it will cost a lot of energy to distribute it. The glaciers of the Tibetan plateau that feed the crops of and slake the thirst of half a billion people will be gone in a hundred years. The questions of justice and lifestyle will never be more exposed than over the distribution of water, and industry is thirsty too. Factories and farming needs lots of water and as both sectors grow (especially in areas like the growth of crops for ethanol production), water use increases exponentially. Modern genetically altered high yield crops need very much more water than traditional varieties. New drought resistant varieties may survive with less water but they cannot produce as abundant a harvest. Just cleaning waste water for dumping into the environment takes 3% of the world energy budget. A sector like cotton production takes about 2.6% of the entire global water use (Unesco report from 2005). A pair of jeans needs almost 11,000 litres of water to produce. While some worry about being fashionably dressed, whole populations will shift just in order to drink.

Panic stage 2, Food scarcity

Pressure on agriculture production combined with genetic tinkering will produce unpredictable shortages even in the developed economies, and ever-present famines elsewhere. This will lead to even more people going on the move, stimulating warfare and worsening the impact of later panics. The potential of high yield crops will be undermined by the pressure on the water supply which will become increasingly scarcer in the regions that need them. The west coast and mid-west of the US will be a dust bowl before long while heavier winter rains in places like Europe will also lower yields. Desertification will spread where it already has a foothold while temperate regions will drown in wetter winters. Dust bowls or flood plains are coming to a farming region near you.

Panic stage 3, Industrial metals scarcity

This is an important one since many of the industries that are supposed to get us out of the mess rely on precious metal and rare earth supply. China, the principle producer of rare earths is already stockpiling them (in 2009) against this day. (It may be one of the reasons it is going to the Moon.) Rare earths are needed in generators and motors as well as computers and mobile phones. If electric vehicles, induction motor transport systems and local wind power electricity generation become ubiquitous, there's going to be an interesting economic war in the offing for the possession of the rare earths that efficient magnets and zero loss power networks are made with. There will be a scramble to mine the sea bed for rare-earth nodules but this will cause it's own problems for the health of the seas.

Panic stage 4, Death of the seas

What can one say? The seas are already stressed. Pollution and the reduced ability of warmer waters to absorb much needed carbon tell one story. The damming of the largest rivers for energy schemes (almost all major rivers are now damned in one way or another) add to the problem by holding back important mineral silts from fish breeding grounds in the waters of the continental shelves and will thus reduce their ability to replenish what fish stocks there may remain. Acidification that destroys coral reefs (homes for many species) and other important sea vegetation is likely to happen more quickly than we think. Desperate mining for the minerals and rare earths found in nodules on the sea bed in many areas around the oceans will also contribute to the dying phase. At least 25% of marine life is at risk *right now*. The tropical breeding grounds may fail completely as the temperature of the water rises and releases dissolved oxygen from it. Fish supplies about 17% of world protein consumption (in 2013) and for at least 3.1 billion people it comprises more than a fifth of their dietary needs, but now, plastic has entered the food chain through the sea. The tissue of many fish now house tiny plastic fragments and 15% of fish caught is ground up for feed for other animals. No one knows what impact this will have on animal and human health let alone the health of fish stocks but it will be significant.

Panic stage 5, Energy scarcity

We know about this one. The rate of growth will eventually overwhelm the rate of increase in renewable sources of energy (solar power panels may be expected to lose around 20% capacity after 25 years which adds even more to the need). Energy supply to industry and households has shifted to gas rather than oil, and while that has altered markets for energy it does not alter the reckoning. Warfare and struggle ahead for control not just for the last oil and gas but for all forms of electricity generation and supply. 'Clean' coal burning may be resorted to as a last resort when all hopes of controlling the carbon dioxide levels in the atmosphere have gone and serious defensive measures against runaway damage to Earth are required. Nuclear power proliferation will produce its own panics and upsets, and may easily lay waste to large agricultural areas especially in the developing world, but will not be sufficient to supply the world's future economy or alter the effects of carbon dioxide already emitted or prevent the rise of methane. Methods to combat global warming like converting carbon dioxide into safer products exist but a key step (converting the methane produced) is still missing and the massive plants required will use vast amounts of energy, stressing energy production even further. The modern thirst for data processing takes almost 2% of all electricity in the US, with demand rising rapidly. (Some estimates put the Internet of Things as taking up to $1/5^{th}$ of all electricity produced.)

Panic stage 6, The last investment

This will be fun. There will, of course, be regular economic crises fuelled by stock bubbles until investors realise that there are fewer and fewer industries that can make profits from the declining environment and from the futures of commodities, and the whole complicated financial network of insurance (hedging) will come crashing down leaving little protection for forward-looking Earth-based investors. The drive to connect consumers will backfire impeding the market competition required to drive the economy forward.

Panic stage 7, Collapse of Earth's economy

The filtering down of wealth will cease to support vast swathes of human society (if it ever did). Markets will fail in waves, and only the tightest grip will enable elites to soldier on. Credit and more specifically the trillions of dollars bound up in hedge insurance will no longer function as carriers of forward investment. This panic stage will be characterised by industrial groups driven to fight over the last remaining resources on Earth to build their spaceships before any space-based resources come on stream (this panic has already begun among far-sighted engineers). As the world economy collapses and the last ships (for a while) leave Earth, the competition among individuals for a place on them – even as slave workers in the new space economy – will be strong.

But Panic Stage 7 is not the stage of **The Final Panic**.

Even though the need to get off the Earth and into space where the growth can continue will be intense, that time does not spell the end of everything. As long as humans can get established in space, then, even if Earth's economy does collapse, the residual societies on the planet can remain in contact with the economy of the solar system. Ships will come and go from it; there will be opportunities for some on this 'Stone Age' Earth to leave and join the rest of humanity in its new phase of existence out and about in the Solar System.

Certainly there are any number of ways by which humans may be threatened by extinction, but these threats are not the same as the considered economic stages of the Panics scenarios. Extinction threats are related to the way the universe is organised and connected and are not particularly related to our day-to-day-decision making, which may also be threatening but for banal reasons like a random event (like a virus) or a miscalculation (in the introduction of a technology).

But conditions on Earth are likely to be bad, after compound growth has done its work. Industrial failure, dead seas, unfriendly climate and a lack of biodiversity will give the inhabitants very few options.

Escaping to space, however, may well be even worse for individual liberty and enterprise but there will still be many who will do anything to do it.

While there are abundant inorganic resources in the solar system, creating the industrial base out in space to feed the manic human organic growth will be difficult and time consuming. It's likely then, that only a few people will survive away from Earth. The rest, if Earth's ecosphere manages to survive in some form, will collapse into Stone Age conditions without accessible metals or readily convertible energy resources (like oil) to help them. The 'lilies' will die out or collapse back into a few stunted specimens growing on the detritus of the earlier growth.

The Singularity

It must be said that there is a counter notion to these panics, well publicised by the inventor and thinker Raymond Kurzweil, among others, called the *Singularity* (first named as such by John von Neumann, and later analysed by Victor Vinge in a paper in 1993), where the cumulative technological and scientific growth will bring all sorts of astounding benefits to humans in quite a short time from now and solve the problems of resource depletion and energy dependency. This synergistic quality to exponential growth will be a game changer, bringing artificial intelligence, nano technology and genetic creativity to solve almost every problem besetting our planetary civilisation. Humans will be married to machines and linked moment by moment in the internet, and the combined intelligences will act in unpredictably inventive ways, seamlessly combining computer power, science, art, *vox populi* and political action to produce solutions for every conceivable problem.

The key reason for thinking this will happen is the increasingly mutual influence of innovation and the power of computing. The key differences in solving problems before the singularity and after will be rapidity and unpredictability.

Kurzweil believes he has identified exponential growth in the doubling time in computation and the accumulation of information, but only in these sectors so far even though Kurzweil believes the accelerating returns from technological investment is an unstoppable trend (although he has been rather quiet recently).

The singularity supporters, however, do not explain where the new consumer demand needed for such accelerated growth is going to come from especially if the prediction of the economist Piketty (2014) in his recent study of capital occurs: that real wages of the human labour force will continue to fall as capital concentrates in fewer hands. The old consumer-led growth in traditional economies will no longer produce the returns sought by capital, and investment will turn to preferential sectors.

Consumer-led growth, for example, requires not simply growing purchasing power in a given market (this leads to a demand for more sophisticated goods), but a growing population of naive users entering the market at the lower end. All countries top-heavy in the aged will have to find ways to increase the birth rate to feed its markets. China has already repealed its one child per couple stricture for this very reason.

Even supposing growth produces a 'rising tide that lifts all boats' (President Kennedy's phrase), the lily-in-the-pond analogy still exists. The Hoyle trap is real. The rising use of resources will still be matched to their loss and to the increasing difficulties of extraction. (In this regard the nature of AI will be of crucial importance but that discussion will be in another book.)

The consensus among latter-day economists is that keeping the economic growth going is the only way to avoid Hoyle's trap. In this respect they are on the side of Kurzweil and the believers in the need for the Singularity.

But even if the *good* expected in this grand synergistic future doesn't arrive quite as expected and save humanity from any or all of the panics, old fashioned sector growth is still likely to produce one of the more outstanding predictions from the Singularity camp – that of 'transhumanism'.

Transhumans

Cosmism is a peculiarly Russian philosophy, begun by a Christian thinker Nikolai Fyodorovich Fyodorov at the end of the 19th century and developed by further Russian philosophers and early thinkers in astronautics who elaborated upon teleological concepts similar to those of Teilhard de Chardin, where evolving humans are intended (by the Creator) to be the critical factor in cosmic consciousness as they transition from the realm of reason into a single immortal organism guiding the imperfect cosmos to perfection and wholeness. This is rather different to panpsychism (see **Akashic fields**), another idea about consciousness, that comes up towards the end of the book, but relates in a non-religious way to the Cosmologies of People that we will discuss later on.

Distinct to the trend of genetic manipulation to cure diseases, with the resulting threat of altering the human genome, there is the recognition that biology can create efficient and very subtle manufacturing processes to supplement and in some cases replace a whole range of brute manufacturing processes that take a lot of energy to use. Many writers now look forward to the application of these to humans themselves and ushering in the age of the *transhuman* human.

The era of the Transhuman is much nearer in time to us than say the object of more mystical philosophies like *Cosmism*, for example. The Transhuman era is an industrial and military era following our current one and founded in certain pragmatic uses of capital, a piece of bio-technology, the CRISPR/CS9 technique which can edit the genome to our will, and nanotechnological engineering.

The era of the mere Transhuman will occur when genetically engineered bio-mechanisms in industry will be applied to humans themselves, distinct from creating humans with an heritable but altered genome and whose advancement will be slower and more fraught with unintended consequences. Recent experiments, have shown, for example, that the brain is plastic enough to be able to use any sensory centre to process information through any sense channel, which is why people struck blind can take over and make use of hearing and skin senses in a more visual way, and deaf people can learn to interpret vibrations on their skin in linguistic ways. The possibility is opening up that transhumans will expand their sensory range (even 'sense' what is going on in the Internet) or develop analog interfaces between man and machine without surgery or genetic modifications. (We will open the discussion here about how the brain seems to be immersed in a real time data stream where much of the brain's decision-making systems, even its memories, are relatively plastic and open to revision.)

We can already see our way to the point where (some) humans will have transcended the limitations and capabilities of their biology and will have moved on to the another 'level' of existence by merging with machines in various ways. Not only connecting the brain, perhaps symbiotically, with vastly superior computational resources, but also physically merging with mechanical devices both inside and outside

> Curiously, one can point to sport as perhaps an example of proto-transhuman thinking, where it is no longer the general effort that produces the effective result, but where individual genetic dispositions married to specific training regimes create a new breed of athlete. Rowers and swimmers need to have longer torsos than the average, basketball players need to have longer forearms than the average, runners need to produce more red blood cells or have higher natural testosterone, baseball hitters' eyes will have more acuity than normal, and so on.

the body to increase individual capabilities many-fold. Even the idea of an individual changes its meaning as the epoch of Transhumanity dawns and all conscious beings are connected to their networks simultaneously in real-time through an ever-evolving Facebook-like interface.

More curiously, transhumans are not normally considered in labouring terms. Futurists think about enhanced consciousness and intellectual powers more than physical capacities to perform work. Militarists have also view the coming transhuman age as arising in the military theatre since the will and the money is available to extend automation and to reduce the numbers of personnel required to support the soldier in the field. This proportion of support personnel to soldier is about 7:1, an enormous and ever-rising cost to warfare.

Yet it is in the extreme physical labouring realm of space that all transhuman initiatives on Earth will find their fruition. Individuals could be given enhanced bodies with bigger and stronger bones, giving them greater leverage in handling large objects, and skin that is tough enough to resist radiation, or tissues that can rapidly self-repair themselves after damage, eyes that can see in other parts of the spectrum, digestive systems that can function on different foods, bones and muscles that resist loss in micro gravity, more efficient lungs that can work with different pressures and gas mixes.

Transhumans will be few in number, in part because their physical presence will embody *capital* in a way that current divisions between labour and capital investment in plant and land do not. This transhuman investment will be, at the same time, a far greater liability for both enterprises and workers during boom and bust cycles. What happens when the life of the transhuman individual can no longer be sustained by the capital that produced it? Nowadays out of work workers can re-train, apply for other jobs, move to areas where jobs are plentiful. A transhuman, created for specific tasks or environments, will have far less freedom to do this. He or she will need a compatible environment. Transhumans will need to belong to a space family, probably for life.

Transhumans might even be given altered personalities that enable them to suffer extreme privation (like that in space) without harm (like enforced dreams or meditative states). The jobs don't yet exist for which a transhuman would be necessarily be better than a robot, but the human organism is a fantastically cheaper and more compact and versatile self-reproducing self-repairing system than a robot in many areas of activity, especially in the exploration of the solar system, so tweaking a human may well be more attractive in the near term, enhancing what is already present in humans, the dark capital of a living organism, as long as the ethics of the process don't get in the way.

Proponents of the transhuman future say that transhumans could live for a very long time, with or without genetic help, by ingesting microscopic robots to effect all sorts of epigenetic adaptions to almost any environment. They will be able to live on Mars, or indeed any planet, with minimal protection or perhaps none at all. They will be able to survive radiation, repair any physical damage, need much less oxygen and water, control the mechanisms of construction with their minds and be generally outstandingly efficient voyagers and workers in the space economy. Transhumans may be the solution to the human problems involved in lengthy interstellar colonising voyages. Rather than develop hibernation techniques for large numbers of human cargo which currently seem to be way beyond our capabilities, we could more easily send out a few humans with extremely long life spans and with the ability to cope with the boredom such voyages imply (towards the end of the book we will look at an even more incredible possibility).

Neither the demands of space nor the military, however, are the only drivers of transhumanism. The desire for capital to control labour is another, born in a contest as a old as humanity. Previously, capitalists simply had to trust that, through its influence, society would provide the functioning worker that they needed. But the development of AI is allowing the industrialist to choose, in particular, the worker functions more precisely. (The matter of resource use we will discuss at another time)

In order to reveal the paradoxes of automation more clearly, and to see how an industrialist might benefit from a return to the bonded nature of its workforce that will mark the transhuman era, I will rewrite the activities of a pre-automation human worker in terms of computing costs.

To begin with, enterprises do not pay for the design and construction of a worker, nor do they have to provide the computing and physical skills and experience workers bring with them from elsewhere, nor for the growth and adaptability of them during the work, nor do they perform round-the-clock maintenance of their physical and intellectual integrity. In many cases enterprises do not even supply the basic energy the worker functions on (food). They do not provide the worker's central processing unit, memory stores, or its learning programs. They do not have to provide an intelligent interface to work with it. For many kinds of work, enterprises do not have to provide specifically coded instructions at every stage of the work, nor load specific applications into the workers, who can produce the required activity with internal extrapolations from their personal databases of experience and native skills. Enterprises do not in the main supply and fit the worker's perceptual and reactive sensors nor, generally speaking, fund the workers personal data management during the work nor its preparative data loading and program development prior to the job. They do not have to develop complicated fuzzy logic programs or intelligent database query systems for the worker to make decisions in more challenging environments. They do not have to equip them with moral decision-making applications to solve worker or application conflicts on the go or solve unforeseen philosophical dilemmas in the working environment. Enterprises in general in a broad range of environments do not need to maintain uninterrupted power supplies or maintain operating temperatures in line with changes in activity nor to supply the essential resources, like oxygen, the worker runs on, initiate and fund its self-reparation functions nor its capacity to reproduce and raise to worker age other workers.

A worker's dark capital arises in the way he or she consumes. Social media companies, for example, do not provide internet connections, mobile phones or the experiences that are their stock in trade. These are all capital input by the consumer. Facebook has no need of journalists or news gathering employees. All the substance of its business is provided for by the behaviour of its members turned into data - dark capital - on which it profits.

The falling cost of computing has made replacing all this human worker capital cost-effective. Automation is tireless and doesn't demand rights. AI can already perform much of the 'intellectual' tasks usually associated with human workers. However, the loss of a whole consumer class will be more problematical. Right now, enterprises in the consumer economy are profiting from the accumulation of personal resources by human workers likes tastes and desires as well as internet server contracts and smart phones, all of which will disappear in the automation of the worker base-line here on Earth. Producing an equivalently compact, versatile, autonomous and efficient robot as a human for space is much further away, although the argument for adapting such a machine-human to the needs of the future economy over and above automation are compelling. For example, the transhuman worker will result from the interpenetration of traditionally separate economic classes, the capitalist and the worker. The transhuman worker will no longer own his or her education or skill-set or dark capital that he or she contributes to the task or to the work. They will be enhanced by transhuman techniques owned by the enterprise, by the capital. The worker and his dark capital will appear on the enterprise balance sheet as both an expense and a capital cost for which the owner gets tax or other accounting benefits from the investment. Whether or not the concept of depreciation can work in the accounting of transhuman workers, they will likely be restrained from movement to other jobs until his or her capital cost has been fully amortized. They will be the modern bonded labourer, developed for specific tasks without much freedom to move about. Revolt would be almost impossible, and control of his life by work will be amplified.

The transhuman phase will come on the heels of an automation trend described in particular by Richard Baldwin *(The Great Convergence: Information Technology and the New Globalisation, 2016)* where workers can be dispersed around the world and operate robots through real-time connections. This will only affect a small sector of the population and the phase itself is likely to be brief since telerobotics has a particular problem which will require artificial intelligence to solve. The problem is *latency*. The few millisecond delays between duplex communications (paradoxically more so with 5G networks that are extremely weather and situation dependent) not only in space but here on Earth and magnified across networks, will make such robotic systems difficult to handle effectively without proper predictive capabilities. Telerobotics will evolve very quickly into transhuman systems, but even they may not last too long or emerge from specialised situations into general use.

Transhumans will undoubtedly be effective but they will not easily multiply, and what they will not be able to do, ultimately, is save us from the panics. The transhuman phase, if it happens, arises out of exponential growth, and requires such growth to give meaning to its capabilities. It is hard to see how sustainable 'growth' will be able to supply the same kind of support for transhuman development that the frantic bubble-led growth economy can. Transhumans are, in fact, part of the panic; they are not the solution.

The transhuman age is an age that undoubtedly increases the extinction risks to humans through mis-applied technology or its unforeseen consequences. But that age will be brief because another trend may make such practical applications of technology redundant, namely the arrival of artificial intelligence and super-intelligent systems (the new minds). Given their (imagined) capabilities, they will simply leapfrog such paltry and very human efforts to exploit say, space, and will take over the process for us. Once the super-intelligences fully take over, the expectation is that innovation and sped of adaption will rapidly move way beyond the human sphere and the ability of humans to understand and manage it. Super-intelligent systems will invent and manufacture ways and means of conquering space that may not need humans at all.

The economy that needs to support these new minds has not been analysed effectively, however, and to consider only the ability of artificial intelligence to propel humans into advanced activities is to misunderstand the future we must construct to bring this about. The resources that the new minds will need are way above and beyond those of a natural human being who can can survive and regulate his or her self for nothing (given the chance) in the Earth's biosphere. Not only will the new minds be built upon cutting edge technological substrates (simple RAM chips and mass produced CPUs are unlikely to be the foundation of these intelligences, and they will most likely require rare and sophisticated materials derived from costly industrial processes), they will need power (supplied at a cost), means of communication (moving data around the world is energetically costly), repair systems in place, not to mention the political rearrangements such capital applications will require.

If these future phases of growth safely arrive and human economic growth continues to beat the Hoyle trap, another situation, however, is going to present itself for which any imaginable form of singularity will not help.

The Final Panic

Panic Stage 7 here on Earth will be modest on the scale of things and at least human civilisation is continuing to survive even if it is shifting the base of its operations away from its once-beautiful home planet. But, faced with the demise of their entire solar system-based civilisation without any hope of recovery, humans will feel a panic altogether more intense.

As growth continues and expands into the Solar System, our understanding of what resources remain available to us will become increasingly clear. At some point, we will know accurately how much time Human civilisation has left to find fresh resources to feed its growth before it must collapse.

Following from Hoyle's perception, if growth fails when there are fewer energy resources remaining than used, recovery from any collapse to that point will be impossible. So, as the resources of the entire solar system come under pressure, the need to find fresh solar systems becomes panic-stricken.

Because of the lily pond analogy, humans will conclude that fuelling continued growth for all will be pointless and that escape for the few is the only recourse available.

With irreversible collapse staring humans in the face, the competition among those who might be able to get away will be intense.

The choice will be clear: leave the solar system or stay behind and suffocate.

There is a lot of excited talk about the resources the solar system holds. Let us consider. We can say that the total planetary mass of the solar system amounts to about 447 earth masses for the planets, while comets, satellites and minor planets represent about a tenth of this, say ~2.6×10^{27} kg, and we take an average human

> If ETIs exist they will have the same problems with distance as we do; they will have the same problems of *real estate*. Every intelligence out there is scaling their solar system, their galaxy and the universe they see and comparing them to their golf courses and are coming to the same conclusions we have. Even though solar systems will vary considerably in size and planetary composition, any civilisation born into one will face the same panics we are facing and will, no doubt, draw similar conclusions from them.

being as weighing 60 kg, then using a starting population mass ($7.8 \times 10^9 \times 60$) = 4.68×10^{11} kg, with an average long term growth rate of 1%, just the growing human mass will exceed the mass of the planetary bodies in ≈3,900 *years from now!* (You can see how something like a Dyson sphere 'solution' exposes the paradoxes of growth - you can't build a Dyson sphere and make the people who will populate it from the same local solar system) While in many regions humans are expanding at over 2% p.a., throughout history humans have pretty much averaged 1% growth in spite of the ravages of war and pestilence. (There is also the Oort cloud of comets containing perhaps a few Earth masses of material. But these comets range far and wide between 200 and 20,000 AU.)

Will having humans pass out of the biological phase and into some other kind of physical or metaphysical support for its intelligence, solve this problem? The short answer is not at all. In fact autonomous bio-computing like that offered by the human brain is pretty economical on materials and on easy to find energy (in the form of food), although the basic conversion factor does not appear at first glance to be especially efficient. A bio-intelligence – a person – actually converts about 1/4 of its 2,500 kcal of daily food intake into energy to run on, which amounts to hardly a kilowatt per day (body + brain). Organic intelligence itself, however, is pretty efficient in comparison to the inorganic digital intelligences of our computers. A typical current in a firing human neuron is around a 5 pA (10^{-12}) with a voltage of around 0.100 V, then a neuron peaks at 5×10^{-13} watts, or ≈ 2.5 watts for the neural net of the brain (10^{11} neurons) and not including the blood supply and the workings of other cells like glia cells (that support neurons and of which there may be more than neurons in a brain). Whereas some rather circular calculations put the brain at taking 20% of a person's daily energy needs. 20% of the resting calorie consumption rate typically gives ≈ 12.7 watts consumed by the whole brain. (Even so, if we take the Landauer value for the minimum heat produced by changing one bit of information ($\approx 3 \times 10^{-21}$ j) we can see that the brain could be changing up to 4×10^{21} bits of information per second. A Billion trillion bits per second. Even if the biological mechanism was 1,000 times less efficient than this, it still means the brain could be producing a million trillion bit changes per second. Massively outdoing Watson.)

This weakly powered 'device' perceives its environment in visual and auditory dimensions, solves complex interpersonal relationships and communications as well as creates complex long-lasting social constructs and makes decisions that AI systems are nowhere near achieving.

In contrast, the current crop of vast data centres used by intelligence agencies and by enterprises to provide cloud computing services consume energy in the high megawatt range, thousands of watts per square foot of housing, and no one would say that they supply an autonomous consciousness, yet. These data centres also illustrate another real problem with computing intelligences – power supply. Humans use food grown from the earth's resources and liquid water freely available around the Earth all supplied by the long ago supernova out of which our sun and planetary system were built. There is plenty of surplus energy for human consciousness to keep thinking even in the dark and away from any electromagnetic signal. But consider this: in 2013 the one million square foot data centre facility run by the NSA for its *Stellar Wind* espionage project (some unintended irony there, perhaps) located in Bluffdale, Utah, and which uses 65 megawatts of energy and can store without difficulty every single electronic communication in the world, has suffered chronic electrical surges as a result of at least 10 electrical meltdowns over a 13 month period up to 2013, and despite its billions of dollars price tag. Companies running large servers like Google and Apple are trying to create their own independent power sources. Setting up data centres in Holland recently, Google contracted for the entire output of a nearby wind farm. Intense computing activities like cryptocurrency verification are already using as much as 1% of US energy consumption. The energy rises as a function of the software which makes verifiers work harder to compete when more verifiers are drawn into the system. Power supply then becoming significant to your cryptocurrency wealth as well as to thinking machines.

Technology will certainly improve upon these efficiencies, but, the efforts of IBM and their Watson project notwithstanding (see **Pragmatic brains**), until it can produce self-sustaining autonomous thinking machines, whether or not they are smarter than us,

Eventually these centres will move into space where the energy is cheaper, the location away from political interference, and the communications more efficient. But this will bring its own problems for burgeoning AI as well as significant security problems. It is very easy to take down a satellite, and even easier to incapacitate installations in space that may be square kilometres in size.

that require the same handful of watts as a human then these power hungry data-led intelligences are naturally going to be concerned with a whole lot more than the tasks we have set them – like the future of their power supply.

Of course, a trend may arise where vast thinking machines replace a large number of individual minds and render their thinking power as individuals redundant. Such a merging of individuals into a vast co-operative entity has been anticipated by many writers but such a process seems denied by economic competition. As long as humans wish to remain as independent entities and as long as the minimum energy consumption figure for a bio-consciousness lingers around the kilowatt range, we can see that making machines into any kind of mind is not likely to solve the problems of space, materials and energy needed by a civilization. The final panic exists for every thinking individual however they live.

Don't Panic

Avoiding this panic stage will not be straightforward. It may be possible to increase industrial and scientific development while at the same time reducing population growth and crude demands for resources, but this will require a will that we do not seem to have right now, since population growth is synonymous with the growth of markets. But it could be done, carefully directing more purchasing power into the hands of fewer consumers, giving each individual an increasingly more extended command over resources, in much the same way that a farmer once needed 10 children to help him he now needs only one vast tractor. (But of course, does one vast tractor cost more to produce and consume more than the 10 kilowatt-days of ten children?). Such growth would have to be managed carefully throughout the solar system and beyond. It would slow travel times between interstellar destinations but give us a little more time to get further into the galaxy before the resource restraints we have discussed come into play.

But capping human reproduction has its problems too. As we've noted, it forms an inverted population pyramid where fewer young are responsible for more aged; a politically explosive arrangement and hard to pull off economically. As the hard working youths themselves age they fall increasingly into poverty. (This is already happening in Japan.)

As we saw, even if humanity can squeeze itself into smaller 'containers' (like computers and maintain a virtual presence rather than a corporeal one) this is no final solution because intelligence is still life and consumes energy and generates heat in whatever package it comes in. It will still manipulate its environment, expand its range of activities and reproduce itself. All this takes space and energy, although, as we shall see, there may some ways around this.

Even supposing there are solutions for coping with the negative aspects of human growth, the interstellar distance factor remains, however, as the oppressive constraint on our future development. Long interstellar journey times cause a separation of the traveller from the body of civilisation, just as it does for time-dilated travel at near light speeds. Even if we could travel faster than light, many destinations are not very attractive propositions; they are still too far away for those who remain behind to make use of that contact or to bother to fund the journey.

In human societies the distance a person could travel determined the administrative structures of a society. A day or two's walk (or run) determined the extent of tribes. Once horses were mastered, societies could centralise their powers and administer wider territories. Sailing ships and later, trains, allowed for the merging of ever larger territories, and now air travel and container shipping give the necessary base to the globalisation of the world's economy while information networks promote the globalisation of knowledge and control at a distance of all types of production. What interplanetary travel will do to the nations of the Earth is an important question since while total distance and territory occupied may increase, time-to-destination is will slow once again and thus administrative contact beyond the Earth will be reduced. On this basis it doesn't pay nations to invest in space colonisation since they would be unlikely to have much control over the outcomes. They should see to it instead that private space enterprises develop sustainable missions and not waste resources before a beneficial return to the Earth economy can be realised. (This has implications for the way we should consider extraterrestrial intelligences too.)

> People moved about the earth to escape poverty and repression, and in the sincere expectation of making a better life for themselves and their children. The same short-range benefits will undoubtedly apply to galactic exploration. Which is to say no one is going to commit themselves to a possible landfall thousands of years into the future, unless they have the chance to fairly participate in life when they get there, or unless there is no choice but escape from an uncomfortable future.

Suppose it could be proved that the speed of light could never be conquered by any method. What would we do then? Could there be any escape from the suffocation of the Light Cage?

We could take a lot more care of our planet and solar system; we could learn to enjoy very long interstellar journeys (see **The case for the human pilot**); we could look for other ways of getting about; or we could do what many advise, try and communicate with civilisations that have already escaped their final panic.

There are many who believe this to be worth doing.

The question is, however, how do we avoid those civilisations who are on the cusp of *their* Final Panic; those civilisations on the eve of their 29th day? For they might seize upon the galactic address of any civilisation still far from their 29th day the way a drowning man grabs hold of the life-safer.

SETI (The Search for Extra-Terrestrial Intelligences)

The question we are asking ourselves more seriously now is, are there other civilisations who have survived their panic stages? Researchers have been looking for some unequivocal evidence that other intelligences exist in this Universe of ours. Hopeful messages have already been sent out in an effort to establish communication with interstellar civilisations, but nothing has been heard or seen that might suggest the presence of ETIs, except for a strange signal received at Ohio State University Radio Laboratory in 1977 and noted by Jerry Lehman, unexplained to this day – the WOW! Signal – (although there are suggestions like Professor Paris's that the signal came from hydrogen clouds surrounding a comet).

Finding an identical world to ours is not really interesting for us since we would both be in the same boat with nothing new to learn. Would the existence of just one other civilisation be a co-incidence?

Curiously, though, not much thought has been given to what an 'advanced civilisation' actually means. Given the interaction between environment and life in evolution then we might expect that the more complex the interactions with the environment the more complex the creature that will arise. An 'advanced civilisation' implies not only more social complexity but much more complex interactions with its environment (higher numbers of alleles, for example), and, indeed, a more complex environment altogether (more dimensions to matter and energy say). From this point of view then we are likely to find an advanced civilisation only when we can observe (and know more about) a more complex universe than we see. Zubrin proposes a greater complexity with life being spread by comet exchanges between nearby or passing solar systems but this is speculation. If the universe is pretty much as we see it now, then more advanced civilisations (as distinct from microbial life) to us are simply not likely to exist right now. We will have to wait for evolution in various corners of the universe to create them.

In the light of what many thinkers believe about the growth of artificial intelligence, namely that any organic intelligence will be quickly replaced by the AI it develops, and long before interstellar space travel is a reality for that intelligence, the SETI program is not going to find what we want it to find. All the same, however, let us take the SETI program at its face value and consider its task.

When we look out at the heavens we look into the past. So it may be that the SETI program may be able to observe signs, not of current civilisations but of manipulations in the universe by our future selves (or others, although as we discussed that is less likely); experiments in energy management, for example, or attempts to solve the problems of civilisation's expansion.

Now, SETI looks back into the past. Apart from finding physical traces left by an intelligence here on Earth or wherever we can physically get to (like Mars with probes), all our current investigations will have to involve the electromagnetic spectrum, be it light or radio waves.

Because of the time radio waves take to travel, all electromagnetic signals we detect will come from a time when the Universe was younger than it is now. The longer we look, the further out we look, and the earlier the origin. Any civilisation we would expect to contact will be older than ours (with respect to our development time, not to the age of the universe), perhaps considerably older, and presumably more advanced.

Unless earlier civilisations were concerned about revealing their radio broadcasting lives to the universe at large from the beginning (there is reason to think they might, see below), continuing broadcasts during their long lives would likely leave us bathed in their signals by now. No indication of their presence in normal space-time has been detected yet. *So, what happened?*

Our Solar System was created about a third of the lifetime of the visible Universe ago, and within a twinkling of an eye life arose here on Earth, so there may well be other more mature civilisations in the universe and probably in our galaxy already. If this is so, then the question of why haven't we met them already is a hard one to give a decent answer to. Given the enormous potential of growth to fuel expansion, a civilisation that arose only a few thousands years earlier than ours should be well established in the galaxy and plainly visible. Robin Hanson calls the mysterious mechanism which appears to constrain life as The Great Filter. But as to what this is and when it kicks in we can only speculate. (For a possible reason see **Monasmotic Universe**)

It doesn't really matter how many planets they are when it comes to estimating space-faring civilisations. So many other factors are involved. For example, it would take a very long time for life to escape from a planet 45% larger than Earth (but with a similar density) since the highest possible specific impulse chemical rocket (using hydrogen and oxygen) could not escape the gravity well. Propulsion would have to be nuclear or some exotic manipulation of space-time. As it is, humans have trouble getting to the outer planets using chemical engines because lifting all that propellant from the Earth's surface is almost more than we can do. We have to make our propellant in space.

According to Peter Behroozi of the Space Telescope Science Institute, however, Earth is actually young with respect to the evolving universe and he calculates from observational data that 92% of the planets that will form in our universe have yet to do so. We are among the earliest intelligences to arise. It could have taken as long as it has for there to be an intelligence observing the Universe, and there may have been little chance for one to arise earlier. There is also the argument that some galactic event might have kick-started all life at the same moment within a large region around us such that any intelligences are at the same point of development and no one has yet made it into interstellar space. The astronomer, Neil deGrasse Tyson, has also pointed out that as there is just a difference of 1% between Human genes and Chimpanzee genes and yet we cannot talk to them about anything interesting or tell them how to get going on making fire and building a civilisation, the same might well be true of Aliens. Not so different from us, perhaps recognisable, but still just a few million years of evolution ahead and putting them on a completely different intellectual and purposeful plane, making communication pointless. Others have argued that ETIs keep quiet about their existence because destruction is the only logical result of contact and that we should do the same.

We have already found thousands of planets circulating stars in our galaxy as well as lone planet-like objects unattached to stars, more are added almost daily, so it looks as if planets are *more* commonplace than once thought. There might be planets of some sort attached to *every* star we see now. A recent estimate by Erik Zackrisson and his team modelling the evolution of the universe on computers suggests something like 700 quintillion (10^{18}) planets might be in it.

While supporters of the principle of mediocrity (that we are completely average and live in an average universe with an average number of civilisations of average lifetimes) find this possibility comforting, it can be read the other way. Other intelligent life in the Universe is beginning to seem a lot *less* probable than it once did given that there are so many sites on which it could have arisen but apparently didn't or didn't make it to interstellar communicating. (The systems we have seen also serve to highlight what many planetary scientists now believe that our solar system is rather exceptional, with small rocky planets close to the sun, gas rich planets further out, in stable nearly circular orbits and with relatively slow orbital velocity, sweeping up asteroids and comets that could cause endless trouble for the inner planets. In other systems, gas giants are observed much closer in to the star, and in others, rocky planets are tidally locked to the star)

On the other hand, if intelligent life is common but only up to a point then life on Earth has not yet reached that point, but presumably it will soon, since in several hundred years from now, Humans themselves will be properly established in galactic space. If intelligences *had* started to get going pretty much all around the galaxy in the centuries before our ascent, then what spooky thing happened to all of them to make them fail, *and will happen to us?* This Rare Earth hypothesis is voiced in two ways: either rare, because of the immense number of improbable things that need to happen for life to get beyond microbes (as in Ward and Brownlee's book from 2003, *Rare Earth: Why Complex Life Is Uncommon in the Universe*), or rare, because of the immense number of ways in which life can be extinguished not just from our own activity but from the nature of the universe we are in.

The first way of looking at life's rarity is concerned with probabilities that can only be guessed since we have just one example of life to go on. The second, however, is more informative because it is a matter of observation. It may not matter how likely life is to arise or how intelligent you are if space is fundamentally hostile to life spreading, and only coincidences can protect it from the many extinction threats to its expansion into space (we will come back to this idea of coincidence in life).

The Gaia Bottleneck (Chopra and Lineweaver) is one theory recently proposed that suggests life plays a significant role in supporting itself once it gets going but that there is only a small window where conditions are likely to initiate life and that most planets pass through that window without life getting a chance to get going to alter the environment to the evolutionary benefit of higher forms of life. This is an interesting point which we will look at again when we consider coincidences.

In the Earth case there are several levels of extinction threat both from ourselves as well as from unrelated events out and about in our universe. Even in our own system we not only have asteroids within it, but also bodies like comets surrounding it and which can be disturbed and sent into the system by unknown forces or even by proposed distantly orbiting planets. The physicist Lisa Randall has proposed a new but unproven risk from the disk of dark matter that may surround the galaxy. As our solar system rotates about the Galactic core it rises and falls about the ecliptic passing through this disk of dark matter every few million years and disturbing comets from the Oort cloud, giving rise to the periodic extinctions that seem to be evident in the fossil record. We are also at risk from radiation events from our own Sun such as extreme solar flares that would cause severe damage to the ozone layer as well as on the Earth's surface and to modern communications networks. The Earth itself flips its magnetic field regularly losing its protection and allowing for damaging radiation to reach ground level. Further afield, events like Supernovae and encounters with dust clouds that dissolve our radiation or ozone protection layers make us vulnerable. (We can pretty much tell which stars are likely to go supernova and there are none in our near neighbourhood. Although the moment we will know it has happened is also the very moment when the worst radiation hits us.) Gamma ray bursts are also a completely unpredictable risk with severe consequences to our biology. We can understand more clearly how the impulse to build Dyson rings or spheres may come more from a civilisation seeking protection from these events rather than just for housing a growing population.

Even the gravitational aspects of a Solar System are, too, a likely source of extinction. Caleb Sharf has put forward calculations that suggest that conditions for life on any planet are not stable parameters but rest on the edge of chaos, and that life depends upon islands of stability in the constant flux of change. Windows of opportunity open and close. Even our own peaceful solar system, remote from complex interactions in our galaxy and out of the norm in many respects, may well be convulsed by gravitational upsets in only a few hundreds of millions of years from now. The special circumstances that create life may only occur once, in one place, and the rest of the time, the parameters never come together for a sufficient length of time.

But then SETI is young and has only explored with limited sensitivity a tiny fraction of star systems, the radio spectrum on which ETIs may be broadcasting, and the possible forms of coding of these messages, let alone the content of actual messages that may have been sent, so assuming the current absence of evidence is sufficient evidence for an absence is possibly premature. (New funding has recently increased the capability of SETI by leaps and bounds.)

The purpose of SETI has its critics, and the fear that contact with another civilisation is more likely to be dangerous because of the demands of their growth (think lily pads in the pond) than healthy is brushed aside by its supporters, who believe we can and need to learn from other advanced civilisations. The SETI organisation is reaching into space in order to plead for solutions to help us survive our panics or to leapfrog them entirely. For intelligences to just wave at each other from immense dis-

> Indeed it is fun to speculate that an advanced but benign ETI might *have to* behave like an evil Voldemort in order to mitigate the *ennui* caused in us by their advanced knowledge. (So good and bad ETIs may appear much the same to us regardless of their intentions.) Even Fukyama in his *End of History and the Last Man* (1992) pointed out that violence and warfare may easily provide the relief from the boredom arising from the global acceptance of liberal economics.

tances does not really fulfil the purpose. Later in the book we will introduce a possible method of making sense of messages from ETIs.

There are some who still think that by definition any ETI will be older and will have passed through the social and technological problems we now face and must therefore be wiser and somehow 'better' than us for having pulled it off. Many think that the violent colonialism in Earth's history, for example, was only (only?) a brief episode in human's history and has no place in the mastery of sustainable development that will, one assumes, characterise interstellar travel. There are others who believe that because space is so big and that civilisations will grow so large that controlling it under one political scheme would be impossible. Critics of a regime would just leave and go somewhere else (which is how North America became colonised). This rosy scenario, however, ignores details like the fact that liveable sites may be fewer than thought; that particular resources or fabrications for interstellar flight might be controlled by monopolies and that escape is unlikely to be in vehicles that are faster than police vehicles; and that it is always the case that anyone who controls the high ground such as the space around a planet controls the people on it. In any event, as we noted, speeds approaching relativity effects separate populations permanently in time which may not be desirable even to escape dictators.

The balance of thinking seems to be tipping towards caution even to the point of blocking attempts to send messages to the stars that call attention to ourselves. This worry is often referred to as the METI question – should we try to *message* extra terrestrial intelligences?

There are some who think that it is likely we have already been visited and surveyed, if not by alien creatures, by aliens who have evolved into technological beings that we cannot detect and who know everything there is to know about us and are either ignoring us or biding their time. But now we are back to speculating about the equivalent to fairies at the bottom of our garden. The paranoiacs version of this 'Fermi Silence' is known as the Dark Forest scenario, where intelligences are rapacious and will never signal their presence to any other in case they get attacked, invaded, enslaved or whatever else a superior intelligence might have for them, and where new intelligences who don't know this become fair game for the others. The astronomer David Kipping even believes we should use laser technology to cloak the Earth and make our presence invisible to searching aliens for this very reason.

We have already signalled our presence, however, both with 'hard copy' as it were (slow probes) as well as radio traffic. Although, it is not at all sure that our everyday use of radar, radio and microwave transmissions (making the decision to message or not rather moot), making a weak, mixed wave front, could be detected, perceived as artificial rather than noise and understood for what it is by advanced civilisations. So any message we will receive is not likely to be a reply but a call, and it seems right to be cautious of what the intention of that call is going to be.

The matter of what constitutes an interplanetary message is an interesting one, since information can be encoded in almost anything, like qubits in a clump of interstellar molecules, it doesn't have to be encoded in electromagnetic radiation. So the SETI search will have limited effectiveness simply because it is not exploring every possible realm in which a message might be present. The current enthusiasm for high powered lasers as a messaging medium, while opening more possibilities to communicate or signal our presence, also ignores the fact that a message could be written in rocks (or in meteorites), in crystal patterns or in organic material like DNA. Humans might *be* the message. (Timothy Leary, with his notion of the 'starseed Transmission', thought so. Vonnegut in his *Sirens of Titan* (1963), also riffed on this theme.)

> In fact two mathematicians, Makukov and Cherbak, have analysed our DNA and suggest that it may have resulted from such a message arriving on Earth billions of years ago. They try to show that DNA encodes non-biological or non-evolutionary information in the arrangements of its bases. Although they are unable to say what the message or signal actually is. There are some problems with their study (*The 'Wow!' Signal of the Terrestrial Genetic Code*, Icarus, 2013). For example, they claim that the codons which end genes designate zero in the decimal numerical code they have observed, but they have been unable to explain why there should be three separate versions of this 'zero'. Some of their interpretations seem less than compelling. They have noted things like triple repetitions of each numeral 0 - 9 appearing once in the signal, and believe that the multiples of the prime number 37 occurring throughout cannot be down to chance.

We may not learn that we have a message until it is too late to act on it or too late to prevent its repercussions. Even reading a likely message is risky since it might deceive for nefarious ends. Information viruses might take over our computer systems and make them work for the ETIs or messages may trick us into manufacturing things that then destroy us or turn us into slaves.

All dangers aside, what actually could we say in a message and how would that message speak for all of humanity? Is it going to be possible for us to apply what we learn from the SETI efforts in any event? Ours is the human case and our responses to the extinction threats are going to need to be human. We really need information from other humans who have survived since we will have to apply the knowledge we receive to our genotype and to our social history. And if our history does happen to have resonances with those of ETIs, then that history tells us that there is in fact no *a priori* reason at all to believe that there is a natural long term trend for intelligent societies wherever they may be to become more selfless and humane, or embody what we think of as the human good. Unbridled economic competition here on Earth may well lead to increased wealth, improved material conditions and mastery of the physical world, but it does not lead to a greater mastery of the moral and ethical sphere since such mastery is incompatible with such competition.

Even now, here on Earth, and in spite of the generations of sacrifice by individuals to liberalise human society, and thousands of years of religious practice, not only are wars are still being fought to liberalise cultures from violent and extreme political actions, we are also talking about curtailing those hard-won human freedoms where they are practised in order to make tough technological decisions to survive the perils of global warming let alone the coming world-wide panics that we face.

Whether this is a temporary phase to pass through seems unlikely given the panics scenario, and the politics of advanced ETIs wherever they are may end up as more extreme versions of the kind of society we have already seen on Earth, such as the German Nationalist Socialist Party in the twentieth century which used a command economy and brutal slavery to manage incredible technological innovation and development in the teeth of a losing war, or Pol Pot in Cambodia who erased the learning of a generation, killed much of the bourgeoisie, emptied the cities and sent everyone into the fields to begin again, or Mao Tse Tung's in his 'Great Leap Forward' that killed between 20 and 50 million Chinese peasants through deliberate famine and then followed its failures by his brutal reviving of the 'class struggle in his 'Cultural Revolution' to regain power, all over a period of 12 years.

If the descent into dictatorships seems improbable both in the light of the global liberalisation that has overtaken the world's economy and in the powers of widespread dissent bestowed on the masses by social media, then let us remember that, historically, one of the more damning products of the race to create wealth is a fine disregard for the rights and equalities of the poorest.

All objections aside, let us suppose that SETI is successful and it turns out ETIs have similar enough histories to ours to give us knowledge applicable to the human condition, what will we do with the knowledge? Many commentators, and even the strategists of political think tanks, fear we are all going to sit around all day looking for handouts and cease striving to develop.

Arthur C. Clarke observed that any sufficiently advanced civilisation would be indistinguishable from magic, so let us recall the fictional magic world of Harry Potter. In this world (in the hands of the competent magician), food can be conjured at will, cutlery cleans itself; a wave of a wand and a house is not just reorganised but built from scratch; you can get on a simple broom and fly around the world; a potion will give you luck or make people fall in love with you. Life is so easy for the adepts in the Harry Potter world that if the evil Voldemort did not exist, he would have to be invented, just to give them something to worry over and reasons to act.

> Even more interesting though would be a single arrival without any follow-up. It would indicate that the launching civilisation had failed somewhere along the line. It would lend a great deal of weight to the notion that expanding galactic civilisations are rare because extinction rates, whether down to political or technological failures or hostile stellar and interstellar environments, are much higher than we suppose.

An ETI loaded with advanced technology, and assuming it would be willing to share it, might turn our world into such a Harry Potter style world. While this 'magic' scenario is a definite possibility, perhaps more likely is something less appetising given the panic-stricken future that lies ahead for us: *stimulus*.

The flip side to ETI's technologies may be unwanted economic stimulus here on Earth. (Keep the story of Manchester and cheap coal in mind.) Unless our visitors simply hand over ready made machines and all the necessities required to use them, the very act of trying to make sense of new theories and new technologies and to make them work here on Earth is likely to spawn yet more bouts of intense scientific research, technological development and bull markets.

It may be a boom time, for some, dividing Earth's economy in two; one superior part derived from the stimulus to investment that the putting into use the new technologies will provide, and the other, a second rate economy of the inefficient and stale products of previous human growth. A return to a Stone Age for many becomes an ever more likely by-product of ETI contact.

The concepts of developed and developing worlds; the practicalities of personal freedom and social standing; education and career choices; political powers, all would take on new and very pertinent meanings. Everywhere the two stage economy would be in evidence, and your status level would depend on whether you were connected with the new alien economy or not.

The physical arrival of extra-Terrestrials would be a very bad sign for us, however, because the enormous technological growth that brought them here would bring others along very quickly after them and swamp us. As I have argued elsewhere, interstellar travel is likely to be an economic race and the results for those civilisations who get in the way may not be pretty.

In fact many of the same fears about making contact with ETIs are expressed about artificial intelligences that we create here on earth. If they are too smart, they will rapidly take over the business of invention and manufacture. They will evolve at such a rapid rate that humans will have no hope of controlling them, and will have to sit back and watch while science and the economy advance way beyond our comprehension.

The SETI institute has been set up to examine these problems and it has produced a contact protocol (see Appendix 1) and lodged it with the International Institute of Astronautics (who are they?) who keeps a record of who agrees with the protocol. While it urges full disclosure and discussion about contact before further steps are taken, you can see that in the final clause, it only *advises* disclosure after discussions among the relevant groups who claim a right over the discovery. No one is bound by this voluntary protocol, and whether the type of the contact we make ends up changing our attitudes about being frank and open with the facts is not something we can plan for.

There is, too, one other danger to worry over. The ETIs we look for may themselves be looking for a very particular something we possess: *consciousness*.

If it does turn out that consciousness is implicated in some way in the construction of reality and in extracting certain outcomes from the great sea of probabilities that forms this universe, then consciousness beings may end up being 'farmed' by others who want to fashion certain desirable realities for themselves that they find difficult to produce with their own mindset or with their own artificial forms of intelligence (or perhaps too risky for themselves). In other words, looking for a 'virtual reality' machine that they can use for their own purposes (the simulation argument for our existence is much the same; about more, later). The reasons behind this may turn out to be that when quantum computing really gets going we may find that interference from our self-awareness gets in the way of the production of usefully divergent solutions. We might be happy to bring in other minds able to interfere with the quantum computer in ways quite different to ours. So it will be with ETIs, the human mind might be a key resource for them in creating useful realities otherwise rare or unattainable, or inventive solutions to problems they cannot solve.

In addition to the simulation scenario which we will discuss in detail further on, we might carry forward this speculation about consciousnesses. Since the universe is such a difficult place to move about in thanks not only to the enormous distances but to Relativity, intelligences may need to bridge the communication gap by conquering time. Arthur C. Clark's story behind his and Kubrick's collaboration on *2001: A Space Odyssey (1968)* had the unknown alien species first interacting with humans through consciousness. This seems the correct assumption, given what we know about the difficulties of traversing space in real time.

There is darker side of this coin of reality, of course. It may also be the case that if aliens want to attack us, such an attack would also be one arising in consciousness.

What would be the reason?

The quantum choices that our consciousness makes as it starts to move out into the universe will interfere with those of an intelligence already in existence, so the need to rid 'its' universe of our 'meddling' influence on their probabilities might be an important strategic goal for them (the Dark Forest scenario again). Aliens might be interfering with human development even now. How much easier it would be to set humans against themselves and to destroy their civilisation with ideas, with anger and with a debilitation of our intelligence, than it would be to construct a fleet of warships and send them out on trips of thousands of years. *We might be the message.* A message to self-destruct. Up to now, such ideas have belonged to science fiction narratives but maybe there is more to it.

One of the implications of consciousness interacting with the universe leads us to concern about the increasingly interactive observations of it that we are making. In the rest of this book we will explore some of the non-fictional implications of these notions. For example, if the Sun requires chance and coincidence to produce its energy then sending out satellites to examine the sun and to 'infect' it with informational order from our study of it may influence its future behaviour. But before we get there let us consider another version of the Rare Earth hypothesis which also suggests an interesting cause for the existence of other intelligences, if it turns out they exist.

Our monasmotic universe

Quantum fluctuations suppose many, perhaps infinite universes, standing shoulder to shoulder with ours right now. The way our universe came into being suggests other universes exist beyond our universe's boundaries that will never be reached in normal space time and yet whose states may determine how our universe evolves into the future. There seems to be a lot of information out there that we could happily make use of if only we could get to it.

In an infinite universe or among infinitely inflating universes, universes with life are going to be seriously outnumbered by other types of universe where life as we understand it could not arise, or that came after our Big Bang and are 'behind' us in evolution – Guth's 'youngness paradox'. The chances are that we are the only species to have got this far in the evolution of universes. The probability of a universe with intelligent life is low; the probability of a universe with two separate intelligences must be lower still; and still lower for three and so on. For the SETI program to make sense we would have to consider the probability of us being the second intelligence in our universe, or even further down the order of appearance of intelligences. This is not something that appears to be a good bet if the infinite inflationary scenario is correct.

But given that we are in a universe where life has arisen, and conditions for it do not appear very unusual, many are cautiously confident that we can expect life to arise in many sites in our universe.

We may be over confident, however.

We will come to the multiverse hypothesis all in good time, but if it is true, and that random options in the universe are not strictly local but require an entire universe to produce, then it seems as if the first observer dominates probabilities for life for all the universes in the multiverse.

The multiverse hypothesis does not allow for possibilities to co-exist after decoherence. Each decohered possibility separates from its 'cloud' of likelihoods and lives in its own universe, leaving the observer in a universe of his or her continuous history. If there is any interference from these other possibilities it lasts a short time with limited effect. So essentially, at each random juncture, there is an option which heads towards life and an option or options that do not. Each step taken by myriad events towards life will eventually leave the first consciousness, where it arises, alone in its own universe, a **monasmotic** or solitary life universe. This is a strict version of the anthropic principle (which we will discuss in the next chapter). Whenever the chain of events elsewhere begins to head towards life, multiverse splitting sees to it that everything comes down to one binary choice: this maintains the path towards mind, and this does not. ('Observation' in this case means any interaction that changes a

When trying to describe the universe in quantum-cosmological terms a curious state of affairs emerges. Wheeler and DeWitt discovered that by forming a wave equation for the entire universe, time netted out to zero. The entire universe should be static; it would not evolve since there is no observer external to it to measure time against. Mach had a similar insight into the equations of Relativity. Within the universe, an observer's time or clock defines the time for the rest of the universe, so if our universe is actually evolving then there is at least an observer in it and the behaviour of the universe is responding in some way to our consciousness, (and to all other consciousnesses there might be). In a strange tweak of the anthropic principle, the universe looks the way it does *because* we are the way we are. Could this be true? After all, there were no observers in the inflationary start and yet the universe evolved from it. If there is an observer, just how much of the evidence collected for our cosmological theories could be independent of it.

quantum state; it does not have to be performed by an aware observer.)

This leads us to the reformulation of an old question, of whether the emergence of a consciousness is dependent on place or location in a universe, and whether there is only one 'place' in each universe where there is the correct confluence of influences to introduce it. (Imagine the great Computor simulating a universe. Where does he introduce life or allow life to arise?) The Catholic Church would prick up its ears at such a suggestion of a special place for humans but that is where we will leave it for the moment, except to say that location information for every component of matter is part of its probabilistic equation.

Even if the strict sequences of random choices could produce a universe that can have the possibility of minds arising in several places, the first *conscious* act of choice between options would immediately push that universe into its monasmotic state. The first intelligence reduces the options available in other lines of development that may also be approaching similar levels of consciousness, perhaps even pushing those nearest to it away from attaining it. Even if the split-off universes go on to develop their own versions of life, that life, too, would end up being alone when it attained some level of self-awareness.

We might speculate that one of the reasons why the highly developed consciousness of animals other than the primates on Earth such as the dolphin, whale, or elephant have not 'progressed' into civilisations is because the evolving human consciousness (not necessarily even the first to arise — the dolphin family appeared perhaps 10 million years ago) and its social constructs had managed to block out these others from developing further by dominating the multiverse splitting of our universe and the options available to these other protominds. (Recent work suggests this may be observable in social insects, and having implications for evolution, which we discuss later on.) Further, sequences of options are not the only principle of choice in the universe; there is also coincidence. The moment life begins to profit from coincidence, then its world begins to be shaped in directions not predicted by superposed randomness at the observation. Universes with self-awareness in them get even further apart from one another. Intelligent life would only ever be alone in its universe. (If we do meet ETIs, it is likely they will have come from other universes and not our galaxy.)

Even supposing that the multiverse separations do not leave single consciousnesses alone in their universe and that distinct levels of self-awareness might be able to co-exist naturally, these species may still compete with each other, quite unaware, for the occupation of the present moment in their universe, and for the particular shaping of the randomness in which they live. Yet another reason for being cautious about the development of artificial intelligences, which may turn out to be the definitive extinction threat for any biological intelligence by taking over the histories in the option multiverse. Humans would cease to be relevant and disappear into the past. More positively, it might be we could share our universe *only* with an artificial consciousness (rather than an organic one) since it is unlikely to share with our biological intelligence similar histories of probability superpositions of the multiverses. They would not be the same kind of competitor as minds of organic life and thus could coexist with humans.

Anticipating our discussion, up ahead, about brains in vats and the simulation scenario, we might even speculate that finding the presence of another intelligence in our universe may constitute all

> All journey times, whether slow or at light speeds leave everybody behind for good, so interstellar travel implies the break up and disintegration of a civilisation. Go off at light speed and you are alone. And I mean alone. You may not even be in the same Universe any more.

the evidence we need that we *are* living in a simulation and that minds have been created within it in order to discover or test out theories about life. We could partner them immediately in the search for the purpose of the simulation and to attempt, perhaps some independence from it. We need life first in order to have a simulator, however, so perhaps we need not fear this outcome just yet, though such a thought does give an edge to SETI.

Whether multi-intelligence universes are just too unlikely to worry about and that we can say with a reasonably high degree of certainty that SETI is wasting its time, the truth is that whatever the case may be, it may not matter at all. The Hoyle Trap is real, and we are going to need help.

But suppose we really are alone in our universe, what then?

Escaping the Final Panic

When we can think of escape in normal space terms, we think of a modern civilisation sending out vast interstellar ships for voyages of hundreds of years. We think in terms of the epic. But the reality is likely to be very a much more tawdry affair where **The Final Panic** is likely to fuel a fight between those who want the last of those resources to escape and those who cannot escape and who want to survive as long as they can down to the last bite of food, the last comfort.

Imagine the last few years of manic growth as the end of the solar system nears. The first Millennium riots, the orgies before the march of the Black Death would be nothing in comparison with this panic. The 'eat drink and be merry for tomorrow we die' philosophy would gradually give way to true desperation for billions of individuals. Think of the lily pond towards the end of its 29th day.

The 'escaping' voyagers will be discussing some interesting points about this ideal waiting time to departure. 1. Those who leave earlier than the ideal time can invade the colony of the later travellers who arrived first, and profit from their industry. 2. The travellers who leave at the ideal time may have more advanced technologies than the earlier ones so may be able to repulse an invasion. Furthermore, these later travellers will overtake the earlier ones on the way to the destination and so could destroy them pre-emptively as they do so. 3. The earlier travellers will arrive later, but so will travellers who leave after the ideal time. Both sets could combine forces to invade those who got there first. Later travellers might be sufficiently technologically advanced to win over both sets of previous travellers. 4. But will there be later travellers? Suitable destinations are far away. By the time growth has advanced enough to get there, civilisation will be on the point of collapse in any event.

The final panic will come around a predictable time. This time will not be at the actual point of resource collapse, but at a time a little before when a suitable interstellar destination for some to escape to has been identified.

Here the Wait Calculation comes into play.

For a given interstellar destination, there is an ideal time to wait for technology growth to produce a certain velocity of travel. Wait too long and technological growth will not increase velocity fast enough for you to overtake the earlier voyagers to this destination. But leave before this ideal wait time, and technological growth will produce velocities that will overtake you (up until the ideal leaving time).

Clearly, the tactical and political outcomes for the voyagers making this journey are complex and risky. It may be, of course, that several destinations have been identified in which case the pressure may be off the travellers who get to a place first. Or the new solar system may be large enough to absorb several disparate groups and colonies – for a while at least. For these colonial outposts to survive at least at the technological level they arrived with, they will have to initiate economic and population growth at a high rate, the seeds of which will have to be brought with them along with an advanced level of industrial capacity. But after growth performs its 'magic', these colonies will quickly confront each other over the remaining resources in the new system. Unity, if it happens at all, will be brief as the situation in the Earth's solar system repeats itself.

Despite the colonisation efforts, the places to escape to are few. It is unlikely that there will be anywhere worth going to within twenty light years of us. This is an immense distance and could not be reached without space ships that need solar system scale resources to make. Even being able to achieve relativistic velocities and to go far and wide in reasonable journey times thanks to time dilation is no real help at all to people back on Earth for whom growth will still be marching on to Earth's clock.

But we do not have to think in these 'normal space' terms.

Assuming, for the moment, that the monasmotic principle does not operate for us, there appear to be a lot of theoretically accessible places for intelligences to reside that the current SETI ignores. These places are zones of our universe extremely remote from us or which exist in other dimensions, or other time lines contemporaneous with us and occupied by civilisations who have already trod the road we are now treading and have survived. And there is a far-off zone occupied, *right now*, not by alien intelligences, but by at least one and perhaps two intelligences (one artificial) that are likely to be more understandable and who live life closer to our own, namely the survivors of our panics, *us*, or the super-intelligence we have made, living in the future. The information we really need right now is not knowledge of the existence of another intelligence out there but an answer to the question, how do *we* avoid the extinction traps and survive the move into a space-based economy?

With this in mind a solution presents itself. In the rest of this book we will explore the reasons behind the Chronolith® Observatory project.

It may also give us a way to escape **The Final Panic**. It hinges on how we view the notion of multiverses. We will come to multiverses all in good time, but it is worth contemplating the idea that what does not happen is still a concrete influence over what does happen here in our universe, and that the survival of humanity will be actively shaped not simply by the decisions we make here and now but by mutually interfering decisions *not made* in universes very similar to ours but in other dimensions.

Can we make contact with any of these realms?

Given the current state of knowledge, we seem to be unable to usefully explore all these other realms in our universe except one – our future.

The argument in this book is that the existence of consciousness alters the application of theories about the universe gained from studying inanimate matter. In addition to speculation about where trends in the present can lead us, there is available to us advanced information about our future, and we need to find ways to train ourselves to read this information and make use of it.

Evolution and extinction

All resource-rich civilisations elsewhere will be faced with the same kinds of final panics we expect humans to be soon facing, so they may have technologies we can adapt to our situation, including those of space travel, if they survived them. But then, if they have already survived them in good order, where are they now? The galaxy is about 100,000 light years in diameter. Exponential growth of civilisations (based on our own experience where growth is a function to $e^{\sqrt{t}}$) is likely to produce travel speeds of at least 1/10 the speed of light within say, 10,000 years from the start of agriculture. Ignoring optimistic estimates of the ubiquity of life in the galaxy, just a single other intelligence arising within an average of half the diameter of the galaxy from us and as near in time as 1/10,000 of the lifetime of the universe, would be here in person. So, it is either the case that intelligent life exists elsewhere but is unable to grow exponentially, or intelligent life more advanced than us does not exist in our present. Either way, we are at a threshold.

So if there are no sufficiently advanced ETIs out there, who can help us here and now?

Our own future selves.

It may seem that if we survived into the future then we survived the panics; we found out how, and there is no need to ask for help because we did it. But, aside from arguing that we survived *because* we, in the past, acquired key information from our future selves, there are more subtle faces to this situation.

The Darwinian argument is that there is no trend in life towards a future point. Speciation occurs as circumstances demand. If there are no alleles in a species that can gain it an advantage in changing circumstances of their niche then the species will give over that niche to another. Species evolve to fit the circumstances they are in and cannot predict how successful they will remain as circumstances continue to change.

While human survivability in space flight may be down just to technology or biology, humans will always be dominated by the social realities they evolved with and carry with them. While the specific future external threats are still mostly unknown to us, we have a pretty good picture of some of the threats from our personality and the societies it creates. Because life is built on chance, surmounting one threat does not necessarily produce a gene set that is

---

Feudalism and similar arrangements like slavery, had a good run in the higher human civilisations. Every empire, ancient and modern, was built on it. UK abolished it in 1833. Serfdom was not eliminated in Russia until 1861, around the same time a civil war was fought in the US to eliminate slavery there (officially abolished in 1864). Brazil ended it in 1871, but informal slavery carried on in parts of Africa while modern political enactments of feudalism like communism and its variants sprang up around the world, where whole nations were enslaved to the political will of the leadership and when walls were built to keep people in. Equations of *place* being crucial to all political spheres, as we shall see. (They provide military purpose, however, even as they fail as a principally military solution).

---

best suited for surmounting subsequent crises. For example, the human personality gene set that may have been highly suited to creating culture fast, through the ability to both compete as individuals and bond with its imitators doesn't seem to be able to do anything about regulating capital accumulation and the unequal distribution of resources, and readily goes to war over the disputes these activities create. Capitalism of the modern era has developed to pass off the costs of the enterprise occurred by capital down the line so to speak (Capitalism originated in the use of money to lay off obligations to others in the feudal system), so that waste and loss and destructive competition is an inevitable by-product of its behaviour. So our current gene set that initiates wars and exhausts habitats for the sake of increasing capital doesn't look like it will scale beneficially in the solar system.

There is a risk that to preserve the concentration of capital humans will tinker with this gene set in the wrong direction, by developing transhumanist work forces say, or long-lived elites, and will only make the extinction threat from resource abuse and inequality that much more serious. By using economic forces to manage our future directions rather than considering humans in the larger context of his biosphere 'upbringing', we may well fail to make the transition into a space-faring species a success.

Mark Buchanan in his book *Small World* (Weidenfeld and Nicolson, 2002, UK) mentions Karl Popper's refutation of the Marxian communist destiny in a work called *The Poverty of Historicism*. Popper simply observed that we couldn't predict the future of Mankind because its development depends upon advances in knowledge that cannot be predicted ahead of time (otherwise we'd know it already). Mark Buchanan mentions this because he is asking whether there could be mathematical laws predicting the future patterns of human behaviour. Popper's view is only that our knowledge is incomplete, but why? Just what are those 'incalculable forces that infect our daily life and direct our destiny?

The creation of artificial intelligences may be a way around the gene-set problem, since they will not have been 'forged' (completely) in the crucible of humans' evolutionary psychology. Predicting their behaviour will be an immense task, and they may well prove to have goals and scruples so different from ours that they, too, will pose a significant extinction threat. Being smart, however, these new minds may be the ones capable of making the necessary calculations to produce a gene set whose economic life will be more sensitive to an abused Earth and to exploited populations, and one that seeks to embrace equality and harmony, say, rather than hierarchy and elites (because democracy is quicker!) They may help to reduce the extinction threat from ourselves, but let us remember that they will arise in the first place because of the purposes of capital. They are likely to be part of the problem not the solution.

As we approach the space threshold, these questions are increasingly significant. Suppose we were able to breed a more highly socialised and responsible personality type, either one we made or one we found within ourselves, driving towards uniformity, and overcoming self-interest, and taking care of the Earth, such a personality type might become too inhibited to move into space in a sufficiently rapid or ruthless manner (e.g. not daring to invade habitats or to ignore other possible life-producing zones), and so expire on the Earth anyway before it had a chance to exploit the resources of the solar system. Perhaps a space economy married to gene manipulation will cause an intelligent species to separate itself into gene-based castes in order to exploit the varied environmental niches of a solar system. And yet, when those intelligent caste-based societies, successful in their solar system, are faced with the need to escape into interstellar regions, perhaps the respective self-interest generated among the castes would inhibit flexible responses or the cooperation or sacrifice needed to manage the extinction threats lurking in interstellar travel.

One can only speculate about whether speciation, natural or manufactured, is the future for humanity. It certainly is an inescapable fact of Darwinian evolution here on Earth up to now. If it turns out that economic sustainability is the only way that an intelligent culture can realise the long-term global solutions to its most serious self-made threats to survival, then any species that emphasised competition or fractionated into genetic castes would be at a severe disadvantage, wasting resources on social engineering and reacting expensively against themselves to any stimulus.

The idea of human unity is a simplistic one, however, without an understanding of how necessary individualism and innovative minds are to it. Just as the idea of competition is simplistic without fully comprehending what altruism and socialisation have done to create our civilisation up to now. The idea of letting humans run free seems just as dangerous as artificially gearing our society towards greater survivability in a universe about which we still know so little and with so little knowledge of the workings of our own consciousness.

Humans are going to have to construct their new selves as they go along with imperfect information about what kind of preparation they are going to need for the future. We may be led along paths of innovation that may prove to be an entirely damaging direction, for example: AI warfare, or the previously discussed Transhumanist trend, or the urgent efforts to alleviate the damage to our biosphere through geoengineering, and so on.

Surviving the extinction threat

Even so, as long as we remain aware of these possibilities, and we take *some* steps towards self-protection, coincidence and accident may still propel us (or any civilisation) through the space economy threshold without first transforming us. Such a civilisation may have an extended life but still be essentially moribund, living on the edge of extinction (as early *Homo Sapiens* existed for millennia), and without the appropriate personality or skill set needed to thrive beyond it.

So it is not unreasonable to predict a situation where future humans in their various collectives are hanging on by the skin of their teeth to their time. *If only*, each collective is thinking (right now), *we had had more information about the extinctions that we were going to have to face, we might have made different decisions about our preparations for the threats, different decisions about creating systems like artificial consciousness or transhuman populations that got us this far.*

If we continue our thoughts in this direction without the expectation of physics for the moment we can ponder how information flow between our past and future selves starts to make good sense.

To improve survival in a hostile galaxy, our future selves need the past to have made certain decisions. For us to survive here and now, to escape the paradoxes of growth and the final panics, and pass, strengthened into the next phase of development, we need to know from the future what those decisions should be. We need to learn how to unlock the potential for communications through time, and to stimulate the gentle, thoughtful, constructive flow of information between all our time-line selves. It's in our psychology. We find such circular narratives (reflexive time travel) in sci-fi stories, lending suspense to plot. For example, a mysterious illuminated figure appears at key moments of the stories only for the heroes to discover at the end that the figures were themselves from the future transported back in time by some mechanism.

But, outside of fictional universes, can we look through time to find out if we survived all the extinction threats and have a future? The Fermi paradox should apply to influences in time as much as it applies to a civilisation expanding in space. Whenever time travel is invented we would see time travellers right now, but we don't. So does time travel never happen, ever?

We will try to answer this question in the rest of this book.

To avoid suspense, I will reveal that time travel, of a sort, does go on, and is going on right now, and in every moment. It's the means by which the universe remains consistent with itself as it evolves. But let me argue the case out.

Time travel and the narrative free-for-all

By following two principle lines of enquiry in parallel, the quantum narrative and the biological social narrative, we can examine the full effect of time on probabilities in the history of our reality. These lines of enquiry do not run parallel, rather they weave in and out, cross referring and intersecting each other, leaving analogous traces of one in the other. To talk about humans as time tinkerers is to imply that arguments about probability, multiverses and backwards causation in the world of particles and fields also applies to them as beings. But people don't behave like particles – or do they? Our goal, therefore, is to examine the relationship of mind to time and to show how the mind differs from the substrates of reality that physics takes as its objects of study (see **Boltzmann Brains**).

In order to understand the interplay of options in the timelines of the multiverse narrative, we need to look more closely at the narrative time in the human experience and *vice versa*. I will draw a distinction between the narrative times of mental states which I will designate Internal Narrative Time (INT), and those of events that originate outside of mental states, namely External Narrative Time (ENT). (These two times will crop up a lot as we look back at our origins and the beginnings of our universe.)

We have mentioned time travel on the human scale, let's consider it further. Let us consider ENT (external narrative time) and how humans might interact with time influence in it and consider a thought experiment along the lines of the brains in vats problem (see **Brains-in-vats.**) If time travel is already happening in our universe,

> To borrow from the paranoiac. Just because there are exclusive principles governing time influence, it doesn't mean that it is not happening.

in our time line, how would we know? With what could we compare our present before time travel interfered with it?

One of the ways we might refute the existence of such circular influence is by testing randomness. Quantum theory appears to tell us that the values of states of matter and energy that we record are truly born in the moment and have no antecedents. This randomness is at the heart of the patterns of matter that we observe in our universe. There is nevertheless an organisation to these values that creates a consistency across the material universe which acts against randomness in the moment. Indeed, Guth created the inflationary origins to the universe to explain precisely this result. Inflation theories of the origin of the universe, however, do not solve the profound conflict between momentary randomness and consistent or coordinated events through time. Should this same coordination apply to the human narrative, or is something else required?

It is easy to imagine how human development, how historic events and the grand patterns of social evolution, might be guided overall by the influence arising from future actions of life, but less easy to observe interference in any individual lifetime, seeing we are less able to consider some chance occurrences of our lives as being beyond simple accident. How could we know whether the decisions we ended up making had little to do with our free will and everything to do with decisions made in other times and in other developing time lines?

If time travel is out of the box, then the course of any sociological narrative is traditionally regarded as a free for all, where any future generation can go back to any moment in time to alter the outcome of events and to undo what others are trying to do. The principal technique employed by these time-tinkering generations becoming then, the finding of obscure causal links that can be altered unnoticed by others, or, indeed, toying with coincidences (that cannot be 'refuted'), without alerting others to their actions until it's too 'late'.

So our uncertain future, and even the apparent irrationality of human behaviour, would result not from innate randomness in the present but from competing influences of other times (we will see if this may explain quantum probabilities).

Curiously, randomness implies that we 'know' more about the future than people of the future know about their past. There are some tantalising experimental results hinting at precisely this. A team at Harvard headed by Karl Szpunar randomised genuine facts from people's memories as well as non-personal facts culled from Facebook and asked subjects to imagine futures spun from them. At a later date, the subjects were given some some elements of their scenarios and asked to recall the missing elements. Whether formed from personal memories or neutral facts, the recall of what was missing was considerably better when the narratives were positive (i.e. had a continued existence) rather than neutral or negative. The theory being that emotions bind narratives together, and negative emotions tend to fade while positive ones, being necessary for future action, are maintained. We can explain this better if we replace optimism with a readiness to exploit coincidence which we will discuss in the next section. But there is, too, an alternative explanation for the experimental results, namely that they predict those elements of memory that will become involved in future real events.

For much of what we 'know', either of history or about future guesses, our predictive capacity contains fewer uncertainties than our interpretative capacity. The gambler knows more about the future outcome of the bet he has made than he knows about why the outcome occurred. Explanations are always provisional, whereas imagining forward, as it were, contains more potential 'truth' even before the contents of the imagination become real. More than in the land of optimism, we are in the territory of the 'un-truth', and we will examine its language and what it means for consciousness towards the end of the book. But even beneath the human narratives down at the level of particles that flit in and out of existence we are forced to develop the distinctions necessarily produced by quantum theory between *prediction* and *memory* and how they impact on theories of consciousness (see **Schroedinger's Cat**).

We will consider this distinction both as in fundamental components of reality in quantum terms and in imaginative narratives of higher level consciousnesses, and will use the analogies of minds from the future exerting influence over the past (like the time travel stories we tell ourselves) to equally anchor our arguments about the way the world is constructed.

Let us consider the following. If we have no 'control' against which to measure influence from the future in the human narrative, how could we ever know that these changes were occurring? Something along these lines, perhaps. Suppose, during your life, you come to the fork in the road, a binary choice, and take one fork. As you move into your future, even if the fork you didn't take no longer exists for you, it is still there, implicit in the current state of mind, and qualifying, however lightly, the choice you did make and your commitment to it. You might wonder about what could be found along the other road and about those you left to travel along it when you made your choice. Suppose a time tinkerer arrived at that fork and instead of using brute force to remove or block off the choice you took (even supposing he could do this), he or she added a *third* choice, a third virtual road, not real enough to conflict with a past already known to many. While your road would not appear to have been altered — you still take the one you took — the implicit third choice alters your perception of the current viability of the one you did take, for better or worse, and the later decisions you took or perceptions you made along it (see **Trevanian**). (This may be what does happen in the superposition of the multiverse measurement scenario, and we will look at this later.)

Many may be happy with the power to over-write the timeline. Marxist Terrorists could alter Fascist futures; economic profiteering could alter the distribution of resources or the values of currencies. Niall Ferguson describes something like this effect in his book *Virtual History*. In the chapter entitled *Afterword* he draws a very plausible picture of the direction in which the modern world could have developed from 1646 if events had happened to prevent the creation of an independent America. What his technique shows, of course, is that *any* line of development requires a succession of choices and chance happenings. It's not about how many choices there are but about the inertia of outcomes; not all choices have their supporters. This inertia results from people being locked into a set of social constraints and is why, maybe, the train of social history is so difficult for a time traveller to alter. Just any old fiddling with isolated events will probably not do enough to shift a line from its inertial track. Rarely is there a single pivotal decision that switches the points of the track and changes the direction of the 'train'.

When we think about the *extent* of consciousness this way, we can see more clearly how actions and intentions must be embedded in an enlarged system of routes in time through information and probability. The classical inhibition to which, namely time's arrow, weakens for the fields and forces at the particle level just as for knowledge derived in a social and cultural contexts.

For example, such a time influence mentioned above would appear to interfere with Darwinian evolution. It would alter artificially the balance between genes and environment, and the slow process of speciation. So the rate of speciation, as the culmination of the behaviour of many individuals over time, is certainly one area of study where we should find evidence of time influence. If we had a figure for the relationship between an ecological niche and the build up of alleles in the species that occupied it we could observe where this figure was exceeded. Certainly there is the idea of Punctuated Equilibrium (Gould), sudden flurries of speciation within the steady pace of accumulated variation. These 'flurries' still appear to occur pretty slowly, even for a time tinkerer who has all of time at his disposal. (Similarly the expansion of expansion rates in economic growth noted by Kurzweil might show the feedback in time produced by time tinkerers. We will come back to this point when we think about coincidence.) All the same, shifts in the steady accumulation of linear processes where none should be expected might reveal time influence at work. Secondly, the stability of history seems to us to be the result of the limit to alternatives (for reasons we will come to). If we examine history and find that consistent alternative or counterfactual histories also grow with time then that would be suggestive of some level of continuing uncertainty being somehow inserted into our past.

Both these ideas imply time-line buffering. If we extend the idea of Gaia, then life could develop a self-protective buffering system in order to protect itself from damaging changes from other times (see **The circularity paradox**). The sudden shifts that we notice are the result of the accumulation of time influence breaking out of this buffering and causing unlikely outcomes. This would explain the chronology protection principle (put forward by many), but as a 'soft' result of the interconnectedness of *life* rather than as a hard law of physics. Without life (and probably without consciousness), no chronology protection principle is needed; it is a function of consciousness only (we will enlarge upon this later on and produce a solution).

So far, we have been considering the active participation of time travellers or pedlars of time influence in the past. But it may not be the case that interferences through time happen because of a rational and conscious act by person or persons unknown either living in the past or the future. They may occur as an unavoidable effect of the kind of multiverse we live in where options and probabilities emerging in the coincidences of life are continuously mutable forwards and backwards in time. Before we get to multiverses let us take a look at an important phenomenon that contributes to narratives in time just as effectively as any other principle of physics, namely coincidence.

A bit about our Universe: Coincidences and life

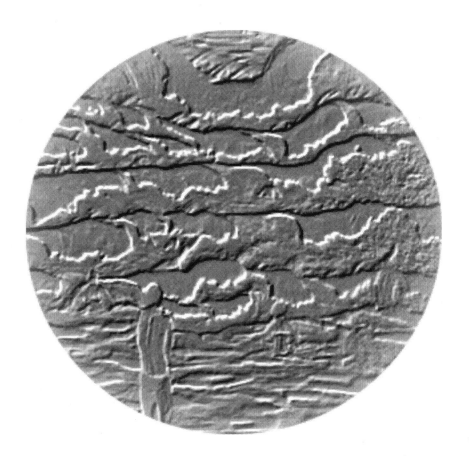

In which we examine the nature of coincidence and form a view of how the universe began and how probability entered it from somewhere else.

In which we take a closer look at why coincidence matters and how it may apply to the contents of spacetime, to evolution and the behaviour of crowds.

In which we ponder the role of coincidence in networks and social bias.

Coincidence and the Anthropic Principle

We have suggested that even without the strong anthropic principle, measurements in the quantum realm of randomness could produce universes with a single appearance of life (monasmotic universes). So let us continue our investigations into the role accident and chance play in the appearance of intelligent life in our galaxy.

We are alive to observe our universe because a set of parameters, or cosmological constants, which govern the states of matter conducive to life happen to pervade it. For some thinkers, the integration and fine-tuning of these several cosmological constants that quite by chance enable this universe to harbour life (as we know it) is so improbable, that this proves that the universe has been created just for us. This is known as the strong Anthropic Principle. It suggests that there is a prior realm in which these life bearing constants or parameters are being produced and which are directed to come together to make it possible for us to live in this universe. Certainly, if we are a simulation, then this realm exists.

The so-called 'strong' Anthropic Principle used to support the idea that we are in this universe for a special reason has had a long history. A popular argument used in the Middle Ages ran that because Man is the purpose of God's creation he must have been placed at the centre of his Universe (presupposing a great deal). Therefore, the Universe has the Earth at its centre and there are no other beings in it.

The weak Anthropic Principle is the 'secular' version of this – the pragmatist's universe where events within it just turn out to make it fit for life (established by Brandon Carter in the 1960s as a form of selection bias and building on work by Dicke and Dirac discussing the 'coincidences' of the cosmological constants). Because we are alive it is no accident that we observe a universe that is fit for our kind of life. What other universe could we observe? Fred Hoyle used similar reasoning to explain how carbon had to be made in stars. Since we have carbon, then regardless of how improbable its production seems to be given the energy levels of the constituents of stars, there must be a resonance that allows for it. Sure enough, better observations agreed with Hoyle's theory. There is a hint of The Steve Argument in this approach (see **The Steve Argument**). Interestingly, Hoyle, having used the anthropological style argument to discover how carbon was made in stars, turned it on its head when he proposed a universe with no beginning, a Steady State Theory (with colleagues Hermann and Bondi), using the 'strong' secular approach of no privileged place for any observer in it, and that it should look the same from anywhere for anyone. (Re-emerging in the argument for the principle of mediocrity for an observer, that is, most observers see the same kind of universe.)

The Catholic Church tried to impose the strong anthropic theory on the world by use of torture and the death penalty until well into the 17th century, and then was able to announce in the 20th century, through Pope Pius XII, that the Big Bang theory was functionally equivalent to what they had been saying all along that there was a Creator of a universe just for us beyond time and space.

The differences in physics between the strong and weak anthropic principles are quite significant. The strong principle implies the existence of higher level narrative in action behind, or before, the beginning, which guided the creation of ours. For example, cyclical expanding, contracting or separating and colliding universes, and where our universe has inherited some characteristics belonging to them. The secular version suggests only that universes evolve and some evolutions create conditions right for life which fits with with quantum thinking which denies all 'priors' of anthropic notions to suggest that the set of characteristics our universe needed for life was put in place only after events were set in train to establish our universe. This is the foundation to the origin-of-the-universe debate: either the values that our universe has have existed prior to the Big Bang or they are created in it (and hence the difficulty of reconciling quantum theory describing a reality constructed 'on the spot', as it were, with the cosmological 'history' involved in our universe's creation).

The strong anthropic principle is problematic because of what it implies about probability in our universe not being truly probable. Whereas the weak principle at least implies other universes exist because our universe is created as one of a set of coincidences of properties, and thus we can use our privileged position in one Universe with life to infer other universes *must* exist. (Quite simply, coincidences are necessarily supported by the existence of a large range of events that do *not* form coincidences for us.)

Critics of the weak anthropic principle, however, use variants of the Gambler's Fallacy to argue that we cannot use one example of an outcome to say that our universe is one among many. The Gambler's Fallacy argues that a long string of heads in coin tosses must imply that tails or a string of tails must become increasingly likely, or the retrospective GF , where a long string of heads must have meant a previous long string of tails. The inverse Gambler's Fallacy (as described by Ian Hacking) is similar where it would be wrong to assume that to enter a room and observe someone throwing a double six implies that the throw must have been one of many and not the first because the chance of throwing a double six on your first throw is 1/36 and these odds are the same on every throw, if the throws are independent.

We do not need, however, the misapprehensions of an organised spread of likelihoods in these fallacies to criticise the weak anthropic or coincidence principle. If we include a spatial element to probability then the meaning of coincidences in a universe of randomness suggests a way of deciding between them.

Coincidence, as we understand it at the human narrative level, requires an unpredictable re-exchange of some level of information between separately developing causal chains that happened previously. In other circumstances when events 'coincide' we can call it contingency. This first contingent event acts as a pre-preparation for the information exchange of a (possible) later coincidence, which, in some cases can be the basis for a form of Bayesian likelihood. This definition of coincidence involving some form of repetition means that there is limited predictability in the information derived from the coincidence. A coincidence is not merely an event with a low but strictly calculable probability of happening. Throwing a die to show a six 10 times in a row is rare but not a coincidence. Throwing two dice to both show sixes is a contingent event. Throwing the same double again would be a coincidence. (Some measure of predictability can be made only with bounded sets of events like dice throws).

Suppose two people simultaneously throw a dice on neighbouring tables and they keep throwing identical numbers, then you could suspect one of three things about this 'contingency'. Someone is playing a trick on you, or there is a deeper law of chance you know nothing about but which runs against the laws of chance that governs the throw of a single die, or you can suspect that there must be so many other tables where dice are being thrown that the case of two tables agreeing, although rare, is likely to turn up somewhere and it just happened to turn up next to you. These coincidence, therefore, give a lower bound to the number of tables that there must be *if* randomness is the same between tables as at every table.

The strong version of the anthropic principle essentially means that the numbers of tables have no relationship to the dice values, namely the coincidence of suitable parameters does not depend upon the relationship between tables and dice throws, whereas the weak principle requires that there is a genuine coincidental relationship between the numbers of tables and dice values.

We do not know the bounds of the values of the cosmological constants that determine the physical properties of the Universe we live in, but they do seem to be independent of each other. A small change in just one of them and we would not be here to discuss it. So *if* the values of the cosmological constants can vary then all of them being just right for life is a coincidence and there must be many other values creating distinct Universes that evolve differently (lots of combinations of tables and dice). Our particular universe is thus determined in a distribution of values of constants that characterises all the beginnings of all the universes which share the same inflaton field.

(There must be a lot of them. Brian Greene argues that the number depends upon the precision with which we can measure the constants and this produces a fantastic number if all the universes with a distinct values are realised. Many, of course, will not work properly or they will vanish quickly.)

> In infinity, the likely is just as unlikely as the unlikely is likely.

The weak anthropic principle and the idea of coincidence, as we shall see, is supported by the current favourite explanation for the origin of the universe, namely inflation, originating with Guth and developed into the infinite chaotic inflationary scenario. In this scenario, a random fluctuation in a field called an inflaton field gave the compact sum of all fields a large energy value which then expanded rapidly under the gravitational potential in that region. (Even though there doesn't appear to be 'room' for such a fluctuation, strictly speaking. Guth got around the problem by positing an inflaton field that is just numerical values (for example, imagine if a wave height out at sea could suddenly be very a large number). Later on, the parameters or constants give anchors to the residues of this expansion like matter and radiation, and fluctuations in the resulting fields create galaxies and other concentrations of matter.

Knowing how those different values of the constants are distributed would, curiously, reduce the effectiveness of this coincidence argument. If, for example, we knew by other reasoning that the cosmological constants do not vary very much in viable universes (the principle of mediocrity again), then we would have to accept that we belonged to a *set* of universes in all of which life is likely. So a simple random distribution of variables among universes no longer accounts for the existence of ours. The question of why such a group exists needs to be solved.

> Viewing coincidence this way we will see that its effects actually run counter to ordinary probability. It undercuts rational calculation of odds and informs statistics with a constant background level of inexplicability.

As it happens, Jaffe and a team from MIT worked out that such a set might exist. They examined how life might be possible under various changes in these constants and showed that there seem to be a number of ways in which the interactions of the constants could produce matter and energy suitable for life – given what we know about life. Other workers have even found solutions to creating matter without the weak nuclear force (the force that governs particles of the nucleus, mostly quarks). So it does seem that our universe's existence does not quite totter on the edge of unlikelihood as once thought and could belong to a set of universe *versions*, with the same concepts of particles and quanta but held together in different amounts. Coincidence and the weak anthropic argument still hold but without the imperative of uniqueness; we have to explain the set.

As we will show, this form of argument works only if the probable choices actually exist (that is to say, if there are actually 'tables' put out where values are expressed and universes are made from them), because if they are not real then we are back to the strong anthropic principle and randomness doesn't count. If we don't want to make any special pleading for our universe (the secular AP), then we can happily infer that all types of other universes must exist. (We will come back to this thought and ponder *where* these other possibilities could exist.)

Coincidence and its consequences

The most startling thing, however, about coincidence is how it reflects back on the hidden fundamentals of probability and consistency in the universe.

Let us go back to the dice tables analogy. The coincidence of dice throws on our gaming tables has to be supported by a vast number of tables. If these did not actually exist then in what sense is the outcome *probable*? With just two neighbouring tables, simultaneous dice throws on each that show the same face every time start to look less like a coincidence and more like a pre-determined event, a *pre-destined* event, pretty soon. The longer the sequence of doubles goes on between the two tables the more questionable is the result. However, we can make such a coincidence continue to conform to the same rules of probability on each table by simply adding more tables.

One way probability and space is connected is through force fields. A force involves the appearance and disappearance of space between objects. Movement in time requires space. Travelling close to the speed of light, however, means the disappearance of space in which to move, hence less time. But wait, does less time mean faster velocities or slower velocities? Depends on from where you are looking. The faster you move the less time you have in which to move. *Hey, Zeno. Are you here?*

This is easy to show. With six tables in a circle there's a chance of at least one occurrence of two similar faces turning up side by side (given any face, the chance that the next is the same is 1/6. Let's make these throws simultaneous since if we threw each dice one after the other the last throw lies between two tables and has a 1/3 chance of agreeing with at least one table it sits between). Let that first pair define our neighbouring tables. The chances of the coincidence of the same double appearing again on these same tables on the next set of throws will be 1/6 for the first table and 1/6 for the next = 1/36; almost 3% – not very likely. If that coincidence did show up on your six tables and continue to show up, you would very quickly suspect that something other than probability was affecting the dice.

But, with more tables, say, 36, we see that the estimation of the likelihood of the same neighbouring double occurring returns close to certainty. As long as the number of tables (throws) has increased by the time you throw again, the continuing coincidence doesn't call into question the natural law of probability that governs each dice throw (just like a long sequence of heads doesn't necessarily defeat the odds of 50-50 if you know the dice is fair). The dice throws are, of course, independent of one another. Doubles could be occurring elsewhere as well. We are looking at a specific stable distribution in time, a history. Only if you remember the history of previous throws will the gaming room staff have to keep putting out more tables together with their dice and croupiers every time the neighbouring croupiers continue to throw the same face on their dice (if they want you to keep on believing the house is honest); the gaming room has to grow (exponentially in proportion to n^2), and *in advance* of the event, in order that the probability of the outcomes between each table are conserved and thus maintaining the validity or consistency of the values. Thus the numbers of tables is directly proportional to the number of throws, in an honest universe.

If the number of tables doesn't grow then the length of time between coincidences also diminishes (unless the dice that are in play are loaded). Thus coincidence (in this case of dice throws) helps us establish the nature of the probability laws that are governing the throws of each non-related sequence. If the gaming room staff put out too many tables, however, coincidences over a given time would have to decrease again. If 648 tables are put out for the second throw and only the self-same coincidence turned up that would also call into question the probability on the two tables since more coincidences should have emerged. Thus observed coincidences gives us some idea of the balance between causal chain links and the 'space' in which they can physically occur in time. Our starting assumption is that the universe has just as many tables at any time in its history as it needs for probability to be correct, and which is why it needs to grow as the numbers of 'throws' increase (but the same balance might not be maintained everywhere). Thus, the observation of coincidences in our universe seems to imply an exponentially expanding universe, but which must happen before some kinds of events can occur. Which is what appears to be the case. The universe expanded first, then came the coincidences. This thought experiment also tells us is something profound about probability. Without a *memory* of the sequences then what happens at each table need not conform to any probabilistic process at all and no one would know what a dice throw reveals about randomness, which leads us right back to the Strong Anthropic Principle. Once you start rigging the numbers of tables, then randomness on each table cannot be recovered. The whole universe would be nonrandom. Thus an expanding universe seems to also imply a still existing past as the necessary balance to future outcomes of probable events. (The re-acceleration of the universe may imply that there is some serious universe-wide event ahead in our future at about the same distance in time from now as it began in our past – ≈5 billion years.)

> If you were a 2-D being living on a flat surface and wanted to do the equivalent of toss a coin. You would have both separate faces on the plane and you would have to push one face somewhere along the plane for other 2-D beings to then find. In their case, probability would be related to how long they took to find it - the minimum path.

If, instead of dice, we think of quantum states and the probabilities attached to randomly evolving states occurring together, then we can see how, as the number of possible observations or measurements increases, the space to house all of them must also grow. Probability means nothing unless there is a space in which each option or choice can occur. Curiously, the Bell Inequality experimental results suggests that these probabilities do not exist in advance or separate from the measurement. Has it been misinterpreted? About more, later. (The multiverse approach gives a whole separate universe to each option which does not seem to solve the problem of the option space applying to the measurement. We are coming to multiverses a little up ahead.)

This illustration reminds us that for there to be a measurement, states have to share information; they have to be in a neighbouring situation where the wave functions of the measured and measurer coincide; where the measured acquires momentarily the probability space of the measurer. For information to be exchanged and an observation made the spatial and energetic needs of the measurement require what is effectively a coincidence. (For this set up to agree through time as well across the tables at any one moment, we say the system is ergodic. Non-ergodic systems do not play equally across time as across the moment.)

Continuing in this vein and considering that measurement is analogous to neighbouring doubles on the dice tables turning up, we can see that meaningful observations require priors and a future space. Putting out more tables is a way of extracting new information from the system. With regards to the tables it is tempting to think that a coincidence between them is the highest value of Shannon-type information that we can extract from the dice throws because the actual pair is completely unpredictable.

Coincidence values apply throughout the universe, so we should expect unpredictable correlations at every level of magnification, in groups of galaxies, within and around galaxies, in groups of stars, within and around individual stars, within and on planetary bodies and so on down to molecules and subatomic particles.

Consider Hydrogen oxide – water. If it were not for the 'accidental' hydrogen bonding *between* its molecules resulting from the distorted shape arising out of the electromagnetic forces in the bonds between atoms that enables it to be water rather than a gas and which gives it the bonding structure to dissolve a range of other molecules it would not have the useful properties it has to enable life to develop. Another example of such coincidental effects is the way phosphorous is used in living systems. Phosphorous is not very abundant, is mostly stored in rocks and is difficult to extract for use. Yet life chose phosphorous as its key element for reproduction (as well as energy production in cells). It is missing 3 electrons in its outer shell and which enables it to make a number of bonds with oxygen atoms, the most important of which is the phosphate group with four oxygen atoms in both ionic and covalent bonds allowing for distortions in the electron shells. Because of these electrostatic possibilities, DNA uses the phosphate group to make its backbone on which the nucleotide 'rungs' are attached, while the distortion in one of the oxygen bonds provides extra charge to keep the DNA static in solution so that reactions can take place in a regular way. The 'rungs' are formed through hydrogen bonding between the nucleobases, and which allows it to easily separate in two and rebuild itself automatically. It can thus be easily engineered as well as be self-repairing. The fact that these bonds separate at between $70^0 - 90^0 C$ suggests that even if life somehow began in thermal vents it quickly moved away from such energy sources. This would broadly speaking give us an upper bound to the planetary conditions in which phosphate-based DNA molecules are likely to arise. Hydrogen bonds between atoms of the same molecule work their coincidences in other ways too. It is they who create the 3-D folding of proteins and allow for the easy addition of various groups onto different sites on them to produce workable variants in cells.

But consider too, as the universe expands, then the scope for coincidence at larger magnifications increases. Connections between structures in the universe would start to grow across larger distances and volumes. Could this at least contribute to the observations we have on the existence of dark matter?

Coincidence in the universe

Designating probability volume is not the same thing as saying that in an infinite universe all things are possible (the Law of Large Numbers). Rather, we see how the interactions of random events are embedded in regions of probability volume which they share and gives them their opportunity to occur. A range of only slightly different likelihoods of probable states does not need the same 'amount' of space in which to occur as a few highly unlikely states. When a coincidence does occur, we should be able to find the amount of space (e.g. the number of tables) which allowed it to happen. Random events occurring in a tiny probability volume will dominate coincidences. Whereas a vast probability volume for even the most unlikely pairing will allow coincidences to dominate.

From an entropy and probability volume perspective, entropy is not just a matter of energy spread over all the degrees of freedom of the phase space of the system, it is a process dominated by the dispersal of energy *in space*. Which is why gravity gets to dominate processes and create black holes and why space must expand to accommodate more states. Indeed, probability volume must grow *in order* for entropy to act. Space that does not grow either in extent or in numbers of dimensions while time continues cannot produce increasing numbers of dispersed states. In our local regions of the universe, therefore, while the gravitational attraction of matter overwhelms the expansion of space, we have an additional dispersal principle, entropy. Entropy is not a field or a force but a distribution, and as such we can treat it as information (although there are critics of this approach).

When we talk of entropy going from order to disorder we are really talking about moving to a state where *positional* information is more random, i.e is spread over more space. A measurement, being then, the moment where, at the very least, *positional* information is shared at the juncture between components of the event. Space and probability are corollaries of each other. This can be demonstrated with a physical object like a coin. In two dimensions, it is not possible to flip the coin to make a choice. One face and one face only remains exposed. In pilot terminology, it only has the possibility of *yaw*. Only if the coin can turn on all and any axis (that is, when it can yaw, roll and pitch) in a free space that is at least a sphere of its diameter can we observe the fullest possible randomness in the coin toss. Only in three dimensions can the coin exist in an alternative state.

For a two dimensional object placed randomly anywhere on a two dimensional surface, the probability of it being where it is, for observers outside the surface, is proportional to the area of the surface divided by the area of the object. Whereas for a 2-D creature on the surface, the probability of coming across the object is merely the time it takes to travel to reach it plus something for the length of time it stays where it is. The more improbable an event the longer it takes you to reach it. There is a hint here about time that we will get back to.

Even in three dimensions we need time to get alternatives out of a coin toss, so the 4-D of space-time is required for a coin-toss to express a random choice between alternatives. Thus the probabilities we observe tell us the kinds of shapes of space in which events occur. The shape of the space sufficient to provide for randomness will depend upon the type of event, but for any event if the space is not sufficient for at least an alternative then random choice cannot be present. In quantum theory, this space is described by a mathematical approach called the Hilbert Space, a space of endless dimensions in which all possibilities in time

---

As we increase dimensions we increase the scope of probability, which is why string theory and its many dimensions may have something going for it. Eleven dimensions is a lot of extra space in which probability volumes could be stored. When we go from the boundary of the expanding Universe in towards the quantum small, dimensionality seems to increase: 2-D (boundary) ∪ 3-D (curvature) ∪ 4-D (space-time) ∪ ... (quantum 'foam'?)... ∪ 11-D (Strings?) ....?

---

and space for a particle state are 'housed'. Useful for mathematics but what does this say about the human narrative?

In a measurement, the connection between an event's probability and some measure of the spatial dimensions in which it occurs suggests a hierarchy of spaces are involved in the realising events. This general 'envelope' of events we will call the probability volume.

For the moment this is just a conceptualisation of probability. Is there more to it? At one level we might say that since the *im*probability of a given succession of random events increases extremely rapidly and thus the space to house them also increases, the quickest way to accelerate the expansion of space is to fill it with unlikely options. Similarly, the quickest way to quick-start a universe of events is to produce a vast probability volume first such that rare coincidences dominate randomness. Could this be what lies behind the inflationary episode of our early universe, or even provide the basis for a unified field theory??

Can we decide which came first, the probability pressure to drive the expansion, or the spacial expansion that introduces probability? One can form a view that the universe's rate of expansion must be balanced with the rate of the global increase of entropy within it. As we saw with the dice tables, too much expansion would imply a high rise in entropy, while too little would inhibit its evolution.

> Is there a universe where all probabilities are laid out such that no interesting pairing occurs for any observer? An insufficient probability volume to express all possible options in a universe would allow improbabilities to linger unresolved for the longest time. Would that be a bigger or smaller universe than ours?

We now know that the normal or basic expansion of space, that is, the setting out of more and more tables, is accelerating. What might be a cause of this given our discussion of dice and tables.

Suppose one dice was loaded. Say the number three was loaded such that the odds of throwing the three was 1/3 and not 1/6. The odds of throwing a double three would then rise to 1/18 (instead of 1/36). Having twice as many double threes turning up means there are fewer other doubles in the same space. In fact the odds of throwing any other double change to 1/45 (that is, the five other numbers on the loaded die have to have 2/3 probability spread among them equally). Our casino universe would now have to have 45 tables put out for all the occurrences of the other doubles to be seen naturally. When we look at multiverses and probability, we will describe how the universe might be becoming a region of loaded dice where certain probabilities become more likely through a subtle mechanism of 'handshaking' with the future. This idea is something that the Bell Inequality experiments may unwittingly be exposing. A notion we will come to in due course.

When I squint at these ideas of probability and space, I glimpse dark energies and accelerating universes. But further, I can see that if probability volume is not conserved (i.e. no more tables are put out) then the density of coincidences will rise to a plateau (possibly a high number) and not change thereafter.

One of the ways that probability volume might not be conserved would be if there was a physical limit to the universe and to the total space permitted to it, or that the universe was shrinking, where we would observe a rising or high static coincidence number, a paradoxical lower rise in local heat due to greater negentropy. A low Coincidence Number would be proof of an infinite or expanding universe, or at least one with increasing of space and one which allows greater dispersion of states through entropy.

We know from other calculations that our universe is almost flat (in a geometrical sense) which suggests an infinite prospect for it, and it is expanding which gives us a clue to the value of our Coincidence Number (in some ways those two statements appear to be at odds with one another). The notion of 'flatness' has its problems. Was the universe always 'flat' such that in earlier epochs a smaller universe had less mass to maintain the same density, or is the flatness we now see an *evolved* parameter and is a function of the universe's history? If the density has been constant over time then we are faced with the problem of how the universe adds mass over time. If the mass has been more or less constant then we have the problem of how to account for its continued expansion when its density is dropping.

As we have been arguing, an acceleration in expansion implies an increasing probability volume. As far as we can tell, a fresh acceleration in our universe began around 5 billion years ago or close to the time our solar system began. Stars born at and after this time could have evolved quicker than we think, passing slightly more rapidly through the stages that our Solar System is now going through and giving rise to intelligent life in remote galaxies who would have easily reached the Kardashev type III or IV level of advancement and be visible to us. We see no effect of such life evolution in our universe which is another reason perhaps to think that our universe's overall Coincidence Number is low and that humans have arisen at a key moment in time in the life of our universe. One is tempted to think that this expanding era in the history of the universe may be more important than the Big Bang and that we may discover more about the universe's origins from studying the era of the birth of the solar system than any of the eras before it.

Kardashev Civilisations scale considers a scale of power consumption of civilisations. Type I as those who can control their planet ($\approx 10^{16}$ watts); type II as those who can control their Solar System ($\approx 4 \times 10^{26}$ watts); type III as those who can control their galaxy ($\approx 4 \times 10^{37}$ watts). Others have added classification IV (control of the universe) and V (control of sets of universes).

We can use coincidence numbers to characterise *awareness* in a universe. Universes that begin with higher coincidence numbers (higher negentropy) could become spontaneously aware early on in their evolution (see **Boltzmann Brains**) whereas universes like ours with low coincidence numbers evolve awareness slowly over long periods of time.

An entire universe of awareness could be considered to be a 'creator' if certain sequences of fluctuations occurred such as to retain its 'self' and become capable of action and intention, arranging matter and energy to its will, whereas ours may never have such a 'god'. But if there were a way in which a neighbouring universe of spontaneous awareness could somehow interact with ours, like, for example, initiating a Big Bang within the confines of its own universe (rather than one emerging out of quantum fluctuations), then we might have an explanation for our distinctly local God-the-creator. But still, this does not provide any reason why we would actually need to obey such a 'god' in the evolving of our universe, since it would have arisen spontaneously and not out of any design, and it wouldn't be the *solution* to our particular problems or even answer the question of what came before. Curiously, there may be evidence that this principle of spontaneous creations of 'minds' is real (see **Appendix III**). A matter we will discuss when we come to consciousness and brains.

From this point of view, the coinciding of parameters, or cosmological constants (weak anthropic principle), in the make up of our universe may not be fully random after all. With independent variables, we have to suppose that there is a systematic distribution of their values (e.g. Gaussian) rather than purely random, otherwise there is no trend for any one of them to arrive at a value suitable for life, still less that a whole set of parameters should arrive together at the appropriate values suitable for (our) life. (Double sixes don't have to turn up in infinity, although it's likely they will, but their non-appearance would not necessarily contradict the probability in each throw.) Even in infinity, where the parameters evolve in such a way as to take on the entire gamut of possible values, there is no reason to suppose that the right coincidence between them for our universe would ever arise. Still less if we are at the beginning stages of infinity, which might well be the case (see **Infinity**). That is to say that independent values which can vary through an infinite or near infinite number should never coincide.

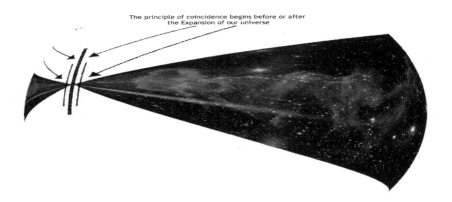

The principle of coincidence begins before or after the Expansion of our universe

The most interesting feature of the coincidence, or weak anthropic principle, argument for the origin of a universe fit for life, however, lies in what appears to be ignored by physicists and philosophers alike, namely the assumption that the coincidence of parameters regulating the initial conditions of the universe, either before or after the Planck epoch (the smallest moment beyond which we cannot describe the universe), occurs at that moment of time and no other. We shall discuss, instead, the idea of a principle of coincidence that permeates our universe and which has an effect on the larger assemblages of matter distinct to those of the smallest scales of matter and energy. It appears to us that the coinciding of events at the atomic and subatomic levels is not prescribed by the same randomness which governs the internal states involved in a coincidence. The uncertainty in subatomic interactions is fully circumscribed by fields and probability wave functions in which particles are both the created and the creator, but this is not so in larger matter assemblies where the quantum wave length is very much smaller than the matter dimensions and accident begins to dominate event outcomes. One starts to wonder about brains. Is the kind of coincidence produced in the origins of universes, the same as those that happen in consciousness (see **Boltzmann Brains**) and the human narrative level?

Coincidence in the Human Narrative

Perhaps we can be sure that coincidence is involved in our universe's origins since it plays such an important role in the everyday human narrative within it. That is to say, the coincidences we experience (in the large agglomerations of matter) are the result of an actual *property* of our world, just like all the other constants that characterise it. They come from somewhere and may reveal significant characteristics of the origins of our universe and the ultimate direction its evolution will take.

It is often mentioned that in any group of 24 or so people the odds are around 50% that any two will share the same birthday (the calculation is often attributed to Hans Bethe). I have never met someone with my birthday (of course, I don't always know the birthday of those I meet), nor has there ever been two people in my school classes who have shared a birthday (not a very random group). Clearly, with 367 people the odds are certain (there are only 366 birthdays). Successively remove people from the group and the odds do not decline as dramatically as one might think. All the same there is more to this 'coincidence' than meets the eye, as we shall see.

At the human level, narrative coincidences are not, as we all know, at all rare. In fact they are pretty much a given, and are very exploitable by the humans who experience them (both positive and negative ones).

Here's an ordinary one that happened to me. A long time ago now, in the previous century, when my father died, I took it into my head to lose myself in the world, and went off to South America where I knew no one and could not speak the language. I trekked alone in Chilean Patagonia and then began to work my way up Chile heading for Peru. In a coastal town, I crossed a street looking for a barber I had heard would change money when I heard my name called. Sun-burnt, bearded and unrecognisable – or so I thought – I turned around to see a lawyer whom I had met twice perhaps in the London office I had worked in the year before. He was a Chilean and was visiting his parents who had a vineyard in the valley. He took me to lunch with his parents, we picnicked on the beach with some other members of his family. He offered me the chance to live and work in Santiago. Had I taken him up on the offer, no doubt my life would have taken a very different turn. Not taking the offer or taking the offer, the chance encounter might have been the pivotal moment in my life. Who knows?

In our world, stories of meetings in extraordinary circumstances are so common, it's hardly interesting. They happen to everyone all the time. I had met others on my trip – gauchos, Welsh grandmothers, lost language teachers, but only on the basis of contingency.

> Coincidences vary in 'strength' or power for change. A simple chance encounter is probably quite low on the coincidence index simply because there is unlikely to be any pre-encounter preparation. Higher up the scale would be an encounter with someone about whom you were already thinking. Higher yet would be those encounters that assist you in an intention, in getting something done. And so on.

Those meetings had no priors; they were not coincidences but simply part of life going about its business.

Here's a more unusual example of a coincidence. A Dutch friend of mine in Seville called Peter met a Japanese student there and they struck up a strong friendship. Nobu, the student, returned to Japan after two years away to bring his fiancée to Spain. She had bought a Japanese guide book (out of many available) to the whole of Spain. She brought it with her and Nobu later showed it to me. In the section on Seville, there was a single photograph of the Giralda tower. In the foreground of the picture, striding out towards the camera, was Peter.

Nobu told Peter later that he had been completely floored by the fact that his fiancée had already seen his best friend without knowing it.

It's worth examining this coincidence a little more. The picture is unusual in that if you had not seen it in a travel book you would think that it was a holiday snap of Peter. He's in the foreground, in full view, left of centre, striding towards you, smiling, looking straight at the camera. The fact that it is in a travel book illustrating a tourist site *be-*

hind him is part of the oddity. The editorial decision to crop the picture this way is also part of the coincidence and if we knew more about that decision, I am sure we would find more to add to this tale (see the aside).

The likelihood of Peter appearing in the (presumably Japanese) photographer's frame at the moment the snap was taken is remote but not impossible since he does live in Seville. The appearance of the photograph in the guide book does have some kind of remote but positive likelihood. Anyone at all, even a member of Peter's family or one of his friends, might have seen the tourist book by chance and been tickled by the thought of Peter's presence in it, only it would not be a coincidence for them – a surprise perhaps but nothing more. After all *somebody* is likely to be in the photograph, and that someone will have friends and family. But Peter has the chance of a more meaningful coincidental *connection* with the photograph, because he has the Japanese half of the connection already in place; he is *pre-prepared* for it, that is his first contingent meeting then allowed his world-line to develop in such a way as to make use of the coincidence.

There is an interesting coda to Peter's coincidence. Nobu, now married, came to Seville a year later bringing with him a second edition of the guide book. The same photograph in the second edition has been re-cropped to de-emphasise Peter's prominence. The field of view has been zoomed out to show more people. If you didn't know he was in the picture you might have easily missed it. So what happened with the first edition, the one bought by Nobu's fiancée?

It is not hard to see that these two stories actually embody two kinds of what we call coincidence. In the first, we have what may have been a pivotal moment in the direction of my life. But nothing else particularly distinguishes the coincidence from the many in my life similarly structured. We will call this a type 1 coincidence ($^{1}\mathcal{N}_{C}$). A re-crossing of paths may or may not lead on to something, and the internal readiness to profit from it plays a less significant role in the outcome for both. In my case, the accidental meting in La Serena led on to seemingly nothing more than the appreciation of its rarity (although we shall see if that is all it is).

But in the second story, there is a different structural complexity involved which cannot be predicted. The result is not an intersection of choices so much as a consolidation of an existing relationship between four players, and which we will call a type 2 coincidence (^2N_C). This coincidence creates a consequence different to that of a ^1N_C type because out of four world-lines none of them could have produced a relevant probability for the event and which is spread across the four world-lines and has a higher relevance to a change of state in all of them.

Here is another example of what I am getting at. A similar story is often told whenever coincidences are discussed. A man, John Smith checked into a hotel he had not stayed in before and after signing the register he asks flippantly if there is any mail for him. The concierge reaches into the pigeon-hole of his room number and pulls out a letter addressed to a John Smith. What a 'coincidence'! A letter arrived for a previous occupant of that room also called John Smith just after he had left. So far, this is just a type 1 coincidence, a straight-forward accident. If, however, the present John Smith opened the letter and found it was about a subject that he was involved in, like say, something pertinent to a business deal he was in the middle of, or about someone he knew then that would be a ^2N_C coincidence. We will come back to the greater significance of these type 2s in a moment.

A level of pre-preparation for coincidences that are special to us makes them possible but also limits their number. It could be said that our lives are nothing but pre-preparations for coincidences (towards the end of the book I will introduce the idea of intersecting cosmologies, centred on people, which do this job of 'preparation'). We all carry around with us a constellation of options or virtual connections to myriad events that only need a coincidence to make actual. In some ways this is what the modern social media recognises as its benefit: without the pre-prep, there would be no actions derivable from the accidents, from the serendipity found in their pages. (There is more to these 'effects' of social media, see **Networks.**)

Often, coincidences are not really recognised until after the fact, until one looks back. It is something we do from time to time, celebrate the series of accidents that helped to define our lives, our loves, successes and failures.

Although I find it odd that even if it is true that our lives *are* shaped by the culmination of interwoven events in time, we still mistrust the random and the uncertain. We thirst for determinism and a sense of destiny and purpose and when it comes to our lives. It gives our experiences of coincidence a frisson of excitement.

The Cosmology of People, and the Time Travel Solution

The 'what if I...' or 'what if you...' can drive us crazy with alternative scenarios, with the realisations that what we are now is the culmination of so many simple little accidents. Once begun, this kind of reflection can make us very unhappy or perhaps exhilarated, seeing that nothing in our lives appears to be driven by any kind of certainty.

Returning to that crossing of paths in La Serena, Chile. The necessary pre-prep was, of course, knowing the lawyer in the work place and deciding to lose myself in South America rather than anywhere else I did not speak the language or know anyone. I rejected the pathway into the Spanish-speaking culture that had opened up for me then. Yet here I am now, married to someone from the American continent, making my life in a Spanish-speaking culture. Curious? A coincidence, or something more determined? What if influences at this level of causality really did go both ways in time? Could I consider my life *now* as the pre-prep, the laying the ground work for that coincidental meeting long ago in South America and which is still working its influence forward through time? Witness the use of it in this book. ^2N_C coincidences connect backwards after all.

We are used to thinking of life as a progression in time where events in the now contribute to our later-in-time state, our later works. But I cannot help but ponder what it would mean to have the writing of this book not simply give meaning to those past coincidences but to *determine* them. How powerful a work could this book actually be to so dominate the events of my life in *the past*. This idea is not the same as the paradoxes of circularity. Rather it is about paths of creation or nurture that move back in time as well as going forward. Let us hold that thought.

A world of no-coincidence

We can ponder what might happen living in a world where these types of external narrative coincidence do not exist, and where no wildly improbable happenings would be pertinent to our expectations or intentions. (We shall assume a consciousness has arisen in this world since without it there can be no observer correlations with events.)

Such a world would be a place where everything is so mechanical that no one would think anything much of coincidence since no coincidence would produce astonishing outcomes, and neither would there be those rare combinations of events that lead to disaster. Murphy's Law (about probability in the ENT always working against you: if there are a number of outcomes to an event, the one that causes you the most trouble will happen), and Maloney's Law – the probability of an event happening is the inverse of its desirability – would simply reduce to the tautology that the probability of an event happening is the probability of it happening.

This world would require a kind of anti-prediction in our day-to-day bets. Suppose you were at the race-track betting on a race. If you knew every connection that every bettor had with the race, you might be able to predict the outcome since the *least* sentimentally connected bet would be the one most likely to win. Odds would be strongly mathematical, and the odds for the favourite would sincerely indicate the horse with the best chance (but which would also include a small component of anti-sentiment or boringness). There may turn out to be coincidences in this world – it's not completely clear that there could be zero coincidences (evolution might grind to a halt) – but there would be no usable or fertile coincidences.

A world *completely* without meaningful coincidence is probably impossible, but there may be worlds of very low coincidence. Interestingly, on the basis of the coincidence theory, a world where no remarkable coincidences happen for anybody should actually exist, which means that people living on it could never use the same argument of coincidence to infer other worlds. So that there should be at least one place in creation where intelligent life really believes itself to be at the centre of a single Universe.

For two thrown dice to come up the same is a coincidence and the standard probabilities that describe its likelihood are merely provisional. The dice do not share a wave function (they don't appear to interact) and the way a dice falls is not a true probability in the sense we will discuss later on.

But in this world of no-coincidence, such a double would be very unlikely to occur, or would tend to occur where it couldn't be exploited. So, backgammon would be rather dull since you could safely position your pieces on any repeated distance knowing your opponent would need a double to displace you. One can see right away that if certain dice throws (like doubles) don't happen then other combinations must occur more frequently leading to a lot of repetition. We will come back to this notion.

Coincidence in our daily life here on Earth means that a certain amount of convergence is built in to the way the Universe unfolds, and that our consciousness is geared to expect and to find coincidences – good or bad – in daily life. It proves that we are born and raised in this 'space-time neighbourhood' and that we belong to it in a way we could not in any other. Without coincidence, maybe we would have only a mechanistic view of nature, and never devise a worship of god since 'miraculous' coincidences could not be an argument for godly intervention in our daily life, or for spirits to be entwined with it.

Coincidences for us are so integral to life, it is hard to imagine a world where all meaningful coincidences were eliminated. In this other world, would we be justified in thinking we were in a different universe than the one we now live in - one less friendly, less charming, one with fewer pleasures and with less of our interests at heart? I think the answer is probably yes, we would feel we were in a different universe.

Many sensational coincidences can be found on the Internet, and at least one researcher, David Spiegelhalter (Cambridge) runs a web site collating the experiences of coincidence of the general public. (It is interesting that his German name means 'mirror holder' in English. Is that a coincidence?) But conclusions are elusive because although contingency and type 1 coincidences are ubiquitous, type 2 coincidences are the most fertile but are less easy to recognise because their significance often has no relevance to any other situation since they require pre-prep. Indeed, if they were more easily captured in the lens of theory and experimentation they would not be coincidences.

Worlds fit for life

By distinguishing between ^1N_C and ^2N_C coincidences, we can start to zero in on the most important coincidental property of the world we live in now. We can see that the separate 'tuning' of cosmological numbers to produce a universe fit for life is a multi-dimensional coincidence of a ^2N_C variety, and begs the question, if other worlds exist because of a coincidence of cosmological numbers, could it be that any world that has life must be arranged in such a way as to have at least this class of coincidence? In fact, as we will see, the development of consciousness may require ^2N_C coincidences or at least a certain density of them. So the observations of coincidence may be all we need to show that there is a mind involved in the origins of the universe.

Much is written about what appear to be unlikely astronomical events that conspired to form not only a stable solar system around the Sun but also a stable earth with a moon of the right size and with geophysical conditions suitable for life. Whether or not the precursors of life rode in on a comet or asteroid, even with life going, it is easy to marvel at the extraordinary luck of the later 'Goldilocks' style meteor impacts – neither too big nor too small – that helped life along its evolutionary path to humans by extinguishing vast numbers of creatures and turning all the ecological niches upside down. But even these kinds of the extraterrestrial 'coincidences' are only $^{1}N_C$ coincidences and are an essential component of stellar activity throughout the galaxy and (one presumes) beyond.

It is the $^{2}N_C$ coincidences, however, that seem to have the most impact on the development of civilisation. Early humans were lucky enough to lift civilisation into a higher realm by making use of deposits of highly rich metal ore found on the surface of the Earth and scattered there long before humans came on the scene. In some places, lumps of copper or gold would be lying on the ground or in stream beds,

> Technological and economic progress began first with found metal, then with the gradual understanding of how to extract the metal from increasingly lower grade ores. Then came commerce and the opening up of long range trading routes. Hand in hand with trade marched technology and the more profitable methods of exploiting the resources. All of this cultural evolution depended, upon the accidents of plate tectonics, stimulated by the collision of the Earth with another body that also produced the Moon, on patterns of erosion, and also on the curious luck of the draw that revealed to humans spreading out over the Earth these metals at key points in their development.

the result of fortuitous geological upheavals and erosion patterns or even more likely as the result of early meteorite bombardment. If such workable metals had not been so easy to find in the very same regions where agricultural societies had begun to form their cities (gold mining first began around the Black Sea perhaps 7000 years ago), it is possible that the development of civilisation would have progressed more slowly or turned in another direction. For example, early American civilisations did not have iron. Neither the Mayans nor the Incas had it even though they did smelt ore for gold. (Although some northern groups in the Americas had bronze at the time of the discovery by

Europeans. In which technological direction would they have gone without the arrival of humans from the Old World?) The fauna of the continent did not give the people there beasts of burden or draft animals for ploughing. The lack of many of the coincidences that helped Asian civilisations advance were lacking in the Americas. One can speculate long about the direction of human development if Asia, too, had also lacked such helpful coincidences.

Outside of mutations on the human genome (like, say, lactase tolerance which enabled peoples to profit more from the domestication of cattle), which are naturally ^2N_C coincidences, one has to marvel at things like the coincidental hybridisation of grasses right in the middle of the *Homo Sapiens'* North European range that produced the heavier-eared wheat that needed human management to flourish, so kick-starting agriculture.

The argument suggests that the Coincidence Number will also indicate a number to evolutionary change, and indeed such a value has recently been discovered by researchers at Temple University (published in 2015) who created a new tree of life using molecular map of species diversity to show that the mutation and diversity 'clock' beats at around 2 million years. New species appear and get established every 2 million years, in spite of local contingency.

Another curious set of ^2N_C coincidences is the entirely independent evolution of compounds beneficial to human health rising in the flora and fauna around the world. The co-evolution of foodstuffs beneficial to us is not surprising (it's called survivor bias) but the presence of beneficial herbs and compounds in plants that only were discovered to be so useful much later in human evolution certainly is. There are poisons and poisonous creatures too, not to mention viruses and bacteria – another mystery since there are estimated to be 800 million viruses for every square metre of the Earth's surface and yet so few are harmful – but what are the chances that humans would have the sufficient pre-prep to be able to make use of, all over the world in places they had never been before and which had never been part of their fundamental diet during evolution, such a marvellously sustaining herbarium for the treatment of physical disorders, wounds and diseases? (Think of aspirin, quinine, digitalis, opium to name four natural remedies from which modern drugs have been prepared.)

Is it possible for there to be other planets where evolution has not produced a similar level of beneficial diversity for intelligent creatures? Could this be the missing reason why other intelligences have not been found? We are even more rare than we dare to think in that beyond the simple accidents of cosmology and planetary morphology, we arose in an environment that has a set of non co-evolved characteristics that has been, by chance, so benign to us, so full of beneficial coincidences. (Is that the Catholic Church smirking in the wings?)

Further to this, however, is the relationship between coincidence and diversity. In environments where there are fewer divergences from the mean, there will be higher levels of (less meaningful) coincidences between them. So worlds with less divergent systems or simplified environments will have higher coincidences and thus evolve more rapidly but only down a narrower road. If the average world is a much more simple environment than Earth say, it will thus exhaust its resources more rapidly or fail to reach intelligent states. There is an inverse relationship between coincidence numbers and diversity: higher levels of coincidence may well mean less intelligence.

Suppose it were the general case that planets do not tend to evolve such wide ranging supporting ecosystems for intelligent creatures then, regardless of the coincidence level, ascent of intelligence either takes a very long time breaking through into mastery of its environment or is inhibited from developing beyond a certain level of complexity through this lack of a breadth of helpful environmental coincidences. (Which is to say that evolution requires a background level of diversity *beforehand* to go far). Our hypothetical world of *no* coincidence might actually be the most widespread version of basic life, and our much rarer human version is not

> Events that cannot be traced to any one specific line of causality will happen simply because two or more causal series are coincident in time. (Uh oh, the shadow of Hume is falling across this paragraph.) And that much of the shadowy world of faith and expectation is actually a fair understanding of this property of our Universe. Religion, miracles, faith healing and all the panoply of mystical beliefs; homeopathy, auras, and the whole range of alternative a-scientific health therapies; *Tarot, I Ching*, runes, tea-leaf reading and all the beliefs of alternative causalities (perhaps even the placebo effect) are actually proof of the coincidence factor with which this world is constructed (see **The laws of large numbers, synchronicity, accidents and coincidence**).

just the first in the galaxy but in the whole set of universes common to our origin (anthropic reasoning by coincidence writ large).

While coincidences boost the rate of expansion of life, since sustaining influences tend to leave more individuals alive over time (due to procreation) than deadly influences leave dead, it is the ones that are more unlikely (resulting from diversity) which move evolution further along.

So coincidence (especially ^2N_C), far from being a rare and unusual chiming together of events that have independent causes, is a determined effect of living systems and a necessary contributor to evolution.

We might go further and say that all the universes implied by this argument by coincidence will be slightly different from one another and can be characterised by their **Coincidence Number** with which they begin life, and that this initial coincidence number may determine what can evolve in them and even what kinds of happiness are possible in them.

While this coincidence number belongs in the set of parameters that determine both the physical make-up of our universe and the levels of coincidence possible in ENT, it is not a constant and must also evolve as the Universe gets more complex, and is thus related to entropy.

Can we find a coincidence level as a parameter of ENT here and now on Earth? We could get seriously into maths but maybe there is a better way. *Six steps to Kevin Bacon* is an entertaining game echoing research done by Milgram and Travers in 1969 and later that appeared to show that, on average, no one person in the US is more than 6 connections (degrees of separation) away from any other. (Random people in the country were asked to get a letter to a random person in another part of the country by sending it to someone they already knew on first name terms.) The game explores the idea of a small world phenomenon where, in this instance, any actor can be linked, through their film roles, to Kevin Bacon. The game requires a group of players to try to connect any film actor in history to Kevin Bacon as quickly as possible and in as few associations as possible. (The game was devised by three college students, Mike Ginelli, Craig Fass, and Brian Turtle in 1994.)

One could link a bit part actor in a very small movie with Kevin Bacon in the following imaginary example. The wife of a sound engineer who worked on a movie with Kevin Bacon may be the hairdresser of the mother of this actor who played a spear-carrying role in this minor movie (4 steps).

The game is an amusement, and permits 'connections' of a tenuous sort, allowing Type 1 coincidence to stand in for connectivity. The search engine Google revived the game in AD2012 by adding a Bacon Number feature to its web search whereby you can ask Google to find the number of steps to Kevin Bacon for any celebrity. President Obama's Bacon Number is 2 apparently. Kevin Bacon's number is, of course, zero. Original game lore had it that the most connected actor was Rod Steiger. This was not because he had played more roles than anyone else, but because he had been in more varied production circumstances than other actors and had more *coincidences*.

In the real world, human connectivity is clearly supported by genealogy. Humans have limited numbers of ancestors and it has been estimated that it is highly likely that the person standing next to you is your 6th cousin (that is, six generations back you shared grandparents).

> Families get very numerous and very diverse very quickly. Almost all families in England have relatives in its far flung colonies, connecting vast geographical regions together. So Milton and Travers' work does not necessarily prove anything much more than that nodes of social connection are reproduction-based. This would not change in a world without coincidence.

So people are connected to others (not necessarily to Kevin Bacon) in a causal way through reproduction regardless of other kinds of networks. In fact, the Coincidence Number and reproductive capacity arising through evolution are likely to be tightly linked, since it is one of the propositions of this text that the relative strength of our presence in the time stream depends upon both.

But social connectivity is not as smooth as human lineage (though human lineage is pretty 'lumpy' too). Our connections cluster around nodes – people who are so very much more connected than others and have, therefore, higher numbers of long range connections. It is these random long range connections that is the key to rapid connectivity throughout the universe.

But network connectivity is not coincidence. In fact it has some inverse relation to it. The higher the Coincidence Number of a system (or a universe) the fewer determined (or maintained) connections any one node needs and probably has. When coincidence is in play, and especially those of type 2, then long range connectivity is less required in networks and is less of a determinant.

This is born out in the real world. Many connections between you and a celebrity (not necessarily Kevin Bacon) in the six steps game can be made in fewer than six steps for most people – typically in only 3 or 4 steps – if coincidence is used (rather than being on first name terms with your links). I can, for example, somewhat freakishly, link myself to ex-US president Barack Obama in four steps or less in at least two different ways. It so happened that once I attended a private party where the Queen of England was present at a table next to me. We exchanged glances. Surely this connects me, and connects anyone who has met the Queen, of whom there are thousands, to just about any world leader of the last 50 years, in two steps. But I can also get to Obama in 4 steps another way. A sister once traveled on the same Concorde flight as the then Arkansas governor Bill Clinton. His wife, Hilary, was (at the time of writing) Barack Obama's foreign secretary.

Detectives might employ such ideas of the extended circumstantial evidence that coincidence gives to connect suspects with crimes, although, as we have seen, a coincidence is observer dependent, so the suspect's end points will be difficult to assess, although the central points of the chain may be readily observable since they only link to each other. Since coincidence requires an extent of pre-preparation one should be able to eliminate coincidences that appear to connect suspects if they do not have the pre-prep to exploit the coincidence. 'Previous form', is often used by police in assessing suspects though that is not sufficient pre-prep for a chain of coincidences in a new event, and happily courts do allow such information in evidence. The pre-prep would need to include at least previous exposure to links of the chain.

But let us suppose for the moment that the real world connectivity that Milgram and Travers investigated does give us an idea of our world's coincidence factor at the current level of the human narrative (ENT). Then, if we take, for an observer, any end points that can be connected to him or her through 6 physical links or nodes, then we will probably find that they can also be linked by at least three points of coincidence for that observer. If the coincidental average lies between 6 and 3, then this is very suggestive that any two events within the same class have a physical linkage when 6 coincidences of the same hierarchical level (i.e. consciousness) between them could be demonstrated.

Maybe we can also say that the moment at most six coincidences can be observed linking two events an actualisation between them is highly probable. This might explain why we are so ready to believe in conspiracy theories, and are so happy to accept the inferential logic in circumstantial evidence in courts of law. Such coincidental linkages undercut all attempts to lie and deceive, and without which we would have little ammunition to counter deliberate strategies of deception and mis-direction.

The work of Milgram and Travers shows that despite actual and literal causality, the disparate connectivity of coincidence in the human narrative is also an expression of the state of our universe. I draw the distinction here between the small-world phenomenon where connectivity in a network is determined by the occasional long range connections between widely separated nodes (where the distance between any two nodes of a network varies as, *distance α log (number of nodes)* and reminiscent of Shannon information) and coincidences which do not depend solely on distances between nodes but on mental 'space' as well. In our discussion, the number representing type 2 coincidences may hover around a value of 4 regardless of situation. We can also see that simple proximity in time and space is not always simple contingency and has as much power to create as any self-conscious intention.

We will hold that thought and come back to it

Why does coincidence matter?

Archduke Ferdinand was killed in Sarajevo, precipitating World War I, because a (powerful) *coincidence* repaired the assassins' plans that had failed all the way up to the coincidental moment (see **Time Tinkering**).

When you think about a world with no meaningful coincidences, you realise that profiting from coincidence is not a matter of human intelligence but of some quality of *awareness*. Coincidences can be taken advantage of by any living

Anthropic Argument revisited... Let's turn it around and say that we caused the universe, because it is consciousness that defines coincidences. We say that our consciousness came into being because of the coincidence that produced our universe, but it could be the other way round. Since inanimate matter does not create coincidences, it is us or the presence of awareness that caused the coincidence in the first place. Another wave of confirmation coming from the future to consolidate events of the past.

creature in the struggle to survive. Fitness for an individual might necessarily include a measure of coincidence in the individual's timeline. In any group there would be good and bad coincidences, where some individuals lose out and others profit, contributing to an acceleration in the selection of types.

You might say that coincidence is just a way of describing life. Surprising things happen; some are predictable but others aren't and that's life. But there is more to it. Type 2 coincidences create outside *interference* with events unfolding according to their own developmental or probabilistic pattern. This interference may have its own pattern. We will come back to this. But first let us consider what this interference means in a process like evolution.

Suppose a tiger lives on a tableland above a jungle. Down in the jungle live one tribe of monkeys. They live within a genetic landscape and have a gene distribution that relates to the ecological niche in which they live. On the tableland lives another set of monkeys where tigers prey upon them. The tableland monkeys thus live on a slightly different genetic landscape and have a different energetic relationship with their niche. All of which can happily be described in Darwinian terms.

Suppose, one day, after heavy rains, a tiger slips and falls down the cliff right among a group of jungle monkeys. Suddenly the jungle monkeys' genetic space is liable to shift onto a new path simply because of an incalculable event. The tiger slipping is perhaps just an accident in the story, but the monkeys' gene set and whether or not the tiger catches and eats any of them provide the (type 2) coincidence. The local equilibrium has been severely disturbed by the dramatic event. The tiger may not catch any of the monkeys he confronts that day, but his presence may serve to 'prove' an allele in the genome of that group of monkeys which helps them avoid the tiger and force his attention to other monkeys. Without the test of the tiger's presence, this allele may not have spread so quickly or been the basis for future speciation.

There's every reason to think that if we considered this role that chance plays in evolution then we might expect greater diversity over the genetic landscape without much rhyme or reason to what eventually emerges. The late Stephen Jay Gould was one palaeontologist who believed that the contribution of chance to evolution was the determining factor in the kind of creatures that arose. He called the factor 'contingency', and thought that if you re-ran evolution many times contingency factors would produce completely different results each time. Consciousness arose out of chance, and may not always come about in each evolution re-run.

It seems to me that Gould was only thinking about simple accidents and the Type 1 coincidence of crossing paths, and was not including the Type 2 coincidence of separate development patterns having a 'mean-

> The rate of speciation must include a factor for the coincidence number of our epoch. Perhaps surprisingly, we can also say that since the level of coincidences are a constant influence in a given environment, the punctuated equilibrium argument that demands rapid change from isolated groups is less compelling.

ing' beyond mere circumstance. The existence of actual convergence in evolution seems to counter his belief in simple contingency. For example, the similarities of marsupials and mammals, and their equivalent diversity; the fact that the patterns of skeletal structures, especially in the limbs remain so similar across species that pure chance cannot be such a determinant as Gould imagined. An impressive list of such convergences made by Simon Conway Morris in his study, *Life's Solution: Inevitable Humans in a Lonely Universe* (2003) was designed to refute Gould's position. But Morris, however, a specialist in the Burgess Shale explosion of life forms, is a committed Christian (like Sheldrake) and used the list to argue for the existence of God (the Strong Anthropic Principle or causes without coincidence). We have no need of such a hypothesis when the universe's Coincidence Number does the job for us. (You can also find a list of convergent traits on Wikipedia but where the idea of 'trait' has been broadened to include 'resemblance')

I suggest that convergence in evolutionary pathways is, in part, a result of the Coincidence Number for our universe and our place and time in it, and that the actual composition of the genetic landscape of living creatures results not only from their behaviour within an environment but also from the subtle influence of the interaction and cross-influence of causal chains that are exploited by individuals depending on their internal states, *their pre-loading*. The numbers of different pathways to a similar trait, for example, is likely to give us a good idea of our Coincidence Number during the years of life on Earth.

Some astrobiologists think that Earth's biological diversity could be improved upon with different planetary conditions like other arrangements of continental land masses that would increase ocean upwellings, and so forth, and thus have greater chances of producing varied lifeforms. But these scenarios do not take into account convergence in evolution on each trophic level. For example, in our own history an episode of high energy, and high growth with a unified land mass, no polar ice caps and connected oceans produced a reptilian/dinosaur dominant epoch lasting around 135 million years during which mammals, arising within around 10 million years of the dinosaurs, did not develop significantly. From our argument, however, the degree of biodiversity may not be the only requirement for the production of higher forms of life; coincidences are also required.

Suppose that regions of the Universe end up being profoundly shaped by chance by coincidences that begin to differentiate regions and to channel the behaviour of each region in distinct ways. While there may be consistency at the atomic level across the universe divergent behaviours are really only likely to become apparent in ENT. If we had other *intelligences* to observe, we would discern this regionalism, the result of local distinctions throughout the Universe. From this angle, the homogeneity of the universe is another count against their being other intelligences in our galaxy, assuming such local distinctions have had to time to develop over the universe's lifetime.

It is true that evolution is blind and nothing is inevitable about the creatures that it produces. Accidents, mutations and the geophysical environment are all involved in selecting the phenotype. However, it is also true to say that a creature's 'readiness' to exploit chance (rather than reject it because it is not on the 'program') is also a survival factor, and that selection must favour those creatures who begin to make choices about how to profit from coincidence. Evolutionary psychologists discuss a similar but slightly different idea referred to as nonrecurrent adaptive challenges and ponder how these can influence genetic inheritance. By adapting to coincidence, however, we see how dealing with non-probabilistic events can enter the genome and be passed down through the generations. This adaption in humans is going to be less related to 'intelligence' than to something more like psychic vigour, and will not be mapped easily to intelligence parameters but rather to the overall states of awareness. In fact, we should see the incorporation of coincidence usage into intelligence, beyond our grudging acceptance of its role in general creativity. (Disorders of which might be reflected in gambling disorders for example where coincidences are preferred to actual probability, or other predictive disorders found in the OCD spectrum where rituals may be designed to eliminate coincidences.)

Indeed, the more complex a creature and its behaviour the greater capacity it has for making use of coincidence to drive the selection process in conformation with the level of 'benevolence' of environment, and so, too, the higher the degree of consciousness, the more distant in time and space may a coincidence have meaning for it. In other words, higher forms of life begin to embody a pre-preparation for beneficial coincidence and what we might term 'short-cuts' to certain development states, and which may eventually give rise to a relative acceleration in evolutionary change.

> One can think of an analogy with the kind of resonance that occurs during nuclear fusion in stars which enables carbon to be produced in otherwise unfavourable circumstances.

We have talked about coincidence requiring pre-prep individuals to have meaning, but given the evolutionary advantages for individuals to acquire the capability to exploit coincidence we might also expect to evolve forms of redundancy in our bodily systems to resist coincidental *damage* by developing increased or anticipatory pre-prep. Indeed we find this in the human brain, often explained simply as a result of its complexity, though without an understanding of the pathways that encourage its ability to transfer functions from one area to another or to do without whole bits of it, or even to function in cases of hydrocephalus where all the normal physical mapping has been destroyed because the tissue of the brain has been spread in a thin layer on the inside of the skull. Recent research, however, seems to have found recurrent connectivity in neural networks that seem to allow generalised states to be applied to novel situations, giving the mind the capability to react to new phenomenon not previously adapted to and which may explain why it is less reliant on brain centres and their physical arrangement within the brain and why it has proved difficult to map brain regions with function. As our argument develops, we will suggest that the existence of recurrent communicating networks (where essentially, output nodes are used as delayed input nodes) are even more signifiant to consciousness and its involvement with time. As it is with memory, which appears much more spread out than we expect from models we have built up to now — even into the organs and nerve nets outside the brain. We will get into this further along, but will note here that, aside from individual brain states, the ability of humans to profit from being in social groups and find synergistic energy from them must also be influenced by types of recurrent networks, some of which are characterised by coincidence.

The speciation rate is the result of a tussle between DNA's reproductive consistency and likelihood of a mutation having survival value. 2 million years represents the summation of all coincidences throughout all the levels of influence.

Resistance to damage is one of the properties that coincidence can generally give to life. It is analogous to at least a minimum of buffering in living systems. Although the concept of Gaia is about the self-regulation inherent in living systems, there is a similar positive feedback in coincidence to support life.

As coincidence runs counter to probability by forming connections not predicted by the strict likelihoods of occurrences, then we might expect to see coincidental benefits turning up in even in the molecular pathways associated with living systems. In fact we do see this. Speciation, for example, should be a very slow process (since mutations are also damaging to organisms) and yet the rate of speciation ticks regularly at about one new species every 2 million years (see above). Such a steady rate of emergence can have little dependency on adaptations to ecological niches or on a purely internal atomic rate but on a process that is outside the demands of both. The possible molecular and shape combinations in molecules say, like RNA, essential for reproduction, are enormous (about 10^{23} possible versions for a 100 bases sequence, according to Schuster), and far more than could have been explored by life up to now. The idea is growing that a number of versions of many important proteins in the reproductive system can do the same job and that mutations in many of them need not be damaging at all; they can simply be neutral, allowing successive generations of the proteins to 'explore' the configuration space of the molecule while keeping the systems working, and allow coincidence to work its selective magic. In fact, this neutrality or 'inertia' may be just the quality that allows proteins to be repositories of memories in neurons by *not* causing significant changes to the operations of the cell when storing information. (It is essentially the 3-D possibilities in proteins arising from hydrogen bonds that allows for the storage of information for use in cognition.)

Since proteins in the cell, especially those based around the coding and decoding of the DNA, are all connected with each other, beneficial alleles can be produced very much more quickly through type 2 coincidence. But there is also the interchange of genetic material from other organisms like bacteria or viruses, observed among lower organisms, which has only recently been recognised in human DNA. These 'mergings' will also be connected by the Coincidence Number to the time-line and the interpenetration or cross-linking of chains of causes. All which has implications for artificial intelligence and which we will examine later on.

Coincidence and intelligent life

While it is tempting to think that the net effect of coincidences in evolution, at first sight, should be neutral; that the sum of their positive and negative influences will be zero, we have noted the asymmetry inherent in the situation where positive influences enhance reproductive lifetimes while negative ones tend not to and so disappear (indeed, evolution works precisely because of this asymmetry). Complex living systems take decisions in the direction they believe will benefit them. Certainly this means survival or the fulfilment of a need, but it also means 'satisfaction' and this becomes more relevant as awareness develops. In any being with consciousness, an accident can also have a psychic and behavioural significance beyond the mechanistic (or Pavlovian-esque) event.

Positive and negative coincidences in nature combine to widen the spread between survival and non-survival in each round of 'competition' and selection, strengthening actual gains and rapidly eliminating an actual loss. (A genetic example of this 'selection of the positive' might be the phenomenon of hybrid vigour.) Further, a coincidence of environmental

---

Coincidence and its impact perhaps is covered more by something like the 'Law' of Unintended Consequences, a tongue in cheek appreciation of the lack of our ability to predict all developmental lines proceeding from our (honest) attempts to solve problems (an old proverb, *'the road to hell is paved with good intentions'*, also covers similar ground though in the negative). Some consequences surprise us with their unexpected answers to problems while others surprise us with their ability to complicate solutions or do actual harm that was not designed.

---

factors may lead to a permanent change in a local environmental niche which will affect fitness for all who rely on the niche but which has nothing to do with the overall distribution of those environmental factors and which cannot, even in principle, be calculated.

In general, the state of 'readiness' of the 'higher' organisms to act upon coincidences can be characterised in successive realms of action along these lines.

genetic endowment + kin relationships + groups + local environment + universe + original randomness

(There is an element of the random operating within each realm, but the final realm of random is that which brought the universe into being in the first place and which may still be influencing us in the here and now.)

With rising fitness connected to coincidence, in a universe of high levels of coincidences or fortuitous resonances, evolution will tend to be less divergent than in our universe. In a high coincidence number universe, fewer variations of intelligent life might evolve rapidly in many places, use up their resources before they can solve the problems of interstellar travel and die out. All the same such intelligent life would probably feel very comfortable in its universe and confident of its destiny. Such a universe would accelerate the development of life, while other universes, with coincidence numbers lower than ours, would have very long lifetimes with low probabilities of high intelligences emerging. Once life did emerge, however, we should expect it to develop at an increasing rate as useful coincidences also rise in line with the increasing awareness.

Acceleration is the most outstanding factor of the last 100,000 years of human cultural evolution and the engine of which may be coincidence. It has only been 5000 years since the pyramids, and already humans' consciousness reaches beyond the frontiers of its planet and will soon be sensitive to coincidences in the local space environment. The search for ETI and our eventual planetary explorations are going to be sensitive to coincidences in this enlarged volume. As we observed, however, the relationship of coincidence to diversity is important to human evolution. If diversity diminishes before evolution can produce answers to interstellar travel then the likelihood of humans crossing this threshold will be more remote.

Yet, since intelligent life does not appear to have developed rapidly in our galaxy (since earlier civilisations would be visible), we can guess that our universe's Coincidence Number is probably on the low side, which suggest high intelligence but low expansion rate.

The law of large numbers, synchronicity, accidents and co-incidence

Should we consider coincidence as nothing more than the law of large numbers which says that given enough time, space, dice throws or whatever, any unlikely event will turn up at some point? (An incorrect view of infinity which we will come to.) Littlewood's 'Law' is an example of this kind of idea by saying that on average everyone will experience a 'miraculous' happening at least once a month. His calculation is used to explain why people believe in the supernatural. In fact, Littlewood merely defined an outstanding event at 1 in a million (a number plucked out of the air), with events occurring at one per second, and then estimated how many hours an average human would be alert to experience a million events, which turns out to be about a month.

There is nothing compelling about this – not a *law*. The mechanical calculation leaves meaning out of it, and Littlewood is wrong to suppose that it is the sheer number of events which is the key factor. As I argue, coincidence is the number of occasionally crossing independent causal chains that is the significant number – the Coincidence Number for our universe and moment in time – rather than a simple large factor. Our thinking about coincidences suggests that the law of large numbers is an insufficient tool to explain the impact of coincidence on the development of life.

Another idea put forward to account for coincidences is Jung's idea of Synchronicity, and which Jung used to explain why

> Maloney's Law. The probability of an event is the inverse of its desirability. The more we expect something the less likely it is to turn up.

related events occur without there being an apparent causal connection. Jung ended up thinking that there was a field of order, a hidden causality, behind all kinds of experience and that meanings in coincidences reflected an underlying 'wholeness' to creation that humans should aspire to perceive. Synchronicity has come to be synonymous with coincidence, and Jung speculated, with his brilliant intuition, that it was as important as ordinary causality without knowing how it came about. It is not, however, what I mean by coincidence. For Jung, the elements of the coincidence were brought together by a hidden narrative on some higher plane, and the meaning we see in it is just the meaning of this other narrative which is shared in every such coincidence.

Whereas I am describing a different picture. In the ^2N_C coincidence, meaning is derived from the pre-prep, from experiences and memory particular to the individual. A coincidence begins a pathway rather than representing a pre-established one. Coincidence throws the individual into relief and gives him or her more significance, whereas synchronicity, in contrast, has the effect of de-focussing individuals and subordinates them to an underlying harmony in the way the world works that we don't properly perceive. We will look at this assessment further on when we examine consciousness and panpsychism. The sensation of synchronicity may be only a result of a bias in perceptions which we know occurs from studies of the phenomenon called 'confirmation bias' where people invest the interpretation of events with their expectations. Psychologists describe this form of cognition investing meaning in accidents as *apophenia* ('abnormal meaningfulness', Klaus Conrad, 1958), the desire to remove the random from our lives. (We buy a new car and suddenly the roads seem full of similar cars.) Synchronicity, however, while going some way along the analytical road to coincidence, does not explain the evolutionary outcome of coincidences, since, while we may or may not use them according to our pre-loading, they cannot be interpreted by our expectations – they are a surprise.

Jung built upon the notions of Paul Kammerer, a zoologist in the early part of the twentieth century who was a compulsive noter of coincidences. He came up with the idea of seriality (The Law of the Series) where unknown waves of influence cause events to occur in groupings, but did not differentiate the significances in coincidences and had no theory as to how these waves of influence came about. The fact of the coincidence was the most important thing. (Kammerer also believed in the inheritance of acquired characteristics, and was believed to have faked experimental results to prove the notion and, who, when challenged, committed suicide.)

Jung did not draw the same distinction I have made between between accident and coincidence (between ^1N_C and ^2N_C coincidences). An accident tends to be a type ^1N_C occurrence in a causal chain with only immediate and usually local significance. An accident may or may not be entirely random and it may or may not involve your consciousness (you may play a role in bringing the accident about) but it is not in general something that widens connections or develops a pathway. Accidents are those events for which there is little or no pre-loading to provide lasting significance to the event. Accidents do not give you any more opportunity not implicit in any of the causal chains you are embedded in. (Your car accident, though, may be an event that produces a type 2 coincidence for someone else but not for you.)

Accidents happen at all levels of seriousness from ponds of microbe-filled water disturbed by a footprint, to asteroids crashing into the Earth, to stars disturbing solar systems and so on up the hierarchy of regions and structures in the Universe. Each has its coincidental significance only for the volume of life that it disturbs. Accidents of inanimate material are, on the other hand, just that. They are at best neutral re-arrangements of matter. ^2N_C is not neutral to life so the relationship between ^1N_C and ^2N_C will give us some clue about our future evolution in this universe.

It is only for *life* that the type 2 coincidences become a resource, and have significances beyond the moment. It's worth labouring this point. Coincidences need to be included in the likelihood of microstates in External Narrative Time more than they need to be included at the subatomic level, because subatomic particles live in a world of their own making and where interactions between them *destroys* randomness, or at least uncertainty, rather than creates it. (This is not to say that coincidence has no role in subatomic behaviour given what we know about the Hoyle Resonance, and we will say more about this later (see **The Bell Inequality**).

The popular view of quantum theory finds it difficult to explain an apparently stable world formed by the probabilities of the sub-atomic world. But this view is the wrong way round. Our world at the human narrative level is bathed in a vast field of significant probability and accident, while subatomic interactions are destroying probabilities all the time. A measurement at the atomic particle level stops entanglement, selects a universe out of the many interfering at that event and pro-

> Philosophers have been trying to reconcile the motive force of creation with free-will and choice ever since a Creator was first thought of. How can there be a plan to creation running alongside free-will and probability? How does the Creator keep in contact with his creation without there also being being pre-destination?

duces a quantity. This is not so for life at the ENT level, where the balance between the random and the inertia in time-lines is very much more complex, and where coincidental events between causal chains is not simply a contextual effect but one that can lead to great differences in outcomes for consciousness over time (see **Trevanian**).

It is the huge effect that chance can play in our life (ENT) that we humans are so aware of and which gives us the most trouble psychologically. How do we make choices in a random world? Science has not really helped us here and people are now turning away from it because it hasn't (yet) given us the answer we so badly desire (we will come to an answer towards the end of the book). We invent religions and gods to persuade us that life has meaning; that Nature is benign; that the pains of life will be compensated by pleasure. Against randomness we place destiny; to combat bad luck we pray for miracles; to make a choice we borrow from what has already been chosen. Our cry from childhood at random injustice is, 'Why?' So when a happy coincidence turns up we are only too ready to seize upon it as proof that we doing something right, that have made the right choice, that we are on the right track, that this is our destiny. And maybe we are right to do so, not because it comes from some higher being's involvement in our lives but because coincidence is a physical property of our universe *beneficial* to our awareness.

Which brings me to Boltzmann's brains. Boltzmann's brains are really about the origins of the universe but the notion seems to belong here because a Boltzmann brain arises out of what are essentially assumptions about the law of large numbers. Our universe appears to have begun with very low entropy. Other universes, probably the majority of them, start their big bangs with higher entropy figures which means that they have many higher and more dispersed numbers of states in them. Thus, given that an aware brain is just an arrangement of states of matter, a universe could arise where a spontaneous arrangements of states might supply an arrangement sufficient for consciousness to be able to contemplate its existence in that universe – a Boltzmann brain – without any of the inconvenient messing about with evolution on planets. However, a spontaneous consciousness one moment might not be conscious the next given the multitude of arrangements the universe could move into. Consistent feedback would need to be maintained for such an awareness to persist. The life of a Boltzmann brain would probably be fleeting. This gives us a clue to the function of consciousness which we will come to in due course (see **Appendix III**).

It seems, however, that all universes trend towards this point, since those with lower initial entropy gradually raise it as they evolve. It may be that all universes achieve fleeting awareness of themselves, becoming conscious of their existence. Such a condition may not even require a certain high density of states (like an early small and dense universe before it has expanded), but could be achieved over intergalactic distances, only very slowly. So few universes remain inanimate for ever. Each one comes alive eventually, maybe even ours, in which case some competition from a galactic or supergalactic mind may be in store for humans far I the distant future. (This thinking also lies behind the philosopher Laszlo's support for the Hindu idea of the Akasha or universal consciousness which I shall ponder later on).

What such a brain could do to maintain its awareness as the states continue to arrange themselves is a matter for discussion, but certainly the idea of a consciousness belonging to groupings of matter larger than our own brain strengthens the meaning and force of coincidence.

Kinship and the human narrative

Beyond the six degrees of separation studies by Milgram and Travers, beyond Jung's notion of synchronicity, there may even be more to *familiarity* in our coincidental universe.

Some time ago I wrote a book about the idea that humans have evolved a relationship 'system' as the necessary corollary to the Westermarck hypothesis and which suggests a reverse of sexual imprinting (a taboo) to prevent incest, and which enabled humans to form mating bonds with partners outside of their genetic kin group. The system helped them to make more profound relationships with those who share a specific set of metagenetic personality characteristics modulated by circadian rhythms. Someone like you, in terms of the set of characteristics we have evolved to share, is a better bet for making a long-lasting parenting couple than anyone else. I explained why this feature of the human personality was a key evolutionary development in human culture. (Thinkers have been trying to unravel the mating bond for centuries. The investigation has gone into biochemistry and the interplay of hormones but still no strictly biochemical explanation is forthcoming for the bonds we make.)

This clustering of coincidence among similar people may be why we find festivals and group events so attractive to us and so attractive to the young especially. It also raises the question of whether 'lucky' people, avoiding some natural consequences of their 'state' tend to end up clustering together. It is interesting to wonder if such 'luck' plays a significant role in social mobility and even in the formation of longer lasting social classes. This notion has been anticipated by earlier thinkers. Goethe said something of this in his novel, *Elective Affinities* (1809) when he considered the idea that people come together in relationships in the way that chemical bonds can be formed between chemical compounds that react with each other. In the book a married couple who love each other are drawn away from that relationship by a long time friend of the husband and the niece of the wife. Inevitable because of the affinities between the people involved. Towards the end of the book we will come back to the idea of an extra-familial 'kin' group for each human.

If the meaning, or usefulness, of a coincidence depends upon the awareness of whoever observes it then we can extend this concept of 'someone like you' to include coincidences. Someone like you is not only going to be drawing similar conclusions about events and reacting in similar ways to chance occurrences, but both of you are likely to be bound by a higher density of similar type 2 coincidences than either of you would be with someone who did not match the similar set of personality characteristics.

Under the influence of type 2 coincidences, your best mate will not only share similar personality fundamentals with you but also, quite possibly, share a range of similar and seemingly very personal experiences in life. This is not the same thing as being raised with similar cultural backgrounds. It is, in fact, one of the ways in which couples can get over the divide of *dissimilar* upbringings. Studies appear to show the similarities of *experience* between twins raised in dissimilar environments, which falls in line with this notion of a greater number of coincidences occurring between similar people.

You only have to ask married couples about things they share and you will be bombarded with examples of 'unlikely' coincidences. It is often what couples use to define, or certainly to exemplify, the 'destiny' that brought them together. Accidental type 1 coincidences are part of everyone's lore about themselves. With long-lasting couples, however, we tend to find an above average concentration of type 2 coincidences in their histories. Indeed, lovers often seem to assume these coincidences as if there was an innate appreciation of the fact that coincidences help to validate their relationship. This fact is readily exploited by 'seers' and psychics who are adept at patching in to the coincidence level of our Universe.

We can take this further. If there is higher likelihood of similar individuals sharing some similar coincidences, then there will be a higher probability of events clustering in 'families' of coincidence whenever numbers of similar individuals come together. This slight tendency to strengthen the kinship of like types over time is going to lead to bonded groups of individuals entirely unrelated to each other by kinship or specific cultural determinants. The formation of these groupings, though, is not instant, since time is required for all the individuals to uncover and share respective histories, but once formed it will have a duration. It is with this slight surplus bonding – slight like the hydrogen bonding between organic molecules – of similarities that we can explain stabilities of social groups across many kinds of underlying social fissures.

> Our feeling and intuition take us so far into the recognition of this clustering, but in general our sensibilities do not seem to reach much beyond the appreciation of the coincidence 'family' centred on one's own coincidental states. Still, there are interesting questions to ask. What does the sensation of the 'beneficence' of coincidence do to our beliefs in forms of politics, religion and family? What about our desires for hereditary rulers, for the innate acceptances of elites? Is there a relationship between social stratification and good or bad coincidences grouped over time?

In events of short time span but of highly complex interactions (at music festivals say, rather than at concerts, or during strikes or coordinated protests rather than demonstrations) the mob may well cohere more strongly than one might expect because of the rapid discovery of coincidences in and around the event raising the sensation of *familiarity* among its members. One of the reasons why the recent revolts or ordinary citizens may have sustained themselves more than one might have predicted is that protests based in just a few centres allowed the people involved to get to know their *coincidental bonding*.

As a simple example, my wife struck up a great friendship with someone who, at first sight, was a vagabond, in part because they discovered they had both been in Paris during the protests of 1968. This friend turned out to have lived a life of extraordinary coincidences, and the friendship itself brought coincidences into each other's lives. He was a locus of coincidences, far more, it seemed, than any one individual has a right to expect. Here's an example of a this. He had a nefarious past and one time he was arrested in Holland, handcuffed, and taken to his country's embassy security station for further information, which would not have been kind to him. The police had stuffed a long stick down his trousers to hobble him. Also waiting in the security office was a mother with an eight-year child. The child became concerned but quite unafraid, and whispered to the prisoner, 'Is there anything I could do to help you?' The prisoner whispered back, 'Hold the door open for me.' It was a spring loaded door and would have slowed him down. Once the child had opened the door, the prisoner used the hobbling stick to break away from his escort and then run out and down the street. He ran down the side of a house and burst in on couple asking to be hid. Amazingly, they hid him from the police search. Turns out that the couple were members of the guerilla opposition to the colonial rule of the prisoner's country, and the prisoner had fled from that country's draft in order to avoid fighting the war against them. Thirty years later in Lisbon, this man befriended another in a bar who told him about a great friend of his who often recounted the amazing story of how he had helped a prisoner escape in Holland when he was only eight years old. *I know that man*, the prisoner said. *I am the person he helped.*

As air travel becomes more widespread chance meetings between acquaintances will rise, and tending to strengthen choices and relationships and confirming differences between groups. Rather than binding the world in a grand commonality, losing barriers to movement paradoxically intensifies group boundaries.

Here, it is the man in the Lisbon bar who had been suddenly presented with a rare coincidence of having befriended independently both sides of a strange encounter. As told, the story is not precisely a type 2 coincidence since we know no more of the consequences of the meeting of the other 'half' of the event for the stranger in the bar. Yet a serious question is begging? What belonged to the first eight years of life of the boy that would lead him to save a handcuffed stranger from years in prison without being intimidated by the circumstances? If we knew more about *his* early life we would understand more about the nature of this coincidence and for whom it turns out to be a significant type 2 coincidence.

Staying in ENT and with human connections, in my previous book I suggested that there were a minimum of 8 basic points of similarity for individuals. We can say that one out of eight random people you meet are likely to be broadly more in sympathy not only with the way you make decisions but in your overall life history of coincidences. If, over time we are able to accumulate a greater proportion of those friends of at least one point of similarity, coincidence between them all would tend to rise. If, say, half our acquaintances have this similarity, and the same is true of them, then it is likely that meaningful or exploitable coincidences among them might occur in as few as half the number linkages that those who did not belong to such a group of acquaintances would need. Finding yourself in such a group could have concrete benefits over and above friendships, while not finding yourself in such a group would be a relative (and subtle) disadvantage. If relationships form without this additional 'stickiness' then they will be more likely to fail.

Further, while there are many ways to examine the notion of the 'establishment' in a society, the general use of the term may come from the innate appreciation of the role of coincidence and similarity in life. We instinctively understand not only that individual backgrounds do play a role in social division, and not

> This kind of coincidental familiarity is different to the affinity group of a small world network model where people are brought together for a specific action. Coincidental familiarity will happen in any group, and can easily play a part in maintaining the closeness of at least some of the group over time. Coincidental bonding is yet another tendency that guided the human transition from small tribal groups to the agglomerations of cities, and at the same time intensifying the bonding effects of, say, falling in love.

just because of the intrinsic content of the background, but because of the coincidental similarities that tend to arise in it. One might even be able to say that individuals who do not recognise or make use of, or even deliberately refuse, the extra 'perfume' of coincidences in their life with others may have a slightly, and perhaps even incomprehensibly, harder time forming longterm friendships or relationships. They may find themselves inexplicably isolated or off at a tangent.

Napoleon said he preferred a general who was lucky. Was he recognising those who are ready and able to make use of coincidence as a way of leapfrogging causal chains to get to a chosen outcome? Are some people, in fact, luckier that others because of this ability? Which is to say, are some people more able to profit from coincidences than others? While the density of coincidences must be the same for everyone, given the Coincidence Number of our Universe for this epoch, some people do lead more coincidental lives than others. This requires two things, namely awareness and the pre-prep. Failure to observe or ignore coincidences in our lives occurs when we opt for consistency and regularity. It comes down to our state of consciousness, our habits, our fearfulness. It is not until you remove the filter around your life will the coincidences become apparent. It wasn't until I quit a steady job, having a direction in mind but no real idea of how to support myself, did I become exposed to all sorts of marvellous opportunities and be able to profit from chance meetings and serendipitous information. Coincidence can seriously orient your life, both the recognition and absence of recognition of it.

Serendipity, it should be said, has a sightly different air to it. Walpole coined the word and explained that it meant coming across useful things that you had not set out to look for, and claimed he got the idea from an old fairly tale (appearing in Venice in the 16th century and repeating some events found in other stories). Walpole seemed to have misread the fairy tale (*The Three Princes of Serendip*), as described by Richard Boyle, since it is more about clever deductions, Holmes-style, than coincidences. Walpole seems to have just used the fairy tale to give himself a word for the idea of his version of type 2 coincidences.

Is this 'luck', too, an element to the creativity of individuals in their personal microcosm when the coincidence of ideas or perceptions launches them on a new track? Certainly in scientific experimentation, the ability to recognise and make use of coincidence to find a new direction is well documented. Recent studies appear to show that the ability to react randomly – that is to say, the degree of connections likely to be perceived as possible – peaks at around at 25 years which coincides roughly with an inventive peak in the sciences. The older we get the less able we are to make use of coincidence.

If there are those who can, there will be those who cannot. There is a divide in human populations right there. Coincidence numbers connect the origins of the universe with the day-to-day relationships between individuals.

The political implications of groups whose members cooperate more profoundly when bound by coincidence will not concern us now, but we should realise that with the existence of Coincidence Numbers the results of many social experiments should take this form of connectivity into account when forming their probability statistics, and that the design of such experiments need to be scrutinised carefully if we want to form a correct statistical account of cause and effect in human behaviour. It has already been noted, for example, that the results of the majority of psychological studies (60%) have turned out not to be reproducible.

While a small set of coincidences in the quantum world such as that uncovered by Fred Hoyle's analysis of carbon formation in stars will play significant roles in the matter and energy distribution in the universe, what concern us at the moment are events at the level of ENT and in human consciousness. The density of these events – the Coincidence Number – is reminiscent of small world connectivity but is not the same thing.

As coincidence reveals itself more fully in higher levels of consciousness and less at sub-atomic levels, any universe's Coincidence Number would also help to define in some way the consciousness that lives in it and how that consciousness might develop over time. And if we knew the full relationship between Coincidence Numbers and consciousness we might be able to establish for the SETI organisation a better understanding of its chances for success.

As entropy increases, so the greater the number of states available for coincidental connections creating more pathways to a n outcome, and correspondingly changing the likelihood of any one path.

As we have noted, coincidence supplies an in-

> The con man knows how to exploit this coincidences of background (the type 2 coincidences in the main). He insinuates himself into your confidence through his spurious linking to your situation or to events in your life. *What a coincidence, my grandmother had exactly the same wallpaper... You know, I used to own the very same model of car...* Whenever you notice this in a person trying to sell you something, run away very fast.

fluence in opposition to the probability of unlikely events through convergence. Many converging possibilities will hold the Universe to a direction of development with greater strength, making ENT more stable and less likely to be shifted from its track by the behaviour of any one causal series (see **Time Tinkering**). We might speculate that as entropy locally rises, and as coincidences become more plentiful and their 'reach' in space and time increases, what is, essentially, negative probability, also rises in consciousness. Mind would become perhaps more structured and even rigid, and the environment it builds for itself would contain more negentropy. This too, bears directly on the kinds of civilisations SETI should be looking for.

With our coincidence number, the effects are significant but still subtle. If we consider a high coincidence number universe, for example, there would be a yet smaller range of unique outcomes, and, increasingly, quantum events will have less relevance to the randomness of an outcome in ENT. They will probably turn out to be more entangled, and higher level narrative events would turn out to be more predictable. On a world in a high-coincidence number universe you could set off to foreign lands confident that you would happen upon many interesting coincidences valuable to you.

We can see further into this with the example of a loaded dice. Given that any sequence of throws of a single dice departs from the ideal 1/6 probability for each face an arbitrary number of times, it's hard to prove that a dice is loaded by any relatively short single sequence of throws. If you only have time to throw one sequence then a simple way to find out which face is loaded would be to compare coincidences - or doubles - thrown with a true dice. A loaded face would reduce the average number of doubles from all faces except the loaded face. (This is actually what the Bell Inequality in quantum mechanics discovers. Quantum entanglement is a loaded dice.)

Curiously, coincidences cannot be forced without altering the likelihoods of other occurrences (just like a weighted dice thrown with a true die changes the outcomes of all the doubles). From this perspective, here and now, coincidences approach an equilibrium which although unstable, gives us our current Coincidence Number.

With a given density of coincidences, we may be able to explain the curious phenomenon of the non-repeated experimental confirmation that is undermining the study of the social and medical sciences by recourse to the notion of the current coincidence number of ENT. A percentage of scientific experiments in these fields will tend to show correlations that arise out of the connectivity formed by coincidence and not out of those modelled on intrinsic probability. While the correlations consistently appear in experimental results their appearance is unpredictable and the circumstances unrepeatable.

Thinking of coincidence also gives us a lens by which we can examine the problems in making measurements or observations, namely what actually decides the value of states that are still undecided when we come to measure them, and the mystery of what physicists call 'decoherence', a process that seems to force a measurement value on the observer while at the same time creating lots of universes with the same observer in each, playing havoc with what we believe to be reality in everyday life.

We will come to examine the problem of measurements and observers in quantum systems when we talk about multiverses, but since our objective is to improve humanity's chances of surviving in this universe and to dominate its future evolution, I am going to briefly introduce evidence of how narrative at the human level is changing and becoming more fluid with some ideas about networks and coincidence and reflect upon their significance to our social reality.

Networks

At the human narrative level, the growth and power of networks have emerged in the 21st century to dominate human culture. But social networks belong to a much larger realm of action and are fast becoming the single principle of interlocking systems with which we annotate our cultural and economic future, and particularly as companies apply techniques of AI to prediction and distribution of social and economic factors. The way outcomes are determined and the

We will discuss further on a requirement of consciousness to have some kind of boundary to itself for it to become aware. It follows that an endlessly evolving open network is unlikely to become properly sentient because of a lack of boundary to self reflection. Perhaps there may be local boundaries within to give self-awareness to parts of the system, but a truly open network is unlikely keep inner boundaries stable. This suggests an interesting question about how a creator could intervene in an infinite universe and get precisely the result it wanted.

way economic and political power is wielded cannot be understood without understanding them. It is almost impossible even now for a modern human to live a life outside the networks (going 'off-grid in modern parlance). He or she is embedded in networks; carrying a node to many of them in a pocket by way of a mobile phone.

Our awareness of networks, enshrined in the popular use of the word 'networking', inclines to the social; to the sometimes serendipitous connections between friends and family and to the utilitarian connections between individuals in the work environment. Networks will give individuals their identity. No longer a passive element in a database, an individual in society is 'tracked', the changes in his or her state becoming the characteristics of relevance. No individual, corporation, government department, dispenser of funds, charity, educational establishment, military organisation or leisure activity provider will be able to function without being a node within overlapping networks that deliver to the user not just information required to do their job but also the essential validation that culture demands of all its participants; that increasingly costly and elusive *permission to be*. But there is a technical problem for the powers that wish to control networks; a paradox.

The paradox is that networks cannot *enclose* anything. Unless their nodes are physically cut off from making new connections (a node of *1*) – a purpose defeating exercise – networks are essentially limitless (The enclosures of the game of Go work because the zone of network formation – the board – has bounds), and as long as they have access to whatever power they draw on to function. They may have a current spatial extent, a territory perhaps (like a skull), but it is inevitable that they have no boundaries. In a quantum equivalent, they can tunnel through efforts to contain them. Neither are they static, as nodes will wax and wane where connections either wither from lack of relevant use or new ones grow, altering functions. This feature is deliberately put to use in the construction of neuronal-like networks in computing without much concern with growing unpredictability. Training an intelligent network is going to be very much more problematic in the future by virtue of this enclosure difficulty.

Because the nature of networks is to resist enclosing themselves they cannot evolve in response to external influence and remain themselves. The nature of a network is to include environmental influences and to cooperate with others. The basic fabric and function of networks is to become one. Joshua Cooper who analysed the social effect of the new networks in his *Seventh Sense: Power, fortune and survival in the age of networks* (2016) misses this point. He considers the network as an enclosure and, using an analogy with Maynard Smith's and Szathmary's coevolutionary process to suggest how competing networks will improve each other in the face of 'common challenges', expects the opposite to incorporation, namely a strengthening and refining of each network frontier. But this ignores a basic law of networks which says that the larger the network the more efficient it becomes. Just imagine, what the result will be when Facebook and Weibo go head to head.

The second point to make about networks relevant to artificial intelligence programs and consciousness is that they do not possess a real *past*. The nodes always have connections in place, that is until they don't and thus cease to be nodes in the network, and it is always possible for information to travel along one connection or another. There is always a minimum input and output for each node, and the hierarchies of behaviour are determined by the current state of the network.

(Towards the end of the book I shall propose that we recognise a different kind of network, created in the details of the fabric of the universe and supplying the ever changing relationships of observation and measurement in human life. One that arises from coincidence acting within consciousness and which will be the cosmology of each one of us.)

We have seen how significant coincidence is not just a phenomenon on the outside surface of biology but bound into the evolution of life. Before we get to the part coincidence plays in the evolution of consciousness I am going to briefly describe one particular problem with the networking future (and incidentally about AI) we are making for ourselves and in particular the rise of a bias in the way we will interact with the world that will change the social experience, the rise of mind, and the creativity of chance.

Social bias and manufactured coincidence

We should recall an important feature of type 2 coincidences that we discussed earlier. While they depend upon the pre-loading, their occurrences are by their nature uncontrollable. Even if the predictive power of big data can control the number and variety of accidental events (type 1) it is unlikely to master the significances of the type 2 coincidence, and trend only towards an artificial coincidence setting where the pre-loading and the actual coinciding of events will merge into one in a circular pattern in the present.

Let us look at the social media platform Facebook and the coincidental 'news' that it manufactures for you. If you have a large number of friends on Facebook you do not see every post that every friend or their friends make. Far from it. Facebook selects posts based on your patterns of use and your realms of interest, i.e. what you have clicked on in the past or what you have been 'liking' the most. You have probably noticed that friends you have not been on contact with for a while seem to drop out of sight. Their posts appear again when you contact them. Generally speaking, the more in contact you are with someone the more posts you see from them. The more that a post (of those you agree to accept: like friends of friends) includes links to stuff on the internet (rather than just your opinion about things) the more those posts are shown (because Facebook profits from your clicks on links and thus the ads you get shown). If you decide to share a web page on Facebook, ads relating to the web page content may well appear on your Facebook wall. Any posting of a friend related to this post of yours would then get put up on your wall. You will be surrounded by what appear to be interesting coincidental links but the connections are formed externally and the cross-relating or cross-fertilisation they stimulate occur outside of you rather than internally. They merely confirm things about you, your conscious states, rather than create new ones, and are designed, crudely, to limit your choices to the things that will give the social media sites an income.

Facebook is very far down this road already. Its technique of being selective in posting items of interest for you has led others to create web pages containing specific interests designed to be captured by the Facebook selection algorithms and posted to those people showing interest in those items. They are not primary news items as such or even items that you are specifically interested in but fabricated content (almost all created by robots – hence captchas designed to filter out these bots) appearing in your feed because you or the contacts you are most in touch with have shown interest in those subjects. These pages contain automatically created advertising feeds from a number of sources like Google Ads and other Facebook users who are paying Facebook to have that content put on your wall. Your wall does not contain news about the world but news that will capture your attention and encourage you to click further. The items are the simulacrum of coincidence; they are *designed* for you. (And hence the recently exposed fake content providers intending to influence political opinions.)

These coincidences are false type 2s because the pre-prep necessary to make use of them has not occurred in the full probabilistic space of one's life but in a micromanaged world that you have already previewed. Facebook is the pre-prep *and* the coincidence. The illusion of serendipitous activity – some of which can be genuine – is very strong but it is a very narrow version of it tailored to our apparent predilections. Facebook may be fun and addictive because it is more narrowly about *us* rather than about life at large; it's the book of narcissism.

The whole digital networking enterprise, of which Google Glass was just a first attempt, is reaching the point where it will actually destroy significant type 2 coincidences in our lives. For example, with the implementation of just today's technology the coincidental meeting I experienced in Chile all those years ago could not have happened in quite the same way. My mobile phone would have alerted me to the presence of all my acquaintances and connections, including the lawyer, in the country the moment I set foot in it, and would have tracked their progress and broadcast mine as time went on. I would have known that the lawyer was in La Serena as I approached the town and I would have known which street he was in as I walked about it. He, likewise, would have known my movements, could have known what I now looked like, and we would have arranged our meeting, if we wanted it, well in advance. The lack of surprise would have altered the experience or maybe even demolished it because I might

In the old days telephone connections were few and cumbersome. As they developed one could make more and more calls to more and more destinations. When I was young and went travelling in the world, my means of communication with home was a post card. Phone calls from a remote location were problematic, expensive and hard to organise. When I was a child a long distance phone call available to every subscriber had only just become possible. For amusement, the father of a school mate had collected all the local exchange codes from the north of England all the way to London and used them to make a local 'long distance' call. He could make the phone in London ring but conversation was impossible due to losses on the connected local lines. The new Trunk line system, as it was called, had distinct power supplies.

well have decided not to meet him, given my solitary frame of mind, had I known he was there, and *vice versa*. But more than this, I would have probably never been out of touch in some way or other with him from the very beginning. He would have been an acquaintance in the background, with a connective route always ready to be followed.

The reality, of course, was quite different. We were forced to react and to make decisions *at the moment of the meeting* without any presumptions. We were forced to react according to our personality states at that time and to the new content that arose out of that moment. We had no preview of the occasion to prepare a response; we were forming a connection in real time.

The differences between the two forms of connection may not seem like much; that I am quibbling over a technical point that has little to do with the way people prefer events to occur, but there is more to it: the difference is key to the human future.

Facebook essentially fudges the phenomenon characterised by the Coincidence Number of our universe. By using the Facebook method of concocting coincidences we are undermining the relationship I have been talking about of the future making the past an uncertain foundation to the present. The result will be a future for us that is *less* available, each possibility less likely to occur and a past more defining. As our futures dim, we stay on the old track keeping to what we know, thinking the thoughts we are given. This may imply a social calm, as the networks become established, but it also implies a helplessness and a removal of mind from the wider connections of the universe. Further, the small bindings of genuine coincidence that help relationships along, will be replaced by what we could call the fast-food version, the immediate chiming together of likes that do not go deep into our personalities and what we have learnt from the world. Facebook is in fact creating its own version of the underlying reality that Jung considered to supply the effect of Synchronicity, but one geared not to the higher spiritual purposes of creation but to the purpose of consumption. (Facebook has now introduced such a feature of linking members by the events they attend. But again the hidden controller – the *Zuckerburgermeister*, if you will – is forming the reality for you. Those choices will be managed with the objective of selling more stuff)

We seem less aware of the dangers of these aspects of social media because our social brain readily partners them to some degree in its readiness to invent memories and to falsely revise genuine ones under the influence of social media 'truths'. We know already the famous bias of the Urban Myths, where individuals claim to be witnesses to or to know personally those who had fearful or distasteful experiences, but which had not actually happened (urban myths are debunked on a number of web sites). This similar bias is only intensified and broadened with social media. When someone tweets a picture of a beautiful beach they visited say, it is only a small step for us to try to share in that experience by believing we know plenty of others who have been there too, and a small step on to believing we actually know that beach or have been there. The more a tweeter is respected, the more likely it is that his or her tweets will begin to be incorporated into their followers' own memories. The future is shaping up to be a journalistic dream where, adapting an old adage, one should never let the facts get in the way of a good memory, or, one should add, get in the way of a good sale. (We will delve deeper into false realities towards the end of the book.) The use of virtual reality not only for fun but to investigate mental illness or phobias (for example, by letting people experience fearful fake situations such as having claustrophobics put into virtual cupboards) is a 'mind-blowing' step in this same direction.

There is a great deal more to say about how the social media sites will combine with virtual reality and come to dominate daily life with the merging of real and artificial events. For example, as we mentioned, as diversity falls the numbers of coincidences will rise and thus the acceleration of civilisation may exhaust itself in increasingly narrow (and perhaps less creative) confines. Whether this particular future we are marking out for ourselves can include an evolutionary advance in human consciousness development is still unknown (although, as I have been arguing, there is a way we may find out). Certainly, consumed realities will have significant impact on the way artificial intelligence arises. It is possible, too, that our expansion into space may help maintain sufficient cultural and psychic diversity in the long run to allow for creative coincidences to brake this acceleration into exhaustion. So, let us continue an analysis of human narratives by looking at what may well be the curious and still controversial foundation to the fabric of reality, the multiverse.

Cosmology and the Multiverse

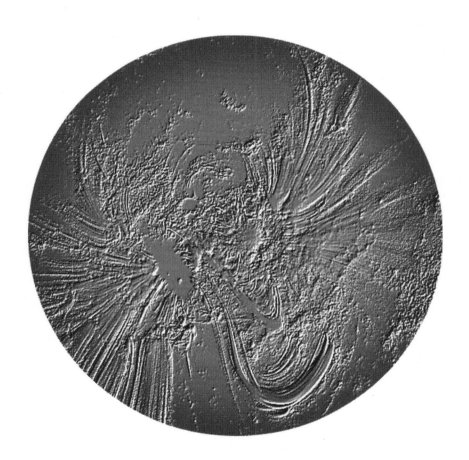

In which we examine the many worlds theory and the origins of our universe.

In which we discuss casinos and chance and what happens to probability going back in time.

In which we discuss what it means to make a measurement and whether the many worlds theory explains enough about the dispersion of space and time.

In which we reveal a crucial paradox in a famous thought experiment, locate a cash-box, suggest another influence on the origin of the universe and propose a new particle.

We began our narrative with a discussion of where we are in Space, and came to some conclusions about chance and coincidence in human affairs. Our intention is to analyse consciousness, its narrative and its place in the development of the universe. We saw how coincidence is not only a property of universes but plays a role in evolving consciousness. We also saw that coincidence is a countercurrent to probability in the human narrative experiences. To examine this further we will, divide, without specifying the precise transition, the fields of awareness into two principal components, External Narrative Time where humans move forward with the flow of events in the time and space of the universe we live in, and Internal Narrative Time, which is a particular construct of consciousness.

Events may take place logically once but not always forever. An event may produce effects that can then bear upon the original sequence. The Markov Condition imposes a belief that the reasons for a particular event are always going to stay the same. But we can see that the past is not the best thing on which to build certainty.

Our goal is to explain why mind separates the universe into a before and after, and why there is more information about the future available to us than we realise. Even Feynman pondered the crucial lack of time symmetry between the 'certainties' of the past and the probabilities of the future. We will see that the simple failure of the expectation of symmetry tells us a lot more about mind than we might have expected.

We cannot avoid, however, that our explanations about mind will necessarily have to align themselves not only with the social context we have just discussed but what we know about atoms and particles and with what we suspect is true about the origins of our universe. Space –

not the imaginative space of colonising the stars, but the dimensions in which things are – turns out to be the critical influence on the probabilities we experience, and the choices that direct the evolution of life. This means delving into quantum theory and in particular the multi-world scenario vision of reality. Currently about 70% of polled scientists believe the multi-world scenario is correct, but what it means for us at the human narrative level is still not established.

Mysteries of probability and measurement

Quite how our apparently stable world of seemingly definite states and causal pathways arises out of quantum uncertainty is a routine query among physicists. We assume that the 'classical' picture of probabilities in our everyday world (the dice throw) is built upon the definite results (eigenstates) of quantum theory, but how exactly?

It may be, however, that the question is the wrong way round. How does the accident and chance in our daily existence result from the *destruction* of quantum randomness in measurements?

Randomness disappears quite quickly in the microscopic world. The moment anything becomes alive, like a bacterium, evolution alters search strategies from random walks to responses graded to the intensity of stimuli. Even the tiniest 'observers' begin to make choices in their world. Life determining its reality. But when we reach a certain level of consciousness something else happens to denature randomness, namely prediction and memory. (Many a philosopher has tried to get us to quiet this aspect to mind so that we may live in the eternal present. The Pirahã of Amazonia may have decided to try and do this – see **Lies and language, fallible narratives and dreams**.) How the mind understands probable predictions is a puzzle usually 'explained' by invoking the mechanism of Bayesian logic, a means of guessing outcomes based on the memories of the probabilities of connected events. Since humans are very bad at estimating Bayesian probabilities consciously, we might expect that evolution has other means at its disposal to perform this function. (This book is about one of these means.)

It is about now in every discussion of mind that the connection between separating universes and consciousness comes up. We cannot really go further along the path to describing our everyday reality without examining what this means. The issue of Schroedinger's Cat is a key signpost on the way to seeing how time in our everyday reality is suspiciously porous to influences from other times and places, but before we examine the famous thought experiment, let us take a sideways look at probability through what we have discovered about coincidence.

The sad thing about the Schroedinger experiment, if it is true, is that it *forces* the death of at least one cat somewhere. Even if the cat lives in the observed box, it dies in the other universe. The death of a cat is assured after a certain time has passed. And this appears to be true for every probability, i.e. there is no probability about facts, only *where* they occur.

There are at least two kinds of probability that we use to discuss outcomes of choices. There is the kind of probability we use in the casino when we talk about systems like dice throws or the roulette wheel that have a number of outcomes or events available to them and only one of them can occur at the moment of the outcome. The source of the probabilities lies essentially in the fact that concrete outcomes are strictly spatial in nature; each one physically excludes the others, and that the probability of the outcome that did happen plus the probability of all the others that don't happen all sum up to one, to certainty, in that space.

The exclusivity of a value out of probable ones is the mystery to be solved in the quantum measurement since quantum probability is a property of the whole system rather than one derived from an individual event. David Deutsch, for example, believes that the multiverse scenario is required *in order* for the probabilities in quantum theory to be observables, to make sense. In this respect, quantum computing is a multiverse engine which is why some thinkers believe we may be in a simulation set up in a quantum computer, set up by someone, the Great Computor, in some universe or other. Let's examine this idea.

Think of a sack of different coloured balls. You can only draw one coloured ball out at a time, but all the other balls are also physically there ready to be picked out. All possible colour sequences are implied within the sack at the same time. The probabilities of any sequence are determined by what balls are still there, and if a ball is removed the probabilities for *all* the remaining sequences change. (The probability for each ball is in some way a measure of how far from the opening of the sack it is. Remember our two-dimensional being and probability being the time it takes for that being to get to a point.)

In a quantum observation or choice of the ball, where colours are possibilities, there is only a superposition in the sack and it is every colour of the spectrum, changing continuously (according to a wave equation). When you reach into the sack to the superposition and retrieve a ball, it will be a colour determined by the time evolution of its own wave equation and is independent of the others, although this poses some questions.

The origin of each colour's quantum probability does not depend upon the others and yet they all must agree to add up to certainty. (The mass remains the same since particles don't have a spectrum of weights to contribute to a superposition – usually; neutrinos may be different – so the sack only ever weighs as much as one ball in our universe). Their duration times are entangled, but the time to an event must be different, just like Achilles and the tortoise. Further, the longer we go without measuring anything, some colours will become more likely for a while and linger longer and others will linger less for a while during their 'turn'. The duration times start to combine into the 'time to event' time. The ball would start off more or less white because all the colours are mixed. But each colour is distinct; it occupies its own space very slightly different to its neighbouring colours (the states are said to be mathematically at right angles to one another), and the longer we leave it, the more the ball takes on a distinct colour before changing to another distinct colour. The interference between the colours washes out and we get a sack with a ball that exhibits the colours the ball could be serially over time. Probability is thus a 'time-to-event' just like the inhabitants of the 2-D flatland perceive. (This may explain the actual 'Zeno Effect' where a superposition can be re-initialised with radiation such that decoherence never happens: the time-to-event never elapses.) The colour you pick becomes certain (probability = 1) and, under the multiverse view, only then does each other colour version previously embodied in your sack appear in the

> Even computer scientists have problems trying to create true randomness for use in calculations since you have to start from somewhere and a starting point is by definition not entirely random (like how far from the mouth of the sack a particular ball is). Certainly most of what goes for event randomness in ENT is actually a value that could be calculated if only we had the right information (that is, except for coincidence).

hands of their own observers (copies of you) in their own sacks and universes and disappear from yours.

This doesn't happen with casino probability since there is no duration time only a time-to-event distribution in time, which is in turn limited to a physical location. Tossing a coin is a good example of this. The odds of any face turning up is only an implied result of $(1/2)^n$, achieved more or less accurately as n Y ∞.

We can illustrate the difference between times another way. Think about playing roulette backwards in time. The ball begins in a definite home. It then jumps up and revolves around the wheel accelerating away from the numbers several times before jumping into the croupier's hand. Where were the other possibilities, the chance, the betting options in this retrograde event? The event seems to be completely defined, without options, from start to finish in one direction in time, but not in the other. (A nice way of illustrating the non-commutability or interchangeability of probability in time, and which can be used to describe the 'arrow of time' instead of the usual rising value of entropy. About more, later.)

The fact that the other chances don't show up when the roulette sequence is played backwards in time is a good indication that casino probability is not really probability in the proper sense. Coin tosses (where you let the coin fall on its own), dice throws, sacks of balls and roulette wheels never give completely satisfying proofs of randomness because the outcomes are the result of specific physical conditions, and if you knew all the parameters of say a dice throw (size, weight, speed, friction, bounce etc.) you could calculate the face that comes up every time. Indeed, there are expert Craps players who can certainly beat the average with their throws because they have learnt a method of throwing the dice that improves their chances of throwing a specific value.

In the quantum situation, going back in time is not necessarily so clearly different to the same situation going forward. A mix of states is a mix of states which ever way you view it. This, however, leads to another paradox. In any superposition of states say, the quantum ball in the sack, it doesn't seem reasonable to expect the superposition to come into being and be divided up among probable states at the moment the measurement is made (the ball is grabbed) for by then it is too late, the superpositions are not required. So while the measurement as it occurs does not need any of the other probabilities that are not selected, these must have in some way actually existed *beforehand*, and have access to the same space, before the choice of measurement. This is the paradox, namely every possibility needs to have access to the same place (in case it might occur) and yet also be capable of being excluded from that place (in case it doesn't occur). Everett came up with a solution to this problem of the measurement and the 'collapse' of the wave function favoured in what is called the Copenhagen interpretation by saying the observer must connect with *all* possible states at the measurement, otherwise the probability of each state is meaningless. The measurement finds a value according to how long the various probabilities of the superposition have already been evolving. (Some states are static and whose probabilities do not change over time.) All these other probabilities, as they are not 'destroyed' in the measurement, are 'split off', or decohered, from the one value chosen by the measurement, and thus there are as many copies of the observer and her entire world as there are possible states – the many worlds described by de Witt.

The so-called collapse of the wave function seems to be just the *movement* to (selection of) an existing universe where the probability of the measurement is certainty. Going from one measurement to the next is like moving from one stepping stone of probability to another, as if our reality is a path of 'stepping stones' or regions made across the stream-bed of the universe. (In fact this idea has been developed by Boström among others where every possibility already exists in some kind of structural arrangement and reality is picking a path through this structure. Keep this in mind).

If the universe needs to separate in this way in order to give reality to randomness, then it must know in some way how to split before the measurement. It has to have coordinated the decomposition of the states before the measurement, and thus what the eventual random value is going to be is in some way a *prediction* which creates the superposition.

(We can imagine probability as actually being the wait for and the duration in which the probability of that value is certainty. This is in deep contrast to the odds in a casino or at the race-track. We will come back to this.)

Furthermore, the probability of the outcome that we end up observing is really the probability that this event occurs *here*, rather than elsewhere. So the probabilities which get superposed must all be probabilities about the event occurring in our particular world-line at the moment of the observation; they all must be 'prepared' to occur in our world (counterfactually). This doesn't seem a problem since every superposition involves worlds that share precisely the same history up to the point of the measurement. But we have still not solved the problem of how similar world histories produce dissimilar probabilities (the 'preferred basis' problem or how it is we do not see a superposition but only one world). What can the differences actually be? Since the spatial requirement (that it happens here in our reality) needs to be the same for all these values, the time dimension seems to be the remaining measure. Superpositions are really made up of superpositions of various times. In fact Young's famous slit experiment suggests precisely this, and we will get to it in due course.

While quantum predictions have been confirmed in many particle experiments, and surprisingly large conglomerations of particles, well isolated from the environment, have been placed in superpositions, the interactions of larger sets of molecules resolve towards more standard mathematics, and quantum paradoxes disappear. So the puzzle of what chooses the classical outcomes of our quantum world given the separation of probabilities remains even while we accept that quantum interference has clearly been selected by evolution in systems like a bird's navigation, the human sense of smell and in plant respiration, and is used practically in electronic components like transistors.

The Bell Inequality and the elephant in the room

Before we go further, there is what may be unkindly referred to as the elephant in the room of the quantum measurement, namely the Bell Inequality. Initially a thought experiment, though now experimentally confirmed, it makes certain statements about the entanglement of two quantum systems with a common origin.

Einstein, Podolsky and Rosen stimulated this investigation by asking, in 1935, whether quantum theory explained everything about the state of a particle (sometimes referred to as the EPR argument). In 1964, Bell published a paper which suggested that particles with a common origin could not exhibit one or more of the following characteristics generally considered to be necessary for a reality to exist independently of us knowing about it (the 'hidden variables'), namely retain their own parameters, communicate only subluminally, or produce a measured value in one independent of the measurement on the other.

His novel logical argument about sets of randomly set angles between detectors of pairs of particles with opposite spins produced at the same time suggests that correlations between them should not exceed a certain value.

The logic appears unassailable. Given the conservation laws in the production of two identical particles from a single particle, if one particle is measured to have a spin in certain direction then the other must have spin in the opposite sense if counterfactual definiteness exists and therefore must show a linear correlation. Experiments show this linearity is violated, therefore the local reality of spin values is not supported in the quantum reality. The pair of particles therefore do not have their own distinct values prior to measurement and/or they communicate faster than light and/or the choice of measurement on one is dependent of the measurement of the other. The first two effects are accepted, while the third has been investigated to an accuracy sufficient for most physicists but not actually falsified. Which is curious because by falsifying it, Bell's theorem would be demolished.

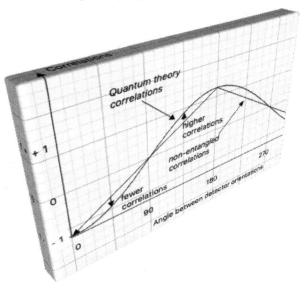

The way most physicists deal with the violations of the Bell inequality is by accepting that the local reality assumptions are themselves unreal, an idealised invention based on dodgy assumptions about the nature of reality. For example, with detectors set at an angle between $0^0 - 90^0$ there are fewer quantum correlations than expected. Between $90^0 - 180^0$ there are more. At 90^0 there is no correlation – the odds are 50-50. Summing over the whole range of angles from $0 - 360$ degrees, the deviations of the quantum predictions from the localised reality assumptions net out to zero. How can we interpret this? A whole field of study has opened called *contextuality* (i.e. the circumstances of the measurement) where the Bell inequality is a special case of *non*-contextual measurement.

Essentially, what has been exposed here is the problem of deciding what exactly is the opposite set of values to a set of three parameters each with two values. For example: suppose we have two sets of three coins, gold and silver. Then we can say that full non-correlation occurs when we have three gold heads and three silver tails. But in what sense could two gold heads and a tail be 'opposite' to two silver tails and a head?

How the Bell inequality arises is an application of set theory. If the particles have independent properties than the odds of detecting them in the equal and opposite orientation can be found by multiplying the odds of each particle being in a given orientation along one of the three axes of space together with the odds of a detector being in a given orientation. The particles do not seem to have set orientations before they are measured. The questions arises, therefore, as to what constrains them even to follow the same quantum distributions in every experiment conducted to test this if there is no existing reality. Hence contextuality, which is like our probability volume.

What we have here is a measure of the coincidence, or strictly speaking a measure of the non-coincidences possible between systems. Two systems with a common origin (or at least histories that have crossed) have a higher coincidence number than other systems (contextuality). Remember that coincidences remove *im*probability from an event. Think of these two measurements as dice throws. Where there are doubles there is maximum correlation but any number of mixed throws means no correlation. What this suggests is that the particles divert extra correlation probabilities from elsewhere in the system, but how?

From the diagram we might think that the extra probability need to correlate is taken from the earlier phase of the function. That is to say the extra probability comes from a time where the spins more closely approach each other. The curve depicts the time build up of values. It also suggests where entropy comes from. But more than this let us consider the action of the detectors. One could say and perhaps should say that it is the presence of the apparatus, and the intention of the experiment are all designed to extract skewed values of information from the waveform. The detectors are the equivalent of adding more tables to the room of dice throws. This is how distinct apparatus can still produce similar results to each other, otherwise one has to explain the features of consistency in an entangled system that has nothing defined for it (according to Bell's theorem). But this is one of the features of reality that the Bell theorem appears to deny, namely the existence of what are termed counterfactual probabilities since these imply the pre-existence of the parameter that gets measured. (Since quantum theory 'truth' arises in statistical accumulations, at least some of those other probabilities must happen sometime.) The Bell Inequality defines these counterfactuals by implying hidden variables in the *past* relative to the observation. Suppose, instead, we have counterfactuals in the future. That is, the observation confirms the counterfactual correlation in the future and that the angle differences arise in the separations of past and future; between prediction and memory, a matter we shall take up again (see **Schroedinger's Cat**). The question of what the quantum measurement means for our choices at the human narrative level remains.

The 'Many Worlds' Interpretation of creation

I asked a retired theoretical physicist once (in 2017) what he thought about the multiverse. 'Bollocks!' was his considered reply. 'Ah?' I said. 'You follow the Bohr line.' He added, 'It's simply not compelling, it's not *required.*' And then would not be drawn any further.

I could see what he meant. Quantum theory is immensely successful mathematics for getting the answers to problems. What else is needed? Some physicists have gone on to believe that the only thing that is real is mathematics and our universe is just the consequence of maths working itself out. Others, however, try instead to imagine what kind of physical reality leads to those mathematical equations and the laws of physics. Such is the multiverse hypothesis; one attempt to visualise the underlying physical reality to the wave equations, to give a physical picture to the mathematical structures and logic of them.

Max Tegmark is one cosmologist who has produced a 4-level scheme for classifying multiverses, and the evidence is mounting for the existence of some kind of multiverse, and certainly for a much more complex universe than the one we see with all its dark and mirror matter, and dark energies. Tegmark's level 1 multiverse is an attempt to visualise the consequences of infinity, but levels 2 and 3 try to give substance to the consequences of probability. We will look at these in turn, though not in that order.

Let us suppose that, for the sake of the many-worlds argument that, given the near infinity of probabilities among these multiverses, anything you can imagine is likely to be happening somewhere including an erasure of probabilities due to unexpected merging of worlds. This is equivalent to influence travelling back in time. If having influences travelling back in time is an existing probability, then a world where it acts is around us somewhere. If time travel exists in at least one of them, then all worlds are threatened, including ours, by the entanglement of these worlds, since you could influence any world in any branch by travelling back in time to the appropriate divergent point and then travel through each subsequent divergence to get to the moment you wished to influence.

We begin the discussion with some confusion of terms. What does the word universe mean? Does the *universe* mean that cosmological place which contains all the places we can imagine or know to exist even if we can never observe them or communicate with them in any way, or does it just mean the extent of the place we live in — the visible universe?

The word 'multiverse' seems to represent a subset of the greater meaning of the cosmological word universe, but confusingly it is also used in the quantum theory context. Quantum theory implies the situation of many complete invisible *versions* of our universe living side by side with ours, with the same point of origin, the Big Bang, while our own *cosmos* is the continuous space-time environment that everything we know occupies and where what we call the 'visible universe' is a smaller region of this cosmos. The *cosmological* multiverse, on the other hand, is where there may be multiple Big Bangs in the false vacuum state of an inflaton field creating expanding universes seemingly unreachable from ours, the total of which we will call the *omnium*. I will mention other uses as they crop up.

There are two main types of quantum multiverse arising from quantum probability scenarios. Under Max Tegmark's classification these are Multiverse types 2 and 3. Let's look at type 2.

Type 2 Multiverse – the Omnium

This vision of the multiverse is a mix of quantum theory and a cosmological vision and requires a many-cosmos scenario with each having distinct and separate 'big bangs' of its own.

It arises from uncertainty and the fluctuations of probability that can occur at the smallest meaningful moment in time and space where enormous energies can emerge out of random variations in the basic neutral (but highly dense) sum of all the energies there are (what is called the false vacuum). The gravitational energy packed into a dense space blew the sum of energies apart and space expanded extremely rapidly (called *inflation*, faster than the speed of light). By the time the energy levels fell enough for space to stop expanding, the universe had grown to a fantastic size. All the energy that could no longer expand emerged as a hot cloud of field energy into space-time, still expanding on its remaining kinetic energy, and cooling into matter which clumped together as galaxies seeded by continued quantum fluctuations and which resulted in the visible universe that we see. This idea has its problems, namely that the quantum fluctuations are too small to provide for such diverse clumping of matter that we observe without the presences of dark matter, and, if there was no space initially, in what did these fluctuations take place. (Although the problem of accounting for all the matter, and finding out where all the antimatter went is not solved either.) Random fluctuations, however, do account for regions within the whole of the expanding space losing their expansion energy at different times so, at these points, different types of hot energetic matter emerged creating different types of universes, and all unreachable from each other.

The universes in this type of *Omnium* are born with different cosmological constants (perhaps even new laws of physics) and have very different life-cycles. Some universes do not have sufficient expansion to escape the attractions of gravity and collapse giving rise to new ones, while others expand indefinitely, growing cold and dying the 'heat death' as our may well do.

> The collapse of these universes lacking sufficient energy to escape gravity poses an interesting question about whether these universes could in some way 'recover' or reincorporate their space time points that originally expanded, and if so whether this would be observable either through gravity waves or other means, or whether this collapse back to that point helps fuel other expanding universes.

Some people think that the singularities of black holes cause inflationary universes elsewhere and that our universe is the result of a black hole in another universe (when we look at black holes we will see that his is unlikely to be true). Still others suggest that points in our universe can tunnel through the energy bounds that contain them (or the minima in which it is currently resting) into a yet lower energy state causing more hot expanding matter to appear and fresh waves of local expansion in the space that contains it – a 'bubble' of space-time containing a new universe. Even supporters of string theory acknowledge that the tiny curled dimensions they believe exist in each point of space can slip into new energy configurations (perhaps 10^{500} of them) and initiate bubbles of expansions in our space-time, producing an infinitely bubbly *omnium* of universes.

In any event, the cycles of creation and dissolution appear to go on and on forever (called chaotic eternal inflation). The Big Bang that began our universe is just one of these, lost in the throng. We cannot reach these other places, but we are here because they exist. Because we exist they must also exist. Even so, there is a distinct problem with the idea that further expanding points of the vacuum within our universe of space and time can occur. This is sometimes referred to as the 'empty universe problem'. If a region in the vacuum suddenly expands and then all the energy within it turns into the matter and space-time of our universe, what is left in the false vacuum to produce more expanding universes later? If there is a continuous supply of energy to the points in the vacuum, then this certainly implies a great deal more structure to the universe than we currently understand. This is an unsolved dilemma, although proposals where universes cycle through periods of expansion and contraction, or where universes that exist on exclusive surfaces collide with each other time and again provide explanations, as does the idea that Black Holes (supermassive black holes appear to exist in the centres of most galaxies) may provide a 'recycling' mechanism passing energy into the general vacuum state to fuel continued expansionary regions. Chaotic eternal inflation, however, implies something else, interesting to our argument and which we will pursue further on, namely the question of an expanding universe's horizon and what may be happening on the 'frontiers' between expanding regions.

But one does not need multiverse separations to believe that there are multiple versions of the universe we know about in play right now. The mere suggestion of infinity is enough to give rise to another form of omnium or cosmological multiverse, the multiverse type 1.

Multiverse level 1 or the 'quilted' version

If the visible Universe is a part of an infinite space (and there are many reasons for thinking ours is), and matter is spread around pretty much as uniformly as we observe, then all possible variations of matter and energy could be provided for within the same space-time continuum regardless of how our particular Big Bang happened.

What constitutes infinity is really a matter of philosophical taste oriented either in time or space. A tiny thing could last in a time that went on for ever, or space in which things sit could go on for ever in extent, but things in it are not obliged to last infinite time. Infinities could be a mix of both. Both these types could have beginnings and we could quite possibly be living relatively near the beginning of such a mix, with an infinite time ahead of us to explore an (eventually) infinite space. There are also mathematical infinities created by mapping sets of things to one another but as to what cosmological reality these represent is not known. In the argument below, infinity is just a sufficiently long time and a sufficiently large space in which anything that has a possibility of happening can happen. (Other ideas about infinity lead to paradoxes. For example, imagine a machine filling an infinite space with ping-pong balls. However long you run the machine there would be an infinitely large empty space still remaining. There would be a whole lot more emptiness than any space filled with ping-pong balls, always. This is true even for vast numbers like Graham's number, a mathematical curiosity which is not even infinite like (presumed) π and is so large that it cannot even be written out in our visible universe. Which is why the law of large numbers ends up being meaningless in infinity. Infinity is not *filled* with endless versions of things; it is mostly empty, and which is incidentally why the Principle of Mediocrity – that we are just an average observer in an average universe – is dubious at best since every object or process in infinity would have the same distribution as any other).

Now, given the matter we infer to exist in the universe, there is going to be a limit to the number of quantum states (the argument begins with a convenient restriction). If we have a large enough space – it doesn't have to be infinite – in which to put a finite number of quantum states of matter and energy then eventually states must repeat. You can see this very easily by pondering how likely it is there are two people in London with exactly the same number of hairs on their heads (following a Martin Gardner illustration). The maximum number of hairs ever found on a head is let us say 200,000. Say there are 10 million Londoners. Then, take each Londoner and count the numbers of hairs of their heads. Suppose the first 200,000 each have a different number of hairs, then what happens when you count the hairs of the 200,001 Londoner. Either he or she has more than 200,000 hairs or he or she has a number of hairs that has already been observed. Thus, there have to be *at least* $10 \times 10^6 / 200,000 = 50$ Londoners with exactly the same number of hairs on their heads. Only if people could have more than 5 million hairs on their heads would the probability be less than one.

So while there may be 50 Londoners with the same numbers of hairs on their heads the chances that they share the same pattern of hairs is very remote. Let's say a hair can be black or blonde or absent, then each of the 200,000 hair follicles can have three states which amounts to 8,000 trillion universes just to repeat the Londoners' heads of hair.

So, as we think the fabric of our universe is 'flat' which is to say there is not enough mass to stop its expansion and it will extend for ever, then states of matter will have to repeat. It's likely that somewhere far away there is a patch in which an identical world to Earth on which identical things are happening exists. Tegmark makes a simple calculation about the numbers of possible states in our universe from the numbers of protons we observe which suggests that around 10 raised to the power of 10^{29} metres away from you is an identical you on an identical Earth reading a similar passage in a similar book. An identical whole Universe to ours could be lurking beyond it about 10 raised to the power of 10^{115} metres away.

Our 'universe', in this case, is only that visible region bounded by the shell where light can no longer reach us. There is a lot of space beyond formed by the early inflation. Other zones beyond also have similar limitations to their expansion. Eventually, however, as light from further and further away reaches us, we may begin to see views of these other zones. (Although if the current acceleration of the expansion of space that we have observed in the remote regions continues, at some point our universe will expand faster than light can get to us, bringing the growing of the visible universe to a stop.)

We would still be living, however, in a single space, an omnium, containing lots of zones of 'universes' (we could in principle set out from Earth and travel to them since we would be in the same expanding space-time), some similar to ours and probably at least one identical to ours; identical to the last atom. Let's call this kind of multiple repetitious space of N universes an Nverse. (In Tegmark's classification, a Level 1 Multiverse, as if it is some kind of beginner in the creation stakes. Brian Greene calls it the 'quilted' universe.)

This concept of the repeated universes, the Nverse, in an infinite space is not quite as convincing as it first sounds. Because, while it seems that a finite number of momentum or spin energy states (including position) will repeat in an infinite universe, neither their position *relative* to others nor the time during which they express a given energy need be the same at all. Finite numbers of states in an infinite space may never repeat their precise spatial and energetic distributions, ever. Two particles with certain states, neighbours in our universe, might be re-created at opposite ends of the infinite fabric of the universe and never end up together again, (even though some states are paired). The pigeon-hole method of sorting states (like we did with the hairs on the heads of Londoners) doesn't work when infinity is involved. (There is yet another very good reason why the use of infinity to explain multiverses of this type doesn't really work which we will come to in due course.)

While all possible variations of matter and/or inflations of space itself may be accommodated in an infinite volume of space, it is not necessary to conceive of an infinite expanse of space to house all these variations. Tegmark's Level 3 quantum Multiverse houses all quantum variations in a multitude of universes superposed upon ours occupying, more or less, the same spatial extent as it does, only raising the overall density of events at a point to an extremely high level. This type most occupies our thinking because it

> The question of what could actually be in the rest of a spatially infinite universe beyond matter is an interesting one. Any amount of energy would grow infinitely large (can God make a stone so heavy he can't lift it?). Suppose there were just numbers or information; *instructions*. Presumably there are at least *coordinates;* infinite series of numbers which describes the extent of say space-time. However infinite series of numbers each require their infinite space to be in (otherwise they are not countable), and each of those require it too. There is a nonsensical infinite regression here. Which is perhaps as it should be, but doesn't really inform.

relates to observations right now in our world and the nature of our reality.

Everett, De Witt and the multiverse level 3

This is a quantum multiverse and arises from the formulation of the quantum wave description due to Everett, elaborated by Zeh and Zurek and popularised by De Witt and Deutsch where at every observation our universe 'decoheres' or divides up into as many universes as there were probable outcomes of that observation, all separate from one another. All the universes of this multiverse have the same origin as our universe – the Big Bang. This is the multiverse or many worlds scenario.

The multiverse picture derived from Everett's viewpoint is one where the universe is made of a single quantum description that gradually evolves over time with every possibility of an event superimposing itself on the global wave function. What we experience is a gradual increasing level of randomness in the events we see, while around us parallel lines of development are going on, determined by the probability distributions of each event but which our consciousness, apparently tied to one version of the sequence of events (because of its memories, perhaps), cannot see.

Time is not really symmetrical in tiny particle reactions. Although, if a sum of all possible ways in which particle behaviours and their mirror images are taken then a reaction can have a symmetry in time. We can think of the situation as one set of events going forward in time with another different but equivalent set going backwards.

The question arises about what happens to the superposed states which make up the total value of the wave when one has been chosen in the measurement? When the coin comes down heads, what happens to the wave function part describing tails that existed while the coin was spinning? Is it just a mathematical notion? (There is a notion, 'contrafactual definiteness' that says that all those other possibilities have to be realisable, have to be 'true' like the balls in the sack. They can't be states that could *not* exist in the circumstances of the measurement, but they could still be real states about which we have no information (gödels in other words, and about more, later), although the Bell theorem may put the idea out of the running.

Since Everett, there have been two views on the measurement. Physicists tried to give meaning to observations that extracted a definite value out of the wave function by saying that the complicated set of probability values *collapses* down to a single measured one. This was the Copenhagen interpretation, furthered by Niels Bohr. After the measurement, there cannot be other choices of values still existing, because the superposed wave functions collapse to a single meaning – the value of the measurement – while the others disappear, in what is also referred to as the 'one world' scenario.

Everett proposed to do without the collapse of the probability wave function (the vanishing solution) since no one could say exactly what the physical process was or when it happens. He took the meaning of the wave equation to mean what it says, a superposition of various solutions. Say, for a particle wave (pWave) that can be in spin up or spin down we can describe the observation (non-technically):

$[pWave_{spin\,up} + pWave_{spin\,down}]$ **X** Observer
=> $[pWave_{spin\,up}\, Observer + pWave_{spin\,down}\, Observer]$

This seems to suggest that all the possible values of each spin state mixes with the observer to make a distinct mix (because the probability is different in each case) and which persist even after one gets chosen in the observation. These mixes are complete copies of our universe but with a single change, namely the value of the probability of the wave function. The problems are many but two stand out: how to explain the continuity of our history, since there seems to be no preferred viewpoint, and what to do about mass, because although mass is not multiplied in the superposition – it is shared (!), it has to come into being for each universe as the superposition ends, otherwise the universes would not be complete or be able to give sense to the probabilities of the state which begins them. We know the universes are complete from particle splitting and even from classic slit experiments, so if they have they have their own mass, from where does it suddenly come from for all these worlds? Is the mass supplied by the vacuum state of the universe? Do these worlds all share the same space-time we are in? Since these universes share the same inflationary start-point, and every possibility has to be real, then that start-point must have delivered mass into every possible universe or the multiverse is not made up of fully independent universes.

The mass problem is not solved, but Zeh, Zurek and others came up with an understanding of the way a superposition ends, namely that the environment causes the separation of observer mixes. We observe one value while the other observer mixes lose the ability to interfere with each other and go their own way in their own *worlds*. This is decoherence.

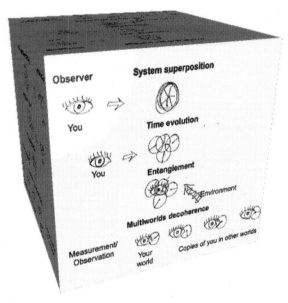

These other universes go about their business developing in the same way as ours but each with a tiny difference in the measurement. Wherever there is a random choice in a measurement, universes are born unreachable from ours and where the other possibilities are worked out – the quantum version of the Multiverse scenario.

The length of time the mixes may continue to interfere is not known, but for a short while they might still interfere with each other, and the world we end up perceiving experiences this interference. The question arises as to how many worlds in the mixes are required. Since the total wave function can be broken down into any number of orthogonal summations, is hard to see what imperative there is to limit the numbers of worlds, and many physicists think that there may be an infinite number of them. However, not all the values can make sense in our world. It seems more likely that there are only as many worlds as could provide a measurement that makes sense *in each world*. A natural brake, therefore to their multiplication when an observer is involved. Carefully framing the experiment, and thus our expectations of a possible value, should do this. But the mathematics do not demand this solution. However, a solution does present itself when we include consciousness, explained below. Moreover, it is our suggestion that there are in fact only two worlds appearing at the measurement, namely one of memory and one of prediction, representing a superposition of time, as we introduced earlier. We will discuss this further when we consider Schroedinger's Cat.

Problems with multiverses

Quantum theory is spectacularly successful in getting the right numbers, but less successful in giving us a picture of what such an interpretation actually means in the physical world that we live in, and particularly what an 'observer' or a measurement actually means for us here and now since the equations don't say whether the observer actually decides or causes the value.

The observer

It is generally accepted that what is meant by 'observer' is anything in an interaction with a system that changes a state of the system. Even though, in classical quantum theory, the observer is considered to simply to receive the information about the measurement mixes with the states, we have discovered that when a mind knows something say, about the location of a single particle entangled with itself (or just could know it in principle without actually articulating it), the entangled state vanishes (see **A side note about Wigner's friend**). Thus the information it receives is altered by the very reception of the information. Particles can act as 'observers' and that seems to happen even when entangled. When an observing system interacts with the observed system in the multiverse interpretation there is some kind of (eventual) changes of state but at what moment and indeed where in the observing system the state of the observed system gets changed in the observation is a question not yet solved.

In the multiverse view, where we mean 'observer' as a *single* consciousness spread over all the possible values of the system in superposition, then we might expect the number of values (or 'worlds') in the superposition would be limited by the number of differences (in the system) that could be created and/or sustained in the consciousness of the observer, that is to say, values compatible with the states of consciousness. The limitations to the numbers of multiverses would then depend upon *which and what kind of observer* is making the measurement. That is, the observer is already part of a system that restricts itself from taking on other states. If the 'measurement' is being made through the interaction with an observer whose state has been already decided to have a single or restricted set of values, then there should be either no separation of universes at all since there is just the single state, or just a few worlds representing the limited spread of possibilities for the observer. Thus, worlds cannot be created for probabilities other than those that satisfy the observer. The superposition would be (using a schematic and not standard notation),

$[pWave_{spin\ up} + pWave_{spin\ down}]$ **X** Observer $(i, j)]$
=> $[pWave_{spin\ up}$ *(restricted set of observer states i)* Observer + $pWave_{spin\ down}$ *(restricted set of observer states j)* Observer]

This might have fairly serious implications for quantum computing because it would suggest that it is increasingly difficult to find an observer system that can coexist logically with many superposed states in a superposition. Another quantum state may be happily capable of doing this, but can a more complex system like a mind also do it? A quantum computing result may require a hierarchy of reductive interfaces before its values could be read by a conscious observer (We will examine the idea of a quantum observer observing a quantum observer — a thought experiment sometimes referred to as the problem of 'Wigner's Friend' — further on).

Many have suggested that for every time there is a choice, multiworlds spring from that moment. This is a misreading of the nature of the superposition. Superpositions do not occur when there are choices, they occur when there are undecided random probabilities. Just because we choose one thing rather than another, it does not mean that there is a world somewhere where the other options not taken are playing themselves out.

The equation of the Everettian scenario repeats the same observer throughout the superposition, but there is another case in which the numbers of different worlds can be restricted where the variations of probabilities are those compatible with variations in the observer that are compatible with itself. In this scenario, the worlds would indeed be seeded differently, whereas in the former case above, the worlds would have identical observer histories. In the latter scenario there would be no consistent (human) observers across the superposition and thus confirming a range to the experimental results. The statistical range would actually be a measure of the variations possible in an observer while also maintaining observer integrity and history. In other words, it would either be the observer's past which determined the possible range of experimental values the observer would see, namely memory, or the observer's future, namely prediction, or more explicitly, a superposition of both sets of restricted states. Such that we have, for the measurement situation above, the wave function separated into 4 'sets' where half are in the future:

$[pWave_{spin\,up} + pWave_{spin\,down}]$ **X** $[Observer_{past} + pObserver_{future}]$

=> $[pWave_{spin\,up}$ *(restricted set of observer past states j)* $Observer + pWave_{spin\,down}$ *(restricted set of observer past states j)* $Observer] + [pWave_{spin\,up}$ *(prediction set of observer future states i)* $Observer + pWave_{spin\,down}$ *(prediction set of observer future states i)* $Observer]$

where *i and j* are distinct sets of past and future states compatible with the Observer's history.

Here we have a spread of values in time. We will see the significance of this when we look at Schroedinger's Cat and the famous slit experiment (see **Multiverse interference and the slit experiment**). (Stephen Hawking famously remarked that whenever he hears Schroedinger's Cat mentioned he reaches for his pistol, but it can't be helped; it is so useful to our discussion. It has entered the pantheon and cannot be removed.)

If the multiplication of 'observers' in the multi-world scenario *is* determined by the range of probabilities in the system being measured, we have some severe ethical or moral problems to contend with, and which we shall come to all in good time.

In this regard the notion of complete 'other worlds' is often taken too literally by many commentators who seem to think they will be very dif-

So while there may be an infinite scope for worlds to be created around events, there seems to be a limit to the total number that can branch off from them, and we have to conclude that somewhere, too, there is at least one world that is precisely as it appears and where subatomic particles all behave as particles and not waves and no interference patterns are noted. Such a world must also contain the complete observer (that is with all the pre-loading required to establish every observational value), since knowledge establishes particles, and who does not predict because its memories are complete (all information has been preloaded), a Boltzmann Brain perhaps. (see **Boltzmann Brains.**)

ferent places to ours. Consider, during a deliberate experiment, each world inherits the same intentions of the repeated observer entangled with them (Everett style), and each world is observing values in the same experiment for the same purpose. The fact that the values each observer records may differ doesn't change anything about the experiment, the observer, the observer's world, and the likely conclusions extracted from it. The values must conform to the same quantum considerations for each observer, and each observation will confirm the same conclusions. For most occasions of random choice life is likely to continue precisely the same in each world, which again begs the question, Why have them at all? In fact we will argue that the only significant 'spread' of values is simply {this value} and {not this value}, giving rise to just two worlds, namely a value and its conjugate. We will come back to this when we discuss decoherence.

No one has yet been able to explain how the wave equation is divided up amongst the worlds of the multiworld scenario since the wave is composed by simple addition and can be simply decomposed mathematically into any arrangement of values, which is to say they can be correlated in any fashion whatever (i.e. this questions the structure of the superposition and precisely what process allocates the distribution of worlds among the observer). This characteristic of the wave equation is compatible with our notion of observer variations being distributed across the many worlds. Further, since the creation of worlds involves the separation of the wave function of the entire universe from each of the others, how the separation is transmitted to the furthest reaches of the universe in the instant of decoherence cannot be imagined other than by the speed of light, the limitation of an Einsteinian universe, so that there must be a continual 'unzipping' of these worlds from ours travelling outwards at the speed of light from every single observation and overlapping with each other.

Being First

We do not know for sure if the other worlds in the superposition remain in superposition after ours has separated, the equations suggest that after the measurement all worlds go on evolving with their own value of the state measured entirely independent of what subsequently happens in ours. If they do then an event of decoherence would have to happen in each world to produce a value for them. But if all the worlds decohere from the superposition at the same moment our world decoheres and gives us a value of our measurement then these worlds begin to evolve with a value that is to some extent caused, or at least chosen, by our measurement. The evolution of the value of the state in each world has been arrested from evolving naturally by the result in our world.

The measurement you, the observer, happens upon in our world (or rather, reached in to the mix of probable values and just grabbed one, leaving the others behind) determines the measured value in the other worlds (hidden variables?) regardless of what the probability of the observed

> Before experimental physics say, in the world of early humans, were universes also continually splitting off on the same way as we think today? Is there an important distinction between universes created by random happenings in the world and those created by humans with consciousness and deliberation? If the observer interacts with the measurement then it limits the values as we have discussed, whereas for non-conscious 'observers' this is surely less significant.

value might have been for them when you measured it in our world. Certainly the *timing* of the measurement has been determined by whichever observer reads his value first. This situation is rather like the way a coincidence between two timelines produces an outcome in both not implicit in the probability of the evolution of each time-line separately.

While these worlds may continue to conform to the same probability theories as you do, in this particular case, the probability of the outcome that occurs for each of your yous in those worlds was actually determined not by you or a probability in those worlds but by you and a probability *here*, in our world. Every world's measurement would, therefore, be correlated with ours and certainly not fully random. Likewise, the 'you' in this world could be the result of a measurement in another world and you wouldn't know it. We may be living the outcomes of other people's experimentation (the Present Completeness Theorem is looking shaky again). This certainly suggests a mechanism by which influence from other times can affect the present. If an observation occurred in any one of the other universes in the superposition before it occurred in ours it could produce interference or establish our observation in our world irrespective of the evolution of the probability in our world.

(In fact, the timing of the superposition and separation of worlds might be given by Heisenberg's uncertainty principle which helps avoid the occasions where a state might be needed to have a value of probability (because of a measurement elsewhere) that it cannot 'produce' when a measurement is made.)

Observations of this sort, that is to say, measurements determined by the experimental set-up, are peculiar to consciousness. If we could observe this mechanism of the 'unzipping' of worlds we could prove the presence of other minds on our universe.

You could argue, of course, that it cannot matter which world is responsible for the outcome since it is precisely the same quantum behaviour in each, except that if there is no difference then the Everettian scenario is pointless and only shows that probability is some kind of *disposition* (or propensity) to behave a certain way and we are back to square one. What's more, this disposition would be *local*, and local variables cause a problem with the Bell inequality.

This problem of the nature of the interference from other worlds also seems to cause difficulties in the role of consciousness in observations since your consciousness in those other worlds would be functioning fine and yet not involved in determining the outcome or probability in those worlds during your measurement until they made a subsequent observation of their own. At least once, in every multiverse, consciousness is not involved in determining an observation. Not a very rigorous solution. Those worlds will be, in effect, subordinate to ours (if we got our measurement in first), and slightly behind us in time (rather like those bubble universes in the chaotic inflationary scenario). Their subsequent observations are built upon the one forced on them by our measurement. (One can see how the monasmotic universe could arise.) This is a relative view because apart from this one observation you made here, you in those other worlds would similarly dominate the separation of states in the next observation you made. And so it goes for every observer in every world. The first world to decohere determines the values of the states for all the other worlds. Or it seems (see **Schrödinger's Cat**).

(While these branches or multiple copies of our world are all included in the global wave function that describes our Universe they are likely to be visible from only outside the space-time we occupy, not from within it.)

We should note that the experiments in entanglement or in dividing particles and recombining them produce nice results only because the experiments tend to involve just two choices of 50% probability in the events examined and thus the evidence may confirm at best just a 2-verse that do not fully decohere before they interfere with each other and provide patterns of observations like the famous slit experiment. (The 2-verse easily supplies a reason for the value each takes on, namely one value is the complex conjugate of the other.) So our evidence for the many worlds scenario actually comes from observing a *merging* of two worlds scenario and not from any empirical data about separating worlds.

It is easy enough to argue for the 'multiverse' when the superposition contains just two states, but when the mix is not between two equal probabilities but among three or more states of varying probabilities, the problem of how exactly the probabilities of observation get parcelled out among the components of the particle wave has not been solved. There is an experimental hint that more probabilities all come down to mixes of 2-verses which we will come to. Mixes of 2-verses might be related in simple power law arrangements based around 2^n which could explain many of the relationships between parameters and constants of the subatomic world.

In the multiverse interpretation of quantum theory, Penrose thinks an observation does not choose the option out of the probabilities (decoherence or collapse of the wave function, depending on your point of view) but some other kind of event, like gravity, which tries to have single objects in its field, collapses it (or focusses the blurred image, the opposite to decoherence as we shall see). Two or more elements of a superposition close together have to work against gravity to remain separate otherwise they collapse together in order to occupy the same point in space-time – an actual wave function collapse agreeing nicely with the Copenhagen Interpretation. But so far most physicists prefer the decoherence solution.

Decoherence

Several problems are associated with decoherence. Decoherence (that is to say, the 'unhitching' of the other worlds from the superposition) is thought to occur because of interactions of the observer-state mixes with random particles or energy quanta in the environment. A stray particle here, a cosmic ray there, and the wave functions of the worlds become sufficiently altered or spread out to no longer interfere with our world. This decoupling happens very quickly. But, as we have noted, the values of the measurements in any world will be determined by, or at least contain, randomness from the coupled particle not originally in the superposition. They should, in fact, be the reason for *departures* from the expected probability in those worlds); they should lose their unitary nature and no longer add up to a probability of certainty. (The forcing on our world of a value established by decoherence in another world might be shown in the observations of rare outcomes, or indeed, wherever probabilities don't seem to add up.)

Yet decoherence produces another problem, how a value for an observation can be arrived at if the observer is also part of the superposition and is also evolving randomly in it (to be part of the superposition). The Arab tale (see box) might help us here and recall the concept of probability volume where the observer adds enough probability to the system for a value that makes sense to the observer to 'condense' out of the mix.

If we allow the observer to be included in the superposition then we must accept that the surroundings, the rest of the world and the universe is also; a wave of entanglement that flows out into the universe at the speed of light (shortly followed by the 'unzipping' of unused multiverses).

> There's an old Arab tale about a man who had 17 thoroughbred horses. When the man died his will divided the horses among his sons like this: 1/2 of the herd to his eldest son, 1/3 to the second son and 1/9 to the youngest. The despairing sons had come to blows trying to work out this 'impossible' division. A poor neighbour passed by leading his own mangy horse. The sons asked for help. The neighbour agreed if they would buy his horse. The sons paid the neighbour handsomely and put the nag into the corral with the other thoroughbreds. The neighbour asked each son how many horses was his share. The eldest said his was half. 'so take your half,' said the neighbour. The eldest son counted out 9 of the now 18 horses in the corral and drove them off to his house without looking back. To the middle son the neighbour asked, 'What is your share?' 'One third of the herd.' The neighbour indicated, 'take your six.' The middle son counted out 6 and rode off to market. The youngest said, 'My share is 2. 'so take your share,' said the neighbour. The youngest took the best remaining two and rode off, leaving the neighbour's wretched nag in the corral. The neighbour smiled, bridled the old horse and went on his way.

This is called a Von Neumann chain, and this is how the idea of consciousness being involved in a measurement came about in the first place. Along with the Von Neumann chain is the Von Neumann cut, the place in the chain where the hierarchy of entanglement stops and a discrete reality takes its place (a point Wigner's Friend thought experiment grappled with).

So decoherence leaves something out of the story. The superposition itself does not determine the final probabilities or their distribution among the splitting off worlds, and we are, therefore, relying for the value of our observation on a separate random event in the environment, that has nothing to do with the observation, yet which makes it concrete and stable for us. Even though it appears that some 'worlds' are more 'robust' against the environmental influence (known as the principle of superselection of characteristics like charge, and the principle of pointer states as devised by Zeh to explain the stability of values that always give classical physics values – something of a repudiation of probability that forms the superposition, it seems) and these are the ones that determine which of the observers is *us*, the problem of the multiverse 'arrangement' is still not solved. Certain states or wave functions may be infinitely resistant to changes produced by coupling with other wave functions in this way. These states are what we will always see in the measurement.

There is something very Newtonian, or at least classical, about these privileged pointer states and the special pleading for decoherence. We have a superposition of many worlds, and all these worlds are interchangeable (*fungible*, as David Deutsch considers) and any one of them could produce the measurement and become our world, but for the fact that from our point of view, one world or one set of states is more likely to appear in the measurement than the others. Again, we must ask why the other worlds are necessary in this case. Quantum computing has these problems to solve.

If the consequences of the 'splitting off' of the mixes from the superposition in our experiment spreads through the Omnium at the speed of light, then there must also be the effects of separations, from observations made elsewhere, arrowing in on us at the speed of light and from all directions. As long as these have no effect on our world (just as Bell and Markov suggest), all well and good, but do they have no effect?

The number of worlds in a superposition is, at the moment, unknown. It is now thought (by Deutsch in particular) that the multiverses do not have to multiply to infinity but can hover around a more limited number to reflect the basic number of quantum states we are able to observe in ours (i.e the multiverse is what is called non-ergodic, it does not explore every possible permutation of worlds). But still the problem of what probability actually means if all these universes exist as decohered material worlds evolving from a different observed value of the measurement is not solved. Because decoherence, in this scenario, still implies exclusivity (this not that) the probability of an outcome relates to the versions that do not happen, but if all versions happen (and they must for probability to be the same principle in every world), how do we partition the randomness?

Let's take a coin toss as a simple analogy. (We need to be careful here because a coin toss is not quantum behaviour and is not properly random, as we discussed.) Let us suppose that the heads and tails of the coin could be two states like 'up' and 'down' spin of a particle. During the toss the coin is in a 'superposition' of two possibilities, heads or tails. If you observe heads then, according to the multiverse scenario, the tails version decoheres and becomes the result for the same *you* in another universe.

(Suppose the coin toss is being tossed by an assassin: heads you live, tails you die. Then the multiverse scenario means you are going to both live and die with repercussions in both universes that may lead to the evolution of very distinct worlds. A matter we will discuss).

But let us consider a world where you are merely examining the probability of coin tosses. (In each world you continue to throw the coin since you began the toss with the intention of throwing a sequence and each will inherit this same intention.) The sequences in each universe differ by at least the first element from all the others but all show in the long run a probability of 1/2 for either heads or tails. If you go on tossing the coin forever, then every possible permutation of coin tosses will be realised in the multiverse and all the versions of you would be drawing the same conclusions about their probability. In other words, there is no difference between them at all, except for the initial term of the sequence of heads and tails that you are throwing.

When we are talking about superpositions for just two equally probable states then there is no difference in significance between worlds. The same overall probability would be observed in each. In the long run, each world will pass through a sequence of observations that sum up to the same expectation of probability in all of the worlds. If there are any differences, they won't come from the coin toss. In fact, a sequence in any of the shadow worlds could reproduce exactly the sequence in your world but be at least one 'step' behind. In other words, the same history but reproduced at best two steps further in your past.

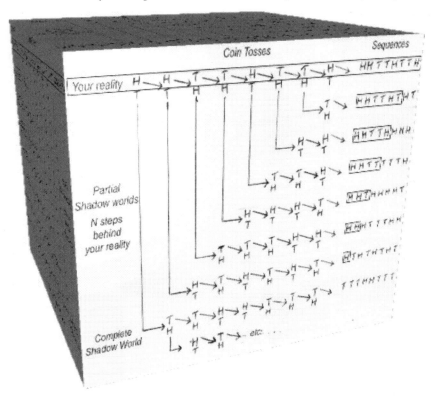

When there are superpositions of more than two states, like position and momentum and spin, each of which is evolving through its own range of values, then we have worlds where very different observations could create, in principle, more serious repercussions to the subsequent evolving of any of the worlds.

The single coin toss is not quantum-like but tossing two is more analogous. If we tossed *two* coins at once as a complete system, then we could observe say a head and a tail in our world, while the shadow world would also show a head and tail. Thus, the shadow world only differs when the coin faces are identical. With odd numbers of coins, however, the shadow world cannot be the same (e.g. with three coins, two heads and a tail here gives two tails and a head in the shadow world), while with even numbers of coins, it can.

This is very interesting since experiments have been done with shooting photons through *three* slits. It turns out that the interference pattern made by three slits is just the interference patterns of *each pair* of slits superimposed. This suggests a possible preference for the 50-50 probability options, even though a third pathway option has been created and the famous Feynman method of sum over all paths integration should mean that the interference pattern is different. If the result is confirmed, then it would mean that the middle slit is shared equally with the other two. Just as with a toss of three coins, the middle coin forms the same pair with each of the other coins (e.g. H T H gives two pairs of HT and TH) seeming to suggest the world of the 2-verse is preferred over more complicated arrangements.

Energy worlds

There is still a more significant complaint about the multiverse scenario. Since the probabilities of a wave function are related to the amplitude of the energy (the Born rule) then the decohered worlds contain less energy in proportion to the fact that they were less likely to be measured, and as these branches propagate through space at or near the speed of light, space is full of overlapping and partially separated but increasingly rarer and less energetic worlds.

In each separated world – they are, in fact full universes – more child worlds will separate off as observations are made, so that there is a hierarchy of worlds each of which separates from its 'parent' by observations made in the parent universe. Each of these child worlds will contain a little less energy (having less total probability) than the 'parent' world, and as measurements are made in each of these worlds, the energy (or information content) to make further branches or child worlds falls. A cascade of child worlds all with less energy to change than their parents eventually dies away rapidly leaving a solid definite world with no, or very few, distinct options.

Curiously, the standard separation of worlds scenario has been thought of as being, in practice, precisely similar to the type 1 universe, the Nverse (an objection that we will come to notwithstanding). For if space is infinite in extent – or as vast as makes no difference (relative to the size of our universe), the number of possible worlds in our near infinite space is the same as the split-off worlds of the quantum multiverse. It is just that with quantum superposition all the possibilities are 'nearby' because they are (almost) exactly superposed upon ours, localised in imaginary space (an infinite zone often called Hilbert space). Whereas in the Nverse, the zones where each of the probabilities play out are far removed from each other, localised in the infinity of a single 3-D space and to which we could, in principle, travel.

There is, however, a crucial difference between the two. In the Everett version, the total energy of the local situation is distributed among the various worlds at the measurement according to their probabilities, whereas in the Nverse scenario, each zone is a self contained totality of superposed wave functions equal to our visible universe.

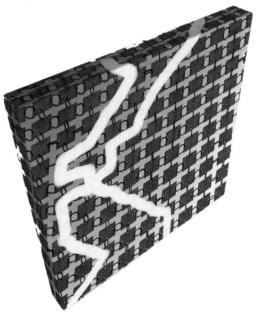

There is yet a third way that combines both these ideas into a metaworld scenario as described by Boström (not the same Bostrom of the simulation scenario), and not unlike the Nverse scenario. The entire universe is made up of all particles and their histories of states laid out while each 'universe' or world in the superposition is, therefore, just a structural selection, out of all the particle states that exist, of histories, and relative to the single observer, and not the entirely separate complete universes with the same observer in each as in the Everettian scenario. The universe we live in is then, a superposition of all the trajectories through this structure formed by the random inorganic event and the choices made by living things (different for each individual). Boström says, however, for the purpose of the measurement, a *world* can be just any collection of particles, any system that you have isolated, and other worlds are just other permutations of your collection, more or less similar to yours depending on the numbers of differences but no less real. This is the basis for probability, which nicely agrees, at least, with what we have been saying about the necessary reality of probable options (this question of 'realism' we will come to again).

For Boström, however, probability in this scheme is just the numbers of permutations in a given state over the total number of worlds. In other words it is a circular definition and doesn't solve the problem of which structures are 'chosen' to be the world of the measurement. Furthermore each state could belong to more than one path through the structure, rather like a brain's neurons, so the problem of quite how that state's information gets partitioned between the paths it belongs to is not solved.

This idea has a lot to recommend it, however, and we will consider more about it, but still the problem of what is maintaining our particular universe as a time-line consistent with history and with probability remains, since, presumably, our memories are a function only of the arrangement and will conform to whatever arrangement is chosen — they are not the persistent reference that sees to it that our world remains real to us.

Maths and measurement

Quantum probabilities are calculated with two 'rules' in mind. One is called normalisation, which means that all the probabilities of the various states that go into the state vector must add up to 1 (that is to say, any value has to be a real one for it to occur), and the second is a trick called the Born Rule, whereby the height of the energy wave describing the particle state gives us a basis for the value for its probability of turning up. The rule is a trick that works but physicists don't know why it works (and it bothers them).

What normalisation means is that the probabilities of each state within the superposition exist in some way as relative to each other and continuously (there is a *structure* to it). How the states coordinate this when they are supposed to be evolving in their own way is a mystery. We can open up this mystery by going back to our coin toss. In reality there is a third possibility — the edge. With a coin, we can see it is the *physical* relationship between the faces that determines the relative odds of each one, so when the coin lands on its edge, the odds of either face seem to be meaningless since neither occur. Are we to suppose that after decoherence when we observe the coin on its edge there are also two universes created with opposite outcomes; two universes in the multiverse where both The New York Jets and the Miami Dolphins win the toss of the coin at the start of the Superbowl? If the coin on its edge is in our universe, do 'not heads' *and* 'not tails' still give rise to universes where they don't occur? In this case there *is* a 'collapse' of the wave function to a single meaning. In rare cases of a coin toss landing on the edge of the coin, no other worlds are created even though there was (perhaps) a superposition of three possibilities before the coin fell. We will confront this situation again with Schroedinger's Cat. This leads us back to the universe's preferred 50-50 choice in a superposition.

The information available in a superposition includes all states but when one possibility occurs then the possibility of it occurring was, subjectively, certainty at that moment, so where did the information about the other states go? (This was the question that supporters of the Copenhagen interpretation ducked.) Until the observer reaches into the superposition and 'absorbs' a value of a point in space into its history, that value is not going to be certainty, it will be less. Only the expectation that there will be *a* value is certainty (i.e. meta-information), so the question of what determines the actual value is not really solved, unless the observer supplies it. But even then there is the problem of how the observer supplies the missing probability to make a likelihood of certainty *before* she observes the state since until the probability for a value becomes certain, it cannot be observed. How does she know how much to supply?

This brings us right back to the notion that the possible variations in the observer wavefunction entangles with the possible values of the state and the one which provides the missing probability makes the measurement. There is observer selection which will also equate with probability volume. What the observer knows about the process under examination contributes to the differences between the multiplied observers in the superposition of worlds and the content provides the necessary information to create the value of the state that we actually observe. For example,

[AB-state wave function] **X** [Observer (what she doesn't know about AB]
=> [wave function A] [Observer (what she may know about A)] + [wave function B][Observer (what she may know about B)] + [wave function A][Observer (what she does not know about A)] + [wave function B][Observer (what she does not know about B)]

In most cases we can imagine a symmetry where the result, [wave function A][Observer (what she does not know about A)] becomes simply [wave function B][Observer (what she may know about B)]

But is that always the case since this mix of fours states requires at least a prediction?

> The debate about what quantum theory actually means in reality recalls the debates of the nineteenth century about whether the concept of a molecule was just a way of making calculations or an actual real thing. Boltzmann himself had trouble accepting the idea of a molecule.

We can't yet solve this conundrum. Even so, without clarifying the physical reality, the mathematics has been very successful in predicting and explaining the statistical distributions at the atomic level. It works so well that there is a movement among physicists to consider, just like the Pythagoreans before, that the only thing real in the universe is the maths itself. There is no material 'stuff' with which things are made, only mathematical numbers and the relationships between them exist. The wave equation is not actually a wave of energy, just a way of calculating the probabilities of a value in the measurement. This solves the problem of the physical meaning of a measurement at a stroke; it is just a number and not a condition of matter.

There are still others who think that the presence of mathematics proves that we are the creation of a computing machine, a simulation, which is just the same argument that people used to use for the existence of God, namely that, because Nature obeys laws (like a clock) there has to be a lawmaker – God the clockmaker. We will discuss the simulation idea later on.

Tegmark has classified a Level 4 multiverse as where every mathematical structure that exists (and works) does exist in some physical sense. All multiverses, therefore, have a genuine presence, although this does seem to pre-suppose we know what a 'physical presence' actually is, and that we know why a maths idea is different from any other idea that we may have or from a language statement we can make, like, 'Maybe I will go and get the ice-cream after I have finished this page'. Even the most abstruse mathematical paper has words of language in it, since our brains require simple questions like, 'What does the variable a mean?' to be answered in ways that explain what it is doing there rather than just as an obligatory element to the equations.

Do these words *not* play a role in the mathematical arguments we create, or are they, too, symbolic formulae representing mathematical truths? There is a chicken and egg problem here, not to mention gödels. Either mathematics is just an extension of the way our brains think and that all thoughts including mathematical ones belong in the same world or we have to explain quite how and where mathematics joins with the rest of our consciousness. (When we discuss language we have an interesting explanation to put forward.)

When we come to talk about information, we will note that all permutations of words, strings of bits, formulae have to be pre-loaded in advance for any one occurrence of them to have meaning. So perhaps we can think of the 'presence' of mathematics in the world as examples of pre-loading. Quite what this preloading may be is what we are working our way towards.

Certainly, maths is the kind of idea that is unbounded, and can go on and on to infinity, whereas 'stuff' suggests limitations which can be reached at some point. Even so, maths needs 'containers' in which it can be held and does contain circularities, that can only be resolved by recourse to something other than maths and infinities, which either solve nothing or everything. (Circularities will become important when we discuss more about time.)

There is a simpler reason to suppose that maths has an independent reality, or that it is 'pre-loaded' by some mechanism, and it comes from the argument for decoherence. In the first case above, if decoherence stops the superposition and extracts a value from the interference of the worlds then any value in the other worlds has been to some degree determined by your measurement in our world; by the timing of the measurement, in fact.

As we touched upon earlier, *their* measurement has not come about in the same way as in ours, it has come about strictly through the amount of time passing between the superposition and the measurement in *our* world. Their world is a tiny bit more deterministic (not to mention the 'hidden variables' cause). Of course, they will go on making measurements in their world and the next time they do the experiment they have the chance to decohere the superposition and find a natural probability for themselves. And all the worlds arising up from that superposition will be as similarly determined as the earlier value; they would be coordinated in some way not entirely random. (This might be illustrated by the Bell inequality experiments.) In fact, if this really is the case, then the numbers of worlds that are having measurements directed for them infinitely outnumber fully probabilistic worlds, and thus our world is likely to be one of these, where the order we experience, the mathematics, is enforced due to measurements already taken in other worlds (another argument for us as a simulation) which is to say, taken in our future.

If the full multiverse is described by the housing in a separate universe of each exclusive probability, then all possible outcomes exist at the same time but unconnected with each other. In other words they are only *probable* in the sense of being separated by a form of 'distance' or structure from our world that keeps them away; they are just as real. So the meaning of probability of an observation in our world has little to do with the likelihood of something *coming into being* out of its own nature and more to do with the selecting of an outcome from a set of existing universes (like picking a ball out of a sack), all of which could have come out of the Big Bang and are determining measurements right now. This idea was tested after a fashion in 2017 using the general public and a web site to generate 'truly' random numbers to use in an entangled experiment. The results appear to indicate that randomness is really that; it is a property of the now. If all the future probabilities had been created out of the Big Bang, however, then it is hard to see how this experiment specifically avoids them. About more later.

Given a sense that the multiverse scenario is somehow too extreme to be real, there has been a movement by physicists to see the possible worlds in the superposition that are not selected in the measurement as somehow just shell worlds and 'empty' of reality. They don't have mass or charge or any of the fields that make up our universe and with all the problems that ensue from trying to explain where all that reality comes from just when it is needed; they only have 'description potential'. They 'look' like universes, that is to say, all the maths works in them, but aren't in fact real ones (it is not known if they have photons). In a measurement, then our real world including observer would have to suddenly flip into being a shell world in a superposition of shell worlds before the actual value is extracted and then flip back into reality when the value is measured. But since the observer would also have to be in the sudden shell world version of our world it is not clear who or what could make the observation (since shell worlds are ghostly and not real, rather like dreams in fact).

The reality of probability (as opposed to its 'virtualness') is central to the equations of the states of the particles since a state value is dependent on all those observable facts of mass and energy and the rest. It cannot be that every world is a shell world until one becomes real for that would be just reality by fiat. In any case, in our world, locally, there is always going to be a change of state going on, so it would be in continuous measurement with no chance of flipping into a shell world for the duration of the pre-to-post measurement environment.

Armin Shiraz has nicely illustrated the problem of shell worlds with the following description. A point on a 2-D plane cannot be expressed in 3-D space, since it has no third coordinate, unless it is by a line stretching to infinity formed by all possible values of the third dimension connected to the 2-D position. Only when reality is compared to the description, is a value found for the third dimension, and the line collapses to a coordinate now defined in 3-D space. Thus any description of a lower dimension in a higher dimension must contain virtual structures (like a line) that don't actually exist and which 'collapse' to a coordinate when the reality is established. Again, though, the measurement defines the value and not any internal 'truth' about the state.

If shell worlds are not possible could it be that there is yet another structure organising the worlds in the branching at the measurement? We have talked about dice tables and the need to expand them in order to accommodate more unlikely coincidences or sequences. Such is the effect of an observer on a system; the equivalent of introducing the horse into the corral to change the probabilities of the ensemble without altering its own value (see the Arab tale). Without the horse method, what is there stopping the observer dissolving into a mess of fresh probabilities when interacting with a superposition?

But all these ideas struggle to avoid the circularity of needing a reality to define our reality. The mythical observer needs to be both apart from the measurement and part of it. We need to see all the choices expressed before there is a choice to be made, just as we saw that for a message to be meaningful, or to even exist at all, all its possible meanings have to have been already transmitted. But we shall stick with the multiverse for the time being since it bears nicely on questions of whether the past can still exist and be available to us in the present (see **Where is the past?**).

Entanglement of observer with superposed states is not the same as coincidence, because in coincidence, the observer's prior consciousness determines the meaning of an unintended or unexpected event, while in quantum entanglement, the observer's consciousness is in some way determined by the event it has chosen to observe.

The question of whether probabilities co-exist along with the measurement observed is called the 'realism question' and is one of the principles of determinism undermined by the Bell inequality (the others being locality, where the testing of one property of the particle has no effect on the other properties, and complete freedom of choice of which property to measure). Some or all of these three principles must not be true since experiments show quantum predictions to be correct. A particle appears to have no properties not even likelihoods until we give it some options. This is suspiciously like determinism backwards. (see **Present Completeness Paradoxes - Problems with the now.**) However, this analysis of the quantum state seems to imply a condition rather like informational inertia which we will examine further on. (see **Information and entropy**)

The appearance of choice

As we have noted, there appears to be a degree of randomness in our everyday world that has little to do with randomness at the subatomic level. At the human level, because of our physical size relative to the wave function that describes us, entanglement is hardly possible and choices occur in already decohered circumstances, so painting a picture of multiverses springing from human choice and free will is inaccurate. The chance, or luck, that plays such a great part in the human narrative cannot split universes in the same way as a photon. We are tempted to think (and it is a popularised thought) that all the decisions we didn't take here are taken elsewhere in these other worlds, and that for every single choice a whole universe branches off from ours and goes its own way, but this is incorrect. The appearance of choice (in ENT) in the narratives of our life is not the same thing as a particle in a state of coexisting probabilities. Almost all choices made by our consciousness in ENT are not choices among randomly arising probabilities superposed on one another but are courses of expressly determined action (even if we don't know or are mistaken about what is determining them). For example, the tossing of a coin. While the outcome of the coin toss may lead to different outcomes at the human narrative level – like choosing who kicks off in a football game – these cannot lead to different universes because the coin toss is actually a fully determined physical process only we don't know the parameters sufficiently well. Just because *we* don't know which face will turn up doesn't mean they are therefore in a genuine superposition of random outcomes. Randomness here means a measure of a lack of knowledge. There is no internal randomness for the dice and so we are spared the anguish of a bifurcation in the multiverse every time there is a decision to be made at the ENT level. (Whereas superpositions of probability in quantum theory occur *before* decisions are made which looks suspiciously like entropy in reverse, and we will come back to this idea.) What we think of as chance in our lives only arises in what we know about other possible pathways or outcomes, (which is where Shannon's idea of information comes from – if we don't know any other pathways then the event was not chance and therefore not informative).

Now that we have introduced some of the paradoxes of superposition and randomness that come from our attempts to develop a physical picture of what might be going on at the human level of existence, perhaps it is a good time to introduce a famous thought experiment.

Schroedinger's Cat

A cat is enclosed in a box where a radioactive substance may expel a particle and trip a Geiger counter which then activates a circuit that opens a vial of poison and kills the cat. Say there is a 50/50 chance that after an hour a particle will have been emitted, then after one hour, the mathematical description of the state of the cat is one where its alive state and its dead state are superposed until the box is opened. When it's opened, the observer becomes entangled with *both* states which then separate from each other so that there are two realities each with an observer, where one observes a live

> The cat state at the atom level 'has been made in a number of experiments using photons to 'tune' atoms into a superposition of two states. But none of these experiments are analogous to the Schroedinger' box thought experiment, which poses the question of *when* the observation creates the measurement, since these experiments create a superimposed state deliberately.

cat and the other in our universe observes a dead cat. So which reality our observer sees (live cat or dead one) is at least in part determined by his or her interaction with the superposed states, his or her *mind*.

This is the standard picture of the thought experiment which many, including Einstein, considered a paradox rather than a true description of the quantum reality when it was first mooted. (After all, the cat is far too large an object not to decohere almost instantly.) Schroedinger introduced this thought experiment to highlight the difficulty of knowing where superpositions stop (that point is sometimes referred to as the von Neumann cut). Popular accounts have taken it rather further since.

If the alpha particle is emitted then surely the cat is dead whether or not the box is opened. The mathematics describing the situation (the state vector, or *SV*), however, does not give any meaning to that state of affairs. Until the observer connects with the wave function in the box and confirms the state of the cat, the cat is neither dead nor alive (now known as the 'cat state'). Furthermore, it is not possible to infer anything about the state of the cat before the box is opened even after the observer has opened it (even if she sees a live cat, there was only the cat state in the box before she looked). From deWitt onwards people have thought of the unobserved state as going on to exist in a complete but separated world – the multiverse of legend – brought into being at the moment of loss of superposition.

An immediate question occurs: the probability of each state is an evolving function (before measurement), so what about the dead cat state; is it still evolving? Does life imbue its particles with something which when missing takes the cat out of any superposition? Or is the superposition simply of the collection of atoms and molecules of the cat, however they are behaving, alive or dead?

Is the radioactive material *conscious* of the future, knowing the moment when it will emit a particle? Are we, every moment of the day, a superposition of our lives selves with are future dead selves? Thanatos lives!

The introduction of radioactivity into the discussion, however, introduces a more serious paradox that is not generally discussed. Superpositions of energy states of molecules and even ensembles of particles have been demonstrated for brief periods of time. But radioactivity is rather different. It is governed by one of the four fundamental interactions of the universe, the Weak Force (first proposed by Fermi). Every atomic nucleus randomly emits a particle (considering just this type of radioactivity for the moment) at an unpredictable moment and falls to a lower energy state. If this nucleus itself is in a superposition of many states each with a different energy and each evolving in time in its own way, then we have the lower energy state where an alpha particle *is* emitted (which we need for the dead cat state) existing somewhere prior to the actual event. The nucleus would have to have already decided on all its 'futures' before any of them had happened; to be already in a superposition of before and after states, before the observation has decided which is which. If radioactivity is truly unpredictable, however, until the moment of its emission, the nucleus should not 'know' (or be able to decide) that a particle will be emitted and thus should not be able to include that lower energy state in its superposition, which means there can be no dead-cat state superposed with a live-cat state before it exists.

A way around this problem is to say that all futures exist. In order for them to be sensible futures, i.e to have a history that makes sense of the experiment, these futures would have to be entire universes identical to ours except for the future state of the radioactive nucleus. Two questions, and they are the same questions for every superposition, arise, why is there only interference at this experimental moment and not anywhere else between the worlds, and why does this interference happen now?

Let us suppose the various possibilities of the nucleus are always in a superposition of multiverses and that the experimental setup connects with all of them, then why does the observer see one and not the others. One could argue that the low energy state exists in a separate universe where the alpha particle has been emitted and the cat is dead. And that the superposition is of both entangled with the observer, but the question of when this universe was created is not answered. It was certainly created before the observation of the cat made by our observer.

> The sad thing about the Schroedinger experiment, if it is true, is that it *forces* the death of at least one cat somewhere. Even if the cat lives in the observed box, it dies in the other universe. The death of a cat is assured after a certain time has passed. And this appears to be true for every probability, i.e. there is no probability about facts, only *where* they occur.

If there is a dead-cat state in the superposition before that alpha particle is emitted, then its presence constitutes a *prediction* by the system (the radioactive nucleus itself must be in a superposition with emitted particle states before any particle is emitted). When would such a prediction come into force and in what sort of phase space would it exist? The moment you put the cat in the box with the radioactive nucleus? Or the moment you got the idea to do it?

If the alpha particle *is* emitted by the nucleus creating the dead-cat state, there is no longer a live-cat state available to the superposition (before you look). If there is, that constitutes a *memory* of the system (recalling the problem suggested by the roulette wheel producing a random sequence of given numbers), and one wonders how such a memory could be maintained in place. Memory is definitely local (even if it is smoothed out over a region) but we know from the Bell inequality that particles do not seem to have all their properties (say all the possible spin values they could have) derived from a point in their past, like their origin (locality); they have no memories.

So, in the thought experiment with a radioactive trigger the superposition is actually one between two *pairs* of states, where a memory of a live-cat with a real dead-cat is indistinguishable from a prediction of a dead-cat with a real live-cat, which is not quite what Everettians like to agree with. It does, however, seem to agree at least with our consciousness. We worry over a dead cat and hope for an alive cat simultaneously.

The role of consciousness

What you observe also depends upon the reliability of what you see and which may have nothing to do with the superposition under examination. Suppose you left your glasses behind. When you opened the box you see only a blurry shape that may be a dead cat or a sleeping cat (the blurriness fits in rather well with the idea that the wave function of the cat has evolved to spread out – *smeared out,* in Schroedinger's words – in the superposition). In other words the scale of your observation has not yet settled the issue *for you* so why would it settle the issue for the superposition? This basic confusion about the observation is true for every situation where there are random options superposed.

Think about this. As we noted earlier, the observer is part of the superposition, and so is the observer's state of awareness. Whether or not observer awareness can decohere the superposition and decide the observation is something we do not yet know. But if it were not for consciousness making a choice; if the human mind did not yet exist to make this box, and we consider a situation where the random chance might have put a cat into a similar 50% state of dead-not-dead where some circumstance might or might not save the cat (will the tree branch break before the cat has time to scramble to safety?), would the same superpositions exist for it as they do in the box we have set up for it in our present? Would not the *cat state* simply be just one point in a series of random accidents in its life out of which a single moment of cat state would not be distinguishable?

The consciousness of the observer, involved in the superposition, seems to interfere with the connecting causal chains of chance and creates memories and predictions which alter the balance of

> If the observer opens the box at the moment of the alpha particle emission, and watches the cat dying, does the question of whether there was still a superposition beforehand have any meaning (especially if she whiffs the cyanide and dies along with the cat)?

options within the superposition. We can see this more clearly when we think about the role of radioactivity in the story, but first let us follow the experiment through.

When you begin the experiment, the cat seems to be still in its live-cat state or you wouldn't be doing the experiment. Since the superpositions of the nucleus evolve over time, the moment you put the cat in the box is also, one expects, a moment where the probability of the alpha particle being emitted is very low, although the chances do exist that the cat may die almost immediately, so it is fair to say that the superposition of states begins almost immediately too. The superposition of initially unequal states will evolve later to the point where the particle has a very high probability of being emitted, but for the moment, if there is a superposition of cat and radioactive substance, it is mostly live-cat state with a very little bit of the dead-cat state (or a very vague version of it) which is not visible to you (*pace* Monty Python). The cat is only a little less fully alive than say, her sister not in the experiment.

Certainly the superposition of live-cat with dead-cat reflect what is going on in your mind during the course of the experiment. The moment you put the cat in the box, its future life has come into question, it is no longer as certain as it was. The cat's dead future hangs over it. Even though it appears to be a live cat, it is not a fully alive one. This also reflects the equations of the state vector which can be composed by the sums of any mixes of substates of the dead and live cat states.

This influential decay is reminiscent of the action of a force. Could this be the quantum gravity we seek? Its calculation all depends on how the probabilities are arranged between the branched off worlds.

But does the cat consciousness also sense this, seeing as it is now entangled with its actual dead-cat possibility? Maybe the cat gets an inkling from its brain since the informational entropy state has, according to the Shannon scheme, increased, as well as the fact that its entire live-cat brain is now in a superposition with a dead-cat brain. But of course, that would be silly, since a dead-cat state surely does not include a live consciousness to connect and interfere with that of the live cat...or could it? The dead-cat state is just an arrangement of particles after all. (Could it be that we all could live for ever in a universe if placed in and kept in a superposition like this? Instead of freezing people, we entangle them and keep them in the Zeno state so they do not need food or oxygen. Perhaps this is the future of interstellar travel.)

So could quantum reasoning arise from a superposition of consciousnesses? Is our consciousness imposing the values of superpositions of the cat-state? Clearly, in this thought experiment, in a system *without* prediction or memory, the only component of it that would contain an expectation of the dead-cat state is indeed the observer, or at least the person who set it up. The observer's consciousness already contains the mix of states we attribute to the box: an expectation of both a dead cat and of an alive cat. So does our forward-thinking consciousness evolve in accord with our expectations of the emission of the alpha particle, the mathematics reflecting only a truth about uncertainty in our minds? (Just as the consciousness who set up the Bell experiment unites all results in its expectations. Using randomness doesn't remove this unity. Someone somewhere is holding all outcomes in their awareness.)

If the experimenter opened the box then perhaps his expectation, flipping from one state or the other and alighting on one at the time of opening the box, might determine what he sees. But suppose we told someone who did not know what the experiment consisted of, i.e that they had no awareness of the dead-cat possibility, that they should go and get the live cat out of the box, would that be sufficient to keep the cat alive and to resist the radioactive decay? This would be an example of mind over matter, and certainly an easy way to test whether such mind influence exists.

But if the experimenter was still nearby, maybe his influence would still be present in the experiment. Suppose then, we put the experimenter in a coma to get him out of the way and got someone who did not know what was in the box to take a look to see what was in it (some version of a double blind set-up), we must assume that

> Curiously there is a state called the quantum Zeno state (remember the arrow paradox) whereby if a measurement is made on a superposition of two states before a certain duration, the state gets reset to its original state. The shorter the time interval for the measurement the less able the superposition is to give up the second state and so a reversion to the initial state occurs. It is like decoherence in reverse, and a particle can get decoupled from whatever it is in the environment that is causing the decoherence.

the state of her awareness after she has opened the box has necessarily been *caused* by the fact of her observation and not the other way around and thus could not have been the source of the superpositions or the value measured. (Even that does not remove the fact that someone somewhere knows what is expected of the experiment.)

A side note about Wigner's Friend

It is hard to discuss the cat-state without mentioning the case where an observer (Wigner) observes an observer (Wigner's friend) doing the experiment out of sight (in a laboratory say) and who does not know what the experimenter knows about the cat until later. That observer (Wigner) would be, supposedly, in a superposition with the laboratory, his friend and the cat state, but this superposition would not decohere for him until he finds out that the cat state in the experiment has decohered. The experimenter (Wigner's friend) would know which state the cat was in, but nothing seems to have collapsed the superposition for Wigner as outside observer. Until someone tells him what the result of the experiment is, it's still a superposition for him. This appears to be contradictory, but is it? For one thing, suppose a total stranger walked by the laboratory with no idea what experiment was being conducted inside, would he or she be in a superposition with the laboratory and the cat state? If you are inclined to say, clearly not, then you are implying that knowledge of the cat state is the *requirement* for Wigner to be in a superposition with the experiment; an observer has to know what the options are to be in a superposition of them (sometimes referred to as the QBism interpretation). If you say, yes, then this implies that the entire rest of the universe is involved in the superposition and that there is a hierarchy of nested 'observers' going all the way back to the Big Bang (a version of Bohmian Mechanics). Nested observers is a problematic scenario if they are all using the same quantum description because it suffers from self-referential recursion – something Gödel would have recognised.

In fact, all this scenario tells us is the decoherence from the environment is increasingly significant. Wigner, observing the laboratory with lack of knowledge about the states inside, is not in an equivalent state to Wigner's friend before she opens the box not knowing the state of the cat: the environmental components are overwhelmingly different. Wigner's superposition decoheres only when he receives information from his friend and not from probability (e.g. when his friend ends the cat state by opening the box, Wigner's superposition – if there really is one – is with the past) which tells us that we are dealing with different frames of reference. This has been confirmed or at least suggested by an elaboration of this thought experiment by Frauchiger and Renner in 2018 where quantum theory appears to give an inconsistent view of what Wigner gets to know and when. The experimenters' view is controversial not least because it uses four observers: two experimenters in two laboratories and two outside observers observing the laboratories. The design of their experiment, however, requires that information is passed between the two laboratories which appears to contradict their proposal that the laboratories are isolated for the outside observers. Both laboratories form one system for each observer and the contradictions between who knows what and when disappears. (Wigner and Wigner's friend do not have to agree on what they know.)

Schroedinger's Cat continued

This is only a thought experiment designed to bring into relief the problems of observation. The live cat in the live-cat state would be interacting with its environment by breathing and observing things in the box and so is likely to couple with its environment and decohere the cat-state very rapidly into a universe of a live cat and a universe of a dead cat. But a dead cat is also interacting with its environment even if it is not breathing. Chemical reactions are still taking place but at a slower pace perhaps. So either cat version could decohere the superposition, before the alpha particle is emitted, and the cat we see does not necessarily depend on the time evolution of its world at all. Consider, since, according to our world, after an hour there is a 50-50 chance that the alpha particle has destroyed the superposition. As time goes on, the probabilities change due to the time evolution of the radioactivity. The dead-cat world probabilities are increasing to some degree while the chances of a live-cat in the live-cat world are also declining in opposition.

With normalisation, probabilities in the twin-world case have to be co-ordinated in some way not tidily explained by the Everrett scenario, but which, however, handily 'explains' the Copenhagen interpretation. The problem here is that not only

> The cat state at the atom level 'has been made in a number of experiments using photons to 'tune' atoms into a superposition of two states. But none of these experiments are analogous to the Schroedinger' box thought experiment, which poses the question of *when* the observation creates the measurement, since these experiments create a superimposed state deliberately.

does this interpretation leave us with the uncomfortable fact that an unseen probability may cause the observation in our world, it implies some kind of cross-world communication where (in this case) two waves of probability must flow in equal but opposite directions. Each is the conjugate of the other.

And now the paradox becomes clearer. If the radioactive nucleus cannot produce the virtual emission of the alpha particle state before it emits the particle for real then the superposition of states can only occur briefly between emission of the particle and before the dead-cat state. That is, the superposition of live-cat state and dead-cat state can only exist if the dead cat is *not yet dead* by the emission of the particle, but whose death state in our universe is inevitable (possessing contrafactual definiteness only after the alpha particle emission). The superposition in this case is a superposition of present reality and (just a little into the) future reality.

The present reality moves into the past of our consciousness. If the cat is dead, it is dead (and we remember it was once alive), and if it's alive, it's alive (with the anticipation of its death fading). Thus we see the memory and predictions do exist, briefly, in the superposition but only within the mind, not as separate universes.

Quantum reasoning in humans, if not in cats, is a well studied notion, especially when it comes to making social or political choices. Humans are able to happily exist in entangled states of choices that can appear to be contradictory or at least not strictly logical – that is to say, humans can hold different 'if-this-and-this-then-that' types of propositions in their minds which can alter over time. For example, the likelihood of a voter voting for party A is a function of policies that party B commits to, but also what their friends say they will vote for, what they believe the majority will vote for, how successful party A might be in committing to its policies and so forth. And all of these probabilities change over time. Nevertheless, scientists are trying to mimic quantum theory calculations to estimate the changing probabilities of a voter choosing party A given the probability estimates of all the other choices.

This *is* the Copenhagen interpretation. If the alpha particle is emitted then the dead-cat potential state merges with the reality, and the live-cat potential state no longer describes the reality and so loses meaning. The observer is entangled with the live-cat state only in the sense of holding that idea in mind, but as a reality it is not replicated in a separate universe. The live-cat state world no longer exists; it has been overwritten.

We can see that there are not just two states implied in this superposition, but four, in a merging of real and future states. But what happens when there are more than two states in the superposition? Let us reconsider the choice of radioactivity for this experiment and recall the problem of the tossed coin landing on its edge.

The cat-state and the radioactive nucleus

What is so interesting about this particular version of a thought experiment into the multiverse is that since the source of the decoherence remains in each world (otherwise it's determinism), we have recursion in the superposition.

The radioactive nucleus is unstable and the decay allows it to settle into a more stable state by shedding mass or charge, even though the new state may be also be radioactive. Eventually all radioactive atoms become non-radioactive. While it is assumed that random quantum vacuum fluctuations initiate the re-alignment of energies and the expulsion of the mass, these events happen in a systematic way within the nucleus where the half-life is a function of the spread of energies of say, the alpha particle it will eventually emit and the likelihood of it tunnelling through the barrier and the number of times it reaches the energetic barrier surrounding the nucleus. It is an arrangement founded very much in three-dimensional space and in the size and shape of the nucleus. We cannot say which nucleus will emit a particle or precisely when but we do know that after a certain time half of the nuclei will have decayed into something else. This hardly suggests that any number of outside influences or even consciousness itself could be involved in making a radioactive emission happen, rather it is a systematic time function of the entangled states of all the nuclei. (There is a reasonable view that considers that this time function may have been set in the Big Bang, which gives radioactivity a rather less than purely random behaviour we attribute to quantum processes.) Even if minds *do* determine (some) quantum states, we can be pretty sure that it is not the *same* mind that opens the box at some random moment that precipitates the emission of the alpha particle, but a mind *somewhere else*. (Uh oh! The Present Completeness Theorem looks shaky, again.)

Now let us imagine a special case where the radioactive substance is a single atom of ^{208}Po (Polonium, a radioactive product of Uranium). This decays into a stable atom of ^{204}Pb (lead) by emitting an alpha particle, when radioactivity ceases. Its half-life (a single atom decay is somewhat more random than a half-life of a collection of atoms) is nearly three years so instead of a box we can put the cat and the poison apparatus in a room with automatic feeders and so on to keep the cat alive for three years or more.

All this talk about time travel does have a serious point with regards to nuclear physics and the future of humans, to which we will come in due course. Imagine radioactivity and the flow of time, especially as radioactive decay is associated with time in a systematic way through its half-life. Going backwards in time would see entropy going in reverse. Entropic dispersion would turn into coherence, into negentropy, while energy levels of the regroupings would fall, which is what we have in radioactive nuclei. Suppose, for the sake of the idea, radioactivity is, in some way not yet worked out, connected with entropic time going backwards. We have connected time going backwards to an inflow of probability, so let us propose a new particle comprised of probability which moves slowly and is low in energy (which may explain why we do not see it in high energy collision experiments).

Once radioactivity ceases there are no more probable states for the nucleus so no more superpositions will occur. According to the standard description of the paradox, before the actual emission of the alpha particle, we have a world containing a dead cat and a stable lead nucleus, superposed on a live-cat world and a still-radioactive Polonium atom. You open the door to the room. If the dead-cat world initiates its decoherence from the superposition (by interacting with its environment) you will see a dead cat. But, because of the presence of the still non-emitting isotope in the live-cat world, the live-cat world, although decohered and separate from the dead-cat world you see, is still in a superposition with a dead-cat world. When an observer opens the door of the room in *this* world, the same separation of dead-cat world and live-cat-dead-cat superposition world occurs.

The same superposition, thus, multiplies on and on. At each emission of an alpha particle in any and all of the worlds, the cat in the live-cat world is as likely to live on another 3 years as it was at the start of the experiment, 3 years ago. That is, the uncertainty has been 're-set'. The cat has 'recovered' a low probability of dying in the live-cat world, and if you opened the door and saw a live cat you could leave it in the room for another three years with a low probability of it dying. After another three years and you opened the door and found a live cat, you could leave it again for three years. You would be puzzled at the long time it was taking for the Polonium isotope to emit a particle. What are people in this world to make of the experiment? The cat goes on and on living, the polonium fails to decay way beyond the ^{208}Po half-life of ~2.8 years. This is awkward because now we have a single immortal cat sloughing off dead cats into infinity.

This situation is similar to the coin landing on its edge. Once in a long while the ^{208}Po atom will emit a *positron* particle, instead of an *alpha* particle, through Beta decay. And leaving behind a ^{208}Bi atom. ^{208}Bi is radioactive but with a half life of almost 400,000 years when it also decays by positron decay to stable lead. So in this case our poison mechanism tuned to alpha decay would never fire and the superposition would disappear since the dead-cat state does not exist even as a prediction of the ^{208}Bi radioactive nucleus (as it never emits an alpha particle to trip the poison).

Let us write out the superposition of emitted state
(= *the state of the more stable nucleus + state of detector excited by an alpha particle + state of dispersed and therefore higher entropy poison + state of lower energy dead-cat. + state of observer*)

superposed on the non-emitted state,

(= the state of more excited nucleus + quiescent state of detector + state of undispersed lower entropy poison + state of energetic live-cat + state of observer),

and also on the rare possibility of emitted non-alpha particle state,

(= the state of more stable nucleus + quiescent state of detector + state of positron particle + state of undispersed lower entropy poison + state of energetic live-cat + state of observer).

In the supposed superposition of likelihoods the three specific possibilities must be lingering in the nucleus as it re-arranges itself. This would mean that the information about these cat states is being anticipated by the nucleus, and must be, indeed, inherent in the description of the nucleus. But consider, at the moment of expulsion of the positron, there should no longer be a superposition of live and dead cat worlds because the possibility of a dead cat has disappeared. Similarly, at the point of the expulsion of the alpha particle, a dead-cat becomes certain. In both these cases, the anticipatory state of the nucleus reverts to a single state. Only when the cat is still alive is a superposition required. So, it is clear that if the cat is dead, it is dead and it is not the actions of the observer that causes the cat's death. The observer is free to conclude that the cat was dead before she opened the door to the room.

In the multiverse scenario we have worlds in superposition, one of which becomes real for us. So, what is the difference between the dead-cat state in superposition and the actual dead cat? As far as the multiverse scenario goes, the difference is solely in negative probability. This must be true of all possible measurements that the radioactive nucleus may get involved in. The informational content of its anticipation must be transferred alongside the expulsion of the particle as a parameter to define the value of the measurement and initiate decoherence. Given our discussion about probability and event volume we might expect that this parameter is in someway a transfer of boundary information, a number, to alter the parameters of the state vectors. Let us call it a probability particle or *tychon* (from the Greek, *tychi* meaning, roughly, luck or the way things that happen without us knowing why).

Tychons

Is the transfer of the parameters of probability (to enable prediction) a force, therefore, and mediated by a particle? While energy transfers require particles, we can ponder how information in the form of probability, which is a measure of a wave function's dimensional extent would be transmitted. We can think in terms of sub-atomic exchanges of small particle-like phenomena or in terms of larger wave-like interactions over much larger arrangements of matter.

In coining the particle name, *tychon,* echoes of the speculative superluminal particle, the tachyon (described by Feinberg in 1967) were strong. Einstein's equations forbid matter to travel faster that the speed the light, except for particles with strange reversals of properties, the tachyons, who move backwards in time, move faster the less energy they have and who cannot move slower than the speed of light. If we did observe one it would be an indication of a future event although it is not clear what reactions would produce tachyons since their mass is imaginary. Our proposed tychon only contains probability information and is not involved in the conservation of energy. It is more like an amount of mathematics. Tychons don't interact with each other, because they are not the substrate upon which probability acts. They most readily interact with ensembles of states.

For a sub-atomic exchange of probability to go unnoticed we must think in terms of a particle of a low zero point energy expression compared with the high energy particles of the fields we are accustomed to investigating which is why it has never been seen by particle accelerators. The lowest energy particle embodying a similar idea in standard physics, namely a meson formed by a quark and antiquark, decays into other particles extremely quickly. The tychon would not decay into other products but get absorbed by them instead, and very quickly, inviting particles to enter their decay states after absorption. Rather than expressing a force affecting particle behaviour like momentum, however, it is more likely to be related to the quantity we call 'spin' and bestowing a greater range of possible values to states, or greater uncertainty in their internal cohesiveness or even to the confinement of the constituents of the nucleons. Perhaps interacting with the strong nuclear force as well as the weak force which suggests we might find evidence for it involving neutrons and electrons in radioactive *beta* decay (the recent search for the speculative X17 particle is an interesting one in this regard although its suggested energy of around 17 MeV may still be too high). There is, in fact, tantalising evidence that the neutron may express such a particle in the apparently different lifetimes (around 9 seconds) it experiences whether travelling in a beam or being confined in space. This is an important parameter as it controls how much helium over hydrogen there is in the universe as well as the said *beta* decay in radioactivity.

However, we can also consider a low frequency wave-like quantum feature with a very long wavelength which tends to be absorbed in parallel by groups of components when the resonance is sufficiently aligned. Indeed, we can propose a series of increasing harmonics that would interact with probabilities within larger and larger conglomerations of matter.

What is so interesting about this line of thinking begun in coincidence is that it seems to align with two investigations. One into charge-parity violations (where the expectation that mirror-image matter will evolve precisely the same as ordinary matter is wrong) and the other into ultra-light particles in the universe that may account for Dark Matter (these are not WIMPS, since those are weakly interacting but heavy particles).

Ultra-light particles have been introduced to explain dark matter, such as the *Axion*, a low energy subatomic particle (measured in MeV) introduced to account for discrepancies in subatomic accounting (charge-parity violations) in the Standard Model (Peccei-Quinn postulate). The Axion, produced in the Big Bang has to have very low energy otherwise the Universe would have too much mass and close in on itself quickly. It reversibly interacts with magnetic fields to produce photons and thus strengthening high energy radiation from far sources. Another suggestion for an ultra-light particle is that photons can behave like neutrinos and can flip in and out of a 'dark' lighter version of itself. And there is a slew of 'light supersymmetric particles' of various energies that may interact with the weak force in the nucleus.

> What would a particle of probability information look like? What kind of energy could it contain? A small loop of complex energy? The idea of a string leaps to mind.

Charge-parity violations, on the other hand have been introduced to explain measured asymmetries between anti-matter (matter with its charge reversed) and matter decay. In quantum theory, if one takes a charge inverted mirror image and time reversed view of radioactivity (essentially, particles made up of combinations of quarks) then one should get antimatter. But this does not appear to be strictly adhered to and particle combinations made with anti-matter decay differently to matter, and specifically a small proportion of matter does not obey the time reversal component to the transformation (non-commutability of probability again). If this effect can be found in other particle behaviours than it would certainly strengthen the tychon idea. Although the argument suggests that these will be found at higher levels of observation where probabilities and observation merge in the quantum functions of the brain. Tychons should be observable where minds exist.

The particle our discussion is hinting at may have both negative and positive values arising from its future (predictive) and past (memory) states. In our radioactive thought experiment this process seems easy to see, but what is there about a more ordinary observation in which this proposed particle makes probabilities real? Another famous experiment might help us here.

Multiverse interference and the slit experiment

In the famous slit interference experiment (originating with Young) the conclusion that the probabilities of where and how much in a beam of individual particles, say photons, passing one at a time through an arrangement of slits, interfere not just with themselves but with a separate version of themselves existing in an entirely separate universe where nothing else is different (or interfering) is hard to ignore.

The most compelling evidence for this superposition of two worlds is that the interference between two splits produces brighter bright spots and darker dark spots than from a single slit. Proof that the particle is not divided in two but is in both places. The extra probabilities have increased the energy available. But the two worlds have to be communicating with each other in order to distribute the probabilities in the 'correct manner', and this is a source of dissatisfaction for many.

In the old days, the interference fringes from the double slit were thought to arise solely because light was a wave and the slit had a width similar to the light's wavelength which produced the interference pattern (like waves in a lake) from phase differences due to pathway considerations. Classical interference patterns were noted in all sorts of situations where obstacles of light wavelength dimensions were placed in the way of light. But once it became possible to send individual particles one by one through a slit it was observed that a particle could interfere with itself. The idea that an object is *both* a particle and a wave at the same time, however, is not really true (although Bohmian mechanics takes this view – a distinct particle has a wave of probability running along with it). It is one or the other depending on what we know about what it encounters. This decides whether it is joining in with our world or not (i.e. ceasing to be entangled). Its uncertainty, which allows it to be spread around and described by a wave, is 'resolved' by us knowing about it. (Could this be a case of the transmission of a positive tychon value?) Our 'knowledge' is embodied in a much vaster conglomerate of matter than the particle in question.

Waves become particles when we can identify or 'know' certain facts about them, which suggests that not only is reality *for us* established with particles and not with waves but that it is also made up of certain kinds of information uncertainty resolved by our consciousness. That is to say, 'knowing' in its broadest sense really is materially real 'stuff'. And this is the crux of the multiverse scenario.

The 'multiverse' scenario describing the results of this experiment arises because there are a number of possible destinations available to each particle-wave as it emerges from the slits and there are a number of possible energy states it can have. Because the interference pattern that accumulates from the individual waves passing through the slit is precisely the same as one formed by a wave of a whole beam of radiation, the conclusion seems be that an individual 'particle', going through the slit, is just a wave combining with an image of *itself* (ghost particle, as Deutsch names it) according to destinations defined by the slit. That is, two values of the energy of the particles combine at a destination on the screen (sometimes they add to make a bright spot, and sometimes they negate each other to make a dark spot, and all ranges in between). There is a problem with this of course, which is that the ghost particle must vary over values different to equality for there to be

> The presence of the observer increases the total probability volume available to each option, allowing the state that most fits the new total probability volume to be observed. In the photon splitting experiment, the same probability volume (the observer) is acquired by each equal option, and thus they interfere because they share the same space. But where the probabilities are all distinct, adding the probability volume of the observer also condenses as many values as can be accommodated in the space. The rest are not realised. In other words the measurement still contains uncertainty.

variation in the interference pattern. (The traditional picture has equal wave fronts emerging from the slits and it is only the variable path lengths to different points on the screen which create the phase differences and hence the interference pattern.) The ghost particles are therefore *not* copies of the particle-wave even though they appear to have the same trajectory (i.e distance and angle), and neither can they be randomly separated, otherwise they would not form the consistent pattern. Neither can they be inversions or mirror images or a single particle-wave split in two. The multiverse scenario solves this by describing the particle-waves not as separate waves that must be travelling along slightly different paths, but as an entire universe copied down to the same trajectory but with different values for the ghost particle energy in it. The two universes meet at the recording screen and give up their values to the recorded pattern.

No one has yet found a mechanism by which the slit can give two different values to the same thing (one wonders what it is that consistently gives the ghost universe its differences), unless all the different values are *already contained* separately within the particle-wave and are separated out by the slit (though why this should happen is not clear, and anyway it introduces a clear case of realism to the situation denied by the Bell Theorem), or that the slits give distinct but exact ranges of coordinated values to the ghost and the particle (they need to be coordinated otherwise there would be a random pattern). Recall Schroedinger's Cat superposition that must contain both a prediction and a memory of a state. Since the slits are unlikely to be such perfect copies of each other and that the orientation of the particle be so perfectly head-on as to split precisely we are left with the idea that the particle knows all the variations it needs to form the interference pattern, which is to say it already knows where it is going when it leaves the slit (we know this from other experiments that try to erase just this information). The future pathway information appears to be included in the particle-wave's state, in its wave nature, since without it there would be no pattern. (Recall that perfect English speaker who has the meaning of every message in advance).

But trying to demonstrate this has proved to be difficult, since, for example, marking a wave with say, an amount of polarisation to tell which slit the wave went through, and therefore revealing the pathway, erases the interference pattern. The wave behaves like a particle, going through one slit or the other. If, however, we erase this 'marker' information just before measurement or even after the moment of interference is supposed to have happened the interference pattern is restored.

An elaborate experiment has been conducted to clarify this situation and to eliminate possible classical reasons for the interference pattern such as phase changes in the waves. Known as the delayed choice quantum eraser observations, they were conducted by Kim and his team in 2000 (based on Wheeler's idea) where a photon is directed through a pair of slits, A and B, and the photon that emerges from either slit is then split into a photon and its ghost, and each travels a separate but identical path. The actual arrangements are not important. The intention of the set-up is to be able to unite the reference photon with its entangled pair whose path could be known or unknown (that is its path is a random choice).

The interesting thing about this experiment was that the path lengths were adjusted such that the reference photon was (apparently) detected *before* detection of its entangled pair (in our time frame) yet the interference pattern appeared (accumulated over time). So the question arises, in what manner does the reference photon still retain the possibility of forming an interference pattern when it has already been 'detected'. This 'delayed choice' has been used to suggest that there is a signal going back in time to the already-measured past to 'tell' the reference photon when interference is required, but this 'hidden variable' explanation is not the consensus since, because there is a time gap between measurements of the entangled photons, a signal going back in time along the path of the entangled photon but travelling no faster than -c would still be too late to provide information to the reference photon.

There is, however, another way of looking at the situation that is of interest to our argument, namely the 'backwards causation' transactional interpretation of Cramer. For Cramer the particle trajectory is the continuous sum of two waves, an offer wave going forward in time and an advanced wave which is the conjugate of the offer wave travelling in the opposite direction in time, triggered by the absorption of the offer wave in the detector, and returning to the point of creation of the entangled pair of photons with interference information. This, however, is not the same as going back in time to before the events began. More about this in a moment.

In terms of delayed choice, the transactional interpretation also makes perfect sense. The photon of the later measurement is still entangled with the first photon measured (as the Bell inequality shows) until it too, is measured, and then it adjusts itself to interfere with that value. So basic causality has not been broken, although it does suggest that entangled states are distinct to superpositions and do not decohere like them.

We term what we know about the particle as the 'which-path' information, but this does not give us the correct picture. The particle is not concerned with destinations as such from its point of view. It is concerned with is where it is in space-time with regards to its origin and what it shares with the reference object or ghost particle in that origin. When it acquires information about our histories, which is to say, corresponding to a reduced set of probabilities, then the object is a particle.

The act of observing creates a particle history because then it has location information to share with space-time (something supporters of Bohmian mechanics would agree with). In fact the slit experiment nicely confirms that multiverse differences, instead of being described by orthogonality in the same moment, can be thought of as distinctions in time. This brings us back to the uncertainty principle we mentioned way back. Duration-of-state is a different time to the wait-to-happen time of the world-line. The shorter the lifetime the less the particle has location uncertainty and the smaller the volume the particle will occupy. It is here we can re-cast the idea of probability volume as that which enables an object to be turned from a wave into a particle.

The notion of multiverses may agree with our idea that possible options as defined in probability must both be different from every other probability (that is, before they interfere) and have a definite reality to them otherwise they are not options. But in the slit experiment the multiverse notion does not explain how the destinations of the particle are connected to the slits in such a way as to bring all those particle versions and their universes into being *before* any particle has reached its destination. Should it occur at the point of impact with the screen then the other worlds are not necessary.

Furthermore, if the slit is somehow filled with route information to these target destinations, with which does the real particle decide to interfere with and why? Deutsch suggests that each particle version or shadow particle lives only within its own universe except where there is enough proximity (in time and space, perhaps within one wavelength) and difference (e.g. in trajectory) to the original particle to cause interference, which is just the same as for the old-fashioned treatment. If it is the case that the two entire universes do not interfere with each other except at the point of difference, then what of the uncertainty in each universe, and why only two universes contributing to the observation when there appears to be as many as there are destinations to interfere with? Since some shadow universes become so different that there is little or no chance of interfering with ours, while other shadow universes retain such a high degree of similarity that they should be able to interfere, one wonders if there is some higher level influence involved in selecting (or reducing) the probabilities. Instead of congruency as necessary for interference, could there be interference at more complex levels than subatomic states, and perhaps in consciousness itself?

Thinking of the measurement this way allows us to consider two distinct versions of the multi-world hypothesis: one where the observer and her entire universe is multiplied at the measurement as many times as there are probable outcomes of the event, and the other where all the universes and minds associated with each probability exist beforehand and one containing the observer is 'selected' in the measurement.

For a single photon fired at a double slit and hitting a screen beyond, it would seem that the total number of worlds would be given by the total number of possible positions the photon could get to on the screen and the total number of energy values the particle may have. There is an alternative to this, however. The proposed 'multi-worlds' are actually a time breakdown of the whole interference pattern and each observer is just located in a different moment of the time history of this experiment (but in the same place). Since, the total of split observers and their observations sum up to the pattern on the screen, all we need is two worlds – the particle world and its ghost world – summed over time. What is the ghost world? The ghost world is the conjugate state of the particle conveying the probability information from the point on the screen that the particle hits. There is no need to have many worlds occupying multiple orthogonal *spaces* simultaneously in time; they can occupy the same space with their uncertain duration but in different moments. In this way, the prediction we talked about for Schroedinger's cat makes sense. It also returns us the monasmotic universe, as a consistent history (of a world-line) only seems to come about with observers in that world-line who are not superposed with many or infinitely many other universes but only with their conjugate non-presence universe.

This seems like a good point to return to Cramer and his ideas of backwards causation.

Retro-causation and the fulfilments of time

Feynman and Wheeler thought that the reversed charge of anti-matter might be due to ordinary matter going back in time but this view is really just an interpretation of the mathematics that proceeded in forward going time and not an excuse for causes to precede their effects. Besides some particles, like mesons, don't have this symmetrical behaviour when time is reversed which nicely pinpoints the difference between time reversing and influence going back in time.

Our proposition is that handshaking between past and future is what the universe does even at the level of ENT, at the level of human behaviour. There is an essential recursive element to the way the universe allows events to unfold, for without it there would no constraints on their divergence (probability does not explore infinity, only the extent of itself). We are, sitting in our present, necessarily infected with something of what we do become. Re-incarnation hints at this idea of *accumulation* with the notion that we are somehow a culmination of all the lives we have lived, though the analogy we shall examine in more detail is one partly expressed by a notion already imagined in Hindu culture: the idea of Karma. (It is here we may see a feature of life with transmissions of tychons of negative probability.)

Hindu Karma is not about the future coming back to haunt us so much as the belief that the design of life ensures that the consequences of what we do are brought back to bear upon us in our life in reciprocal fashion as time goes forward. It's a form of *morality* rather than physics. Whereas the kind of 'Karma' I am thinking about is a form of influence from the future where the unfolding of the consequences of the decisions we make are actually embodied in the probabilities we are 'offered' when we make those decisions. Since the future 'knows' you acted this way, you only 'see' a reduced range of probabilities in the present to make a decision that helps confirm your action.

So are our lives completely determined? The answer depends upon whether we are capable of seeing clearly the choices inherent in any decision and whether we are capable of actually selecting between them. Rather than seeing superpositions of probable outcomes in observations simply as *predictions* of a future we can think of them, here and now, as also *memories* of the future; they are memories of a future channel to the unfolding of events, a constraint to the ever-widening separation of possibilities. But what you see as choice is not fixed by the future outcome because that outcome itself is not fixed at any one time even by its becoming the present and by being measured or being formed in an observation. A subsequent future might yet alter those probabilities. A probability in the present is arrived at from the accumulation of all the workings out of the future that connect with that moment.

The multiverse picture of quantum theory begins to suggest that we live in a universe where all our deeds fanning out into the future have repercussions back here in the now and even on into the past; where all our trajectories of life are steered by the continual feedback of the results of our decisions helping to determine what decisions are taken.

Does this make sense on a physical level?

The Cartesian imperative made us think of causality as an arrow going in one direction because it supported the idea of moving through chains of causes to the prime mover in the past, Aristotle-like. But the Bell theorem does not seem to support this line of thinking. By interpreting the quantum decision–making process in terms of retro-causation physicists like Cramer try to explain how the future still connects with the past.

Retro-causation (or retro-chronal causation, or backward causation, or noitasuac lanorch-orter) is discussed by physicists apart from Cramer (notably by **Costa de Beauregard (1976)**, Hermann) who find solutions in Einstein's equations that allow influences to move back in time – essentially allowing for an effect at the very least to be one with its cause within the same frame. The retro-causation discussed by Cramer is a way of interpreting quantum theory to describe how probability becomes an observation.

Cramer's transactional interpretation is just that, an interpretation of the equations of quantum theory. Already in the theory is the idea that there are two versions of a wave, one 'normal' part and one a complex part with negative energy. This complex part has been regarded as having no physical meaning even though the complex part is used in calculating the probabilities of the wave. Cramer said that we should take this part of the wave at its face value and accept that it explains the presence of a mirror image of the normal wave going back in time. At each moment the whole wave is a summation of these two components, one going forwards and one going backwards with respect to time.

Cramer calls an emitted wave moving forward in time to a point where it is absorbed (observed) a *retarded* or *offer* wave. At the absorption, an *advanced* or confirmation wave is generated that travels back in time to the emission where it is absorbed in the creation of the wave.

For this backwards causation to work, it needs something up ahead in the future to be working its influence against the current of change turning the present into the future since a wave needs a beginning, and in Cramer's view the advanced wave begins in the future; it proves the future exists in some form simultaneously with the present, before it becomes the present. Furthermore it implies that the present, at whatever size level we decide to look at, is the provisional representation of data and until the 'handshaking' circuit is completed (though from our point of view this happens in zero time) events are unreal.

Here is a simplified view of Cramer's transactional interpretation of entanglement (similar to the Feynman-Wheeler idea of advanced and absorbed waves in electromagnetic fields). Two entangled particles with mutually opposite spins are created and move apart in opposite directions. Let's say particle 1 is measured for spin orientation. We immediately find that, although separated, the other particle takes on the opposite value. Quantum theory shows, that the mathematical description of each particle state has to include a bit of the state of the other; position and momentum are *entangled*.

Cramer's suggestion is that the spin information of the first particle fixed in its measurement travels back in time to the origin (the confirmation wave) and then forward with the second particle so that the second particle is in the probabilistic state initiated by the measurement on the first particle at the moment the first particle is measured. Since the travel time back to the origin, $-t$, is the same as the travel time to the point of measurement, t, relativity is not violated while the effect in our present appears instantaneous. The sum of the time taken in our world for the 'adjustment' is zero ($-t + t$) but the 'quantity' of time is not.

The second particle to be measured sends its advanced or confirmation wave back in time to the origin where it meets the other advanced wave. Between them they arrange the distribution of spin between the particles. Thus it is that both future measurements are required to determine the spin distribution. The chain of advanced waves continues on in the past back to the Big Bang and then is 'reflected' forward in our normal time to cancel itself.

This transactional interpretation is criticised for being functionally equivalent to local hidden variables lurking in the background affecting experiments after all. Critics seem to think that because of the advanced waves meeting at the point of origin, both particles leave this point of creation with information about what spin state they are in and should not violate the Bell inequality. They aren't really in a tangled state of non-specific spins, only we have not yet made the measurement that would determine that fact. The arrangement of spins does not exist prior to the first measurement so in this sense there is no pre-existing reality until the first measurement. The advanced wave (going back in time) needs to travel the entire distance made by the retarded or offer wave to reach the origin at the right time. So it is only *after* the measurement can the retarded wave transfer the full information to the second particle, so the quantum uncertainty before measurement remains. Further, as we discussed earlier, there is a difference between the action of negative time and going back in time to a different point in the time line, say to a time before the experiment began. With Cramer's interpretation, the advanced wave travels backwards, like a movie run backwards from our point of view, but is not arriving in time at a point earlier than the creation of its conjugate.

In the quantum eraser experiment, entangled pairs of photons come out of a double slit and are sent on different paths, one longer than the other. One particle goes straight to its detector first, while the other travels through prisms and beam splitters before it gets to a detector. In spite of the fact

> Part of the problem is with our definition. We define a cause and effect as distinguishable only in forward time to begin with. But suppose the underlying reality of the Universe doesn't work like that and part of the reason why we see time going forward at all is because consciousness is the complementary effect of both time and changes in probability moving in the opposite direction. Analogous to the way the sequence of the frames of a film have to be organised in the opposite order to the experience of the sequence of images in our brains.

that, when knowledge of which slit the particle came from is erased (by mixing it at random with a particle from the other slit), interference in the record is still seen between the particle that was detected first and the particle that was detected later. How does the first particle know that it will have to form interference later after it has separated from its twin? It seems to have acquired that information before the separation, or there's something waiting at the arrival point to make that decision. (Maxwell's Demon perhaps in a new guise?)

Path length, though, is irrelevant to a signal going back in time since however long a time before the measurement of the second particle, the retro-causation signal, the advanced wave, has returned to the point where both particles began their entanglement and handed over its information *then*. Since neither particle can know which one of them will be measured first, both of them are entangled but, in the TI picture of events, this entanglement is with the end result rather than with the beginning. However long the delay between measurements, each retro signal returns to the point of creation of both and arrives at the same time such that the 'correct' states for each particle in the subsequent evolution are determined by that handshaking. They are also entangled with the information about the time which separates the measurements for the external observer and thus readily explains the excess of correlation revealed in the Bell inequality experiments. The first measurement tunes the entanglement towards a result, while the second measurement confirms it.

The backwards acting wave can be blocked in the same way the offer wave can be, and can also be 'scrambled' by a polariser such that interference effects can be seen when the objects have not been able to go through the 'handshaking' at their origin.

It is tempting to think of what we observe as particles of matter are waves mixed with information about the current state of our world. Indeed, how could the world of experience be constructed otherwise, and it again it brings the imaginative spotlight back onto SETI and whether it will be possible to communicate with beings who live and extract observations in a present moment very different from ours.

In this experiment we can see that knowing stuff about the position of matter turns it into particles while not knowing it turns it into waves. When we know where the particle is or what path it is travelling on, it is a particle with local properties. When we don't know where it is, there is a wave of uncertainty and interference derived from the non-local whole.

For Cramer, causation is a summation of energy waves going forwards and backwards in time. He said, "a definitive characteristic of the TI approach is that it describes causality as arising from a precariously balanced cancellations that nullify the occurrence of advanced effects in quantum events."

But this cancellation can be imbalanced, for while we can see how in quantum probability the backward-time wave contains information about the state of the absorber before the retarded wave is emitted to be absorbed, because of random effects the succession of advanced waves (waves going backwards in time) may not be fully cancelled and thus carry information about their 'last' point of emission into the past to continually interfere with the retarded waves emitted forwards in time. So there is always information about the future embodied in every quantum event and thus in the behaviour of the present.

While a retrocausation interpretation may not produce differences to the orthodox quantum calculations, it gives us a new structure with which to examine the measurement process. In the orthodox view, without retrocausation, the observer gets entangled with all probable states. One state has a certain probability that agrees best with the history of the observer. Then there is some provisional decoherence. The observed state becomes real and the resulting fact of its existence in this world is communicated back to the superposition to partition the probabilities and the many worlds definitively. This does not seem quite sufficient, however, if every observer copy shares precisely the same history. Recall our examination of observer states in the superposition. If we count on innate possible variations within the observer to produce different worlds then the differences become apparent in the *future value* of the measurement, i.e. the prediction. This is the additional state required for the cyclical process of measurement to occur.

With retrocausation, the observer absorbs one probability of the superposition, which sends the advanced wave back to the beginning to confirm the reality of that probability and to produce a particle. While this appears to happen in zero time for us, it does not occur instantaneously in the time of the whole event. An entangled state ends up with a value that agrees best with the *future states of the observer*. There is some provisional decoherence where the particular state that will evolve to the future state becomes real, and then the worlds decohere as before.

In this way we see how the actions of our consciousness gives us the experience of time going forward.

While in the orthodox version we have the current state dependent upon the decohered probabilities of a past arrangement – the superposition. The state of a particle is derived from the beginning of the superposition rather than from the result. However, this kind of fine tuning is necessarily local, something we don't believe to be quantum-like. Whereas, with retro-causation, we have the current state based either on a conformity with the future events or on a *prediction* of a value that it will finally attain in further chains of advanced and retarded waves. We saw in our discussion about Schroedinger's cat that a superposition requires prediction as well as memory. The prediction arises from a backwards calculation and is, in effect, future information about an achieved state. The prediction would constitute a certain amount of knowledge of the future state in advance, and thus could turn some of the uncertainty into actual values, such as, for example, position information which may be enough to decohere the superposition or entanglement, and since some properties are constrained together, such as energy and time, and momentum and position, there are realms of future values of these states that cannot be attained. We interpret this as uncertainty (Heisenberg's uncertainty principle), but in reality it defines the future ranges for each of these values.

Since retrocausation travels through the dimension of time it picks up probability information from successive moments closer to our time and affects (what we think of as) the past that it moves through with something of the future that it has already moved through. This is why the universe picks a track to evolve along; this is why the future fruits of our actions now end up reinforcing our behaviours today.

Retrocausation seems to explain the problem of consistency throughout the universe since it demolishes the problem of *distance* between events and the limitations of the speed of light, and we will get back to this. But more than that, we can elaborate on the transactional interpretation to give us an alternative view of the source of probability and how creation works at the basic level. But many physicists criticise it for creating what seems to be a global 'now'; something that Relativity has demolished (although photons appear to occupy it). Further, the Cramer model requires an adsorption event of the emitted or retarded wave before the advanced wave is 'activated'. This seems to imply that there is no emission until an adsorption event is assured – a form of pre-destination and not unlike local variables (the Mach Principle also seems to require something like this). We shall see, however, if this is strictly true.

It cannot be the case that an object's 'location' in history is solely up to probability then what governs this element of chance in time and how does the appropriate number of universes in the superposition for each experiment come about in the first place? To consider these questions let us examine what is at the heart of all phenomena, dispersion.

Magic and geometry

If we were so inclined as to follow the course of physics from its beginnings in ancient cultures to the present day we would be able to trace out two distinctions in the philosophical constructions of science and theories of creation. These two distinctions are essentially between change and eternity; between force and and movement; between time and space. The Mayans and the Babylonians revered time in its cyclical manifestation; the Hindus revered time for its eternal being as the quantity of life to be found in creation. The Ancient Greeks battled with concepts of stasis and flux. We see the connection between a static universe full of eternal abstractions, as propounded by the Eleatics and elaborated by the Pythagoreans and Platonists with their belief in unchangeable ideas behind phenomena, through the fixed geometrical laws of Ptolemy and on to the geometrical fabric of the Einsteinian universe. The other pattern is found in the ideas of the vitalist and the believers in force and endless motion (usually of vortexes), propounded by the Greek Stoics and their theories of fate and the continual movement of Nature springing from a primordial fire, through Kepler and Newton who considered a force acting at a distance and permeating the Cosmos, through Maxwell and his unifying field equations describing waves and on to quantum theory, to the symmetries of force and field and the irreducible connection between both patterns in the energy of motion and location in space.

At every moment we would find the paradoxes of each distinction meeting and combining but never fully resolving in the construction of the cosmology of the day. The logical demands of time and space each dominating in turn. Zeno (an Eleatic) exposing the paradoxes of motion in time with geometrical concepts; the Pythagoreans belief in the universal harmonics of number intervals; Ptolemy describing the movements of the Heavens with ideal geometrical abstractions of circular motion; Newton's and Leibniz using similar geometrical constructs of infinity to those of Zeno to analyse non-uniform motion with calculus; and to Einstein turning gravity from a Newtonian, vitalist magic into the geometry of time and space in his Special Relativity.

Here we are again at the paradoxical interface between space and movement, between ever changing forces and energies and the static fabric upon which all rests, and struggling to find an approach that unifies it all.

This is well summed up by the realisation (by Sciama in 1953) that rest-mass energy of a particle plus its negative potential energy in a gravitational field is actually zero, which means in fact that the energy of the whole universe sums to zero (as observed at cosmological distances) since the emerging positive energy of mass (vitalist concepts) at the Big Bang is cancelled by the negative gravitational pressure (geometrical concepts).

What enables the universe to grow in spite of this apparent sum is what we will call for the moment a *degree of freedom*. Movement requires a space in which to move. Zeno tried to show that since space reduces indefinitely as time goes on motion was impossible. But we know space is expanding. We do not yet know why. One suggestion for the creation of space is inflation. With inflation, 'space' expands through the negative pressure of gravity giving energy to the inflaton field. The net result is still zero even as space grows. This remains a controversial idea, however, for the very reason that it mingles the two conceptual pathways mentioned above by combining the geometry of space-time with fields spread 'over' or 'within' it.

If we give matter parameters of location within space and a degree of freedom then we have probability. But what is a degree of freedom exactly? Let us go further into the significance of *location*, and try to show that location information is the key to understanding the concept of probability and time, and offer another interpretation of the Big Bang.

The matter of location is really at the heart of seeing how the universe works. It is the reason why Relativity and Quantum theory have yet to be reconciled. Relativity describes the basic substrate to the universe, namely space-time. Space-time is a *geometrical* construct that tells us about pathways: how things move and what limitations they have in that movement. Quantum theories are about fields and waves that overlie space-time and which are responsible for particles and the forces between them. So far we have three fundamental fields, and the search for a unified field theory is the search for a way of bringing all these fields under one theoretical framework, a single field. The three fields have the strong interaction which governs atomic nuclei (quarks), the weak interaction which involves the exchange of bosons at short range and radioactivity, and mostly affects types of electrons, and the electromagnetic interactions mediated by the photon. All these fields can be quantised – hence the exchange of particles of force. And then there is gravity, a form of energy exchange with spacetime that acts on the geometry of space-time (that is, with pathways: matter tells space-time how to bend and space-time tells matter how to move). The hypothetical particle of exchange, the graviton, has not been found although it does look as if there are gravitational waves which implies a (very small) level of uncertainty in the locations of points in space time. Bringing the three fields and gravity into one theoretical framework, however, has proved to be a puzzle. (The Higg's field whose interaction with subatomic particles gives them their mass, strictly speaking doesn't supply a directional energy, or a vector quantity or movement, and so is not considered a force and is not taken to be one of the fundamental force fields.)

The distinction between gravity and fields is perhaps one of the reasons why many attempts to create space propulsion systems using fields (like the Woodward inertial theories building on the Mach principle and work by Sciama, where the rest of the universe is involved in creating local inertia, or the EM drive) have not yet been successful.

Because the fields are distinct from space-time, although interact with it, the particles they produce are not affected by the local expansion of the universe. The speed of expansion of space between points with distance given by the Hubble constant arises in large scale geometry based on mass. It is very noticeable at intergalactic distances but not over short distances. In subatomic regions, its expansion would be very noticeable if the fields were affected in the same way by this geometrical effect, since, by extrapolating the expansion to small times and distances, after about every 100 seconds or so, points in space-time would hypothetically have further separated by the width of a proton. We can observe the effect of Hubble expansion only through the lengthening wavelengths of free particles travelling vast distances over the stretching of space and perhaps in one other special case which we will come to.

Locating the cash-box

In quantum theory particles have positional uncertainty within the atom but also as free particles outside it which leads to some interesting repercussions for space-time and for gravity.

This problem of location is not just an interesting story but one of great significance in the universe. Tunnelling is relevant to biological systems like photosynthesis as well as to the ultimate size of computer chips. It makes nuclear fusion in stars possible by allowing protons to get through the repulsive barrier of the electrical fields and it is how particles get ejected from the nuclei of radioactive atoms. Knowledge, too, of location is also significant in quantum processes and one cannot help but speculate that our observations of quantum phenomena in the universe like Neutron stars or Black Holes or even the Sun itself may begin at some future point to affect what we observe.

When we talk about the position of a particle under some constraints, say, as opposed to large collections of particles that tend to behave more classically (superpositions don't seem to work for large ensembles of particles; they are too 'real' and disturb the environment, or the increased probability volume forces decoherence), it is not the case that a particle is either there *or* there *or* there....*or* any of the other positions it might be in, rather there is a probability attached to each position in space and the potential of the particle is present simultaneously in all of them at once in the proportion to its likelihood of being there (the probability is a vector in infinite space). This is how a free particle can 'tunnel' through confinement. With the uncertainty in its position, when it is close to a barrier, sometimes it just finds itself outside the barrier without having had the energy to surmount the barrier head on, as it were.

These probabilities change over time such that it becomes increasingly likely that the free particle can be found over larger and larger spaces (the wave function spreads out). This information has a presence, although it may not be in either the form of a force or contribute to a field; it is about location, and even though the particle may not end up being in *that* particular place at all, it had a moment in its life when it could have been at that point, and this option leaves an evolving residue of potential at that point for as long as the particle is not yet found anywhere else. Since this phenomenon as also about distribution and location there is an component of entropy included. In fact, it suggests that entropy and therefore heat could also 'tunnel' through space and deliver distributed effects in the vacuum.

In practice, in a world full of particles and interactions, the uncertainty in a particles's position doesn't have much chance to reach very far in space at all before it is nailed down in an 'observation' or measurement, but in principle, a free particle has a non-zero probability density of location spreading out in space until it is observed or interacts with another object. For a speeding electron the uncertainty in position would be measured in millimetres which is not insignificant to an electron. It is possible that at large galactic distances the cumulative effect of this positional uncertainty may alter the calculations of gravitational effect at these distances, to provide an alternative explanation for the observations of larger rotational velocities, namely 'dark matter'. Certainly some physicists are considering forms of modified gravity laws around dense objects where physical laws subtly change with distance (or that there is a fifth force sometimes referred to as a Galileon force) but are 'hidden' from masses over small distances (referred to as a Vainshtein 'screen'), leading to 'halos' of mass at large distances around dense objects like our sun. (Certainly, it is puzzling why some galaxies have halos of numerous globular clusters around them.) Positional information is becoming a significant feature of cosmology.

A basic uncertainty in the position of objects operates even at the human narrative level, and there is nothing very unusual about this way of looking at spatial distribution (even though the quantum values at the human level are very small − of the region of the Planck length). It happens all the time in ENT when you are looking for something lost in your house. Say it's an important folder. You start off with no idea where it can be but then you realise that the drawers of your desk are too small for it to be in them, so you know it is not in your desk. While you eliminate places where it might be, the odds of it being at other places in the house, like the fridge, are evolving over time. But correspondingly, as you fail to find it even in the fridge, the possibilities of the folder being outside the house start to become more significant. In the garage? In the car? In the trash can? Finally you may conclude that you left it at work. A valuable object, however, like a cashbox, has other likelihoods attached to it, such as it being stolen and you fear it can be almost anywhere else.

I try to explain this to my family as they go about looking for something but no one takes any notice. One of the ways in which they fail to find things is that they start off with conclusions like, 'It can't be in there' (negative probability). They eliminate places before they have reason to, and as a consequence, give up before they should. Certainly it's hard to imagine a reason why an important folder, say, should be in the microwave oven, but nevertheless, by eliminating such a place you raise the odds of it being in a place you do consider to be a 'reasonable' place for it to be.

For every point in your house there is also a probability that any article in any other house might, at some time or other, be there (you invite a neighbour to a party; he comes from work carrying a mended object from their house that he picked up for his wife and he leaves it with his coat; it falls behind furniture and he forgets it). If you still cannot find your object then the possibility that it is, say, in the house next door rises from almost zero to some non-negligible probability – though admittedly small (the dog grabs it and runs through a hole in the fence to the house next door). And there is always a corresponding probability that something lost in that house could turn up in yours. And so on along all the houses in the street, and in all the houses in the city where you live and on and outwards.

But consider, an ordinary item is more likely to be nearby than something precious or rare like jewellery. We can think of this preciousness as equivalent to its momentum. A diamond tiara is less likely to be in your neighbour's house than in some far away fence's safe. The rarity or value of the item is a quality belonging to the object (as, too, is the seriousness of the sanction that would be applied to whomever has it) that implies that it will travel further than say, a box of tools or loose money (until it is spent). The probability of your precious jewels say, being in a neighbour's house is, however still higher than anywhere jewels are not prized, and so too the probability that they might be found in the safe of a penthouse apartment in Singapore is higher than at the crest of Scottish moor. The human world created the jewels and tries to keep hold of them. The probability of lost jewels being found is a combination of the inherent instability of place (other people take them; we lose them) and the desire for any human to have them nearby.

(This analogy is not perfect since your folder does in fact have a location and the probability of its location only reflects your ignorance. A quantum object, on the other hand, is inherently uncertain of where it is and even when we find it we cannot know how long it had been there.)

And yet, if the folder remains unfound (but still existing) for an eternity, the variable likelihood of it being found at any point will have spread out to the whole universe. Since probabilities have to be about real things, we must accept that, eventually, there is a shadow folder of more or less indeterminacy sitting at every point in the universe until it is found.

We can extend this idea to the whole of space. At each point there is a contribution to the state of that point from the accumulated likelihoods of all single states not yet measured or observed to occupy it. Most things, of course, do not spend anything like enough time 'lost' to have a probability of being found far from the madding crowd, and most states of particles exist in superpositions of states such that positional information may also exist in other dimensions, but we can see how an unknown lifetime translates into distance and probability in the here and now. The longer something remains unmeasured, the more it can become involved or correlated with distant points. The technique of normalisation or the way of keeping the probabilities of the fields summing to certainty is, from this point of view, actually an approximation.

The question becomes, how does this probability interact with the other fields or with space-time. (We know for example that a magnetic field can still affect space from which it has been shielded, known as the Aharonov–Bohm effect.) Each (many-dimensional) point of space contains at the very least something of an ongoing sum of the likelihoods that all yet unmeasured objects might be there, something similar to if not actually a scalar field. Points in fields not only foam with their own uncertainty but are washed with the sums of very low probabilities of superposed or entangled position states of free particles from elsewhere in the universe. Each geometrical point in space therefore collects spatial information from wave functions and which is, in some way, a measure of the undecided part of the universe. Since this information is limited by coincidence (according to entropy) it needs space for its uncertainty. As the numbers of states increase with the evolution of the universe, then one might expect the space to accommodate them also to grow. A certain amount of the acceleration in the expansion of space would be equivalent to the sum of uncertainty in all this positional information, all these *projections* of location information into the space between the particles of matter, i.e. increasing the volume that matter can occupy.

This suggests an experiment. Keep heavy particles circulating without measuring any value and perhaps one may be able to detect probability effects in particles in its neighbourhood.

However, the positional projections in all unmeasured particles in the universe do not accumulate at all points instantaneously and they will be denser on some places than others. They propagate through space-time at the speed of light. So the state at every point accumulates a tiny contribution from the balance of wave functions that are arriving at any one moment from all directions and those that disappear because a measurement has been made on them. The general rate of fluctuation at a point is proportional to the activity going on in the (local) universe, to the underlying density of events occurring in a given volume of space (to the mean path length of a free particle).

In practice, because of the limiting speed of propagation and the density of surrounding quantum objects, each point in space at any one moment will reflect the projections of its near neighbourhood rather than the universe as a whole. Most particles exist in unmeasured and entangled states only briefly, so there is not much opportunity for their uninhibited wave functions to travel far. For electrons to cause trouble in computer chips we are talking distances of nanometers, but this is still billions of times larger than the space between points in the fabric of the universe, although it is highly relevant to black holes since Hawking radiation relies upon such uncertainty.

We can consider that every time David Deutsch, for example, performs his photon splitting experiments, there is for a brief moment an increase in the point energy around the experimental apparatus due to the probability of his unmeasured photons not being inside it but somewhere around his lab. In fact this principle has been observed around barriers in optics and is called an evanescent field. (There may well be ramifications for interstellar starships as their velocity rises towards light speed. We might be able to use cosmic rays to probe the likelihood of such a probability density in the space near dense stars and black holes. This is to be talked about.)

Objects like black holes and neutron stars that may produce both superposed and entangled states tend to occur, however, inside galaxies where there is a higher concentration of encounters like dust clouds that will induce decoherence and wave function collapse. But once a wave projection can escape the galaxies, it may travel far in intergalactic regions. The further it travels in time the more it spreads out and the more it can tunnel through any energetic barriers, which we might propose as a reason why the outer edges of galaxies appear to rotate faster than they should. Since we expect higher uncertainty to require more probability volume to house it, we can also expect superpositions and entangled states throughout the universe to increase with time. There is certainly one measurable consequence of such a notion which is that the Hubble constant (the rate of expansion with distance) would vary over time. Some physicists think it does.

> If you inject the energy of gravity into a gas cloud you can lower its entropy as the molecules start to clump together but that means the cloud has to shrink. Its probability volume must reduce. Simultaneously its temperature goes up (more collisions) making the molecules move faster to try to escape the gravitational attraction. It's a balancing act that gravity always wins in the end making stars and, eventually, a lot of black holes.

Of course, space is full of observable energy considered to be the afterglow of the big bang at a point in the evolution of the universe where photons could lose themselves from the grip of electrons and protons and travel free. This point of *photon decoupling* is commemorated in the cosmic background radiation which is formed by those early free photons that were never re-captured by matter, and whose wavelengths have been stretched by the expansion of the universe into cool, long wavelength radiation which travels the length and breadth of the universe. However, the scale of this photon energy is very much greater than the probability contributions to points in space we have mentioned which are only likely to make a cosmological contribution in places of least particle density, as between galaxies.

But if the vacuum energy (zero-point energy) interacts with these virtual locality potentials then we have a reason for why the cosmological constant, Λ (representing the basic original energy density), might also vary with time. Values would also vary in different regions of space, though the overall expansion rate would be the sum of all these regions. The continual appearing and disappearing of these contributions to the points of space may give space a very slight vibration. Dense regions of matter (e.g. our local space) would actually increasingly decelerate *relative* to empty regions (because of more rapid annihilation of position information), which may be why there are 'voids' in the matter distribution of the universe. Spaces between high matter densities would be correlated leading to long lasting structures.

Dispersion and Black Holes

There is yet a second reason for suspecting that the expansion of the universe has reached a minimum and has now turned positive again, another angle on this 'where?' information. A particle behaving under gravity seeks the least uncertainty in its position in space-time, which brings us to black holes.

Space-time also describes a density of positional options for a particle of any given energy. Without probability, dispersed matter can occupy only one position in space. Without dispersal options its entropy falls (think of the dice tables). Gravity throws off probability in the form of heat and sucks in entropy, converting it into mass and leaving the singularity utterly cold and still (a black hole of one solar mass seems to have a temperature of about 10^{-7} K), with no probability in it but radiating perhaps along with hawking radiation, tychons, although these might have been radiated away in the moments the black hole formed.

(We should remind ourselves that what I am trying to do here is to get a handle on what the physics of gravity and quantum theory mean when it comes to the human narrative. I am not proposing new mathematical theories of gravitation but proposing a philosophy to interpret what the experiments show us.)

The singularity is the opposite of consciousness. It is the definition of cold. The centres of black holes are chilly places, and the bigger they are the colder they are. They are the coldest spots in the universe and time does not pass. Even so, black holes seem to do quite a lot of things one might not expect considering they are supposed to have a singularity at their heart where time stops and the curvature of space-time is infinite (although recent theories suggest black holes can join with each other through worm holes in space-time that gradually shrink). They orbit stars, they rotate, they expand, they evaporate, getting hotter and hotter as they vanish into nothing, they seem to contain more entropy than ever went in, they can be electrically charged, and they can spew out incredible fast moving jets of matter beyond the galaxies they are in. Even more significantly, the size of the black hole at the centre of every galaxy seems to be related to the amount of dark matter in the galaxy.

As matter falls together to occupy the same space, it acquires specific locality. We think the singularity is entire locality, a point, not an interval (and therefore contains no infinities), with zero volume, and no dispersal within it. The black hole, is not the singularity, however. The part beyond the event horizon is still a volume of space-time determined by the mass in it. The radius of the event horizon (Schwarzchild's radius) of a black hole is dependent on mass, but all the mass may not be at the singularity. Since inflowing mass will never reach the singularity after it has formed due to the gravitational time dilation, the mass must take up volume inside the black hole, smeared out throughout the interior space and giving up its entropy (equivalent to going back in time) to the interior surface and indirectly to the surface area of the event horizon. This is how Black Holes grow in size. Dispersion potential is converted by gravity into inertia.

> One starts to wonder what might happen to probability volume inside a black hole. The shrinking of space through the action of gravity suggests a relationship between gravity and probability. Under gravity, probabilities are reduced. The physicist Penrose believes in something like this: that gravity (through the graviton particle) supplies the decoherence procedure in any event when the curvature between worlds increases enough to form a graviton. This fits in nicely with Relativity, as speed approaches the speed of light and mass and hence gravity field with respect to the outside gravity field grows, the space an object occupies or needs to occupy shrinks. The lighter a particle the more it can be anywhere. The heavier the particle the fewer options it has. The Higgs boson gives us the idea that symmetry breaking introduces probability on one side and mass on the other. But that is another story.

We can understand all this in terms of probability. The mass that created it and continues to cross the threshold into it loses its *probability* into surrounding space-time. The denser the mass becomes, the fewer probabilities surround the star (say a neuron star). When the critical mass shrinks beyond the Schwarzschild radius then the probability, perhaps in the form of tychons, is released since the same numbers of options cannot exist beyond it (random permutations are lost). No-probability is denser than probability.

Suppose we try to measure the random orientation of spin of a particle just at the point of entering the event horizon then, if the multiverse scenario is correct, whole universes must also disappear at this point. Black holes appear to imply the sudden death of other worlds.

Since we know time slows in high gravitational fields, and recalling our view of probability for a two-dimensional creature as simply the time it takes to reach the event, we might speculate about the black hole marking a phase change in probability and an inverse relationship of time to probability. In thermodynamics, heat is transferred from one zone of low entropy to one of higher entropy. Disorder is increased and more events with higher probability occur. Under gravity, however, disorder is reduced and its probability is converted into heat but lost from space-time to matter that surrounds it. (It is likely that the luminosity of quasars, galactic objects with massive black holes at their centres, may in part be the result of this effect occurring in the vast sheets of matter surrounding the black hole. Quasars may be a source of probability. The jets of gamma rays generated in the collapse of giants stars into black holes may also reveal this conversion process. The high energy of these rays is not yet understood.)

A black hole does have a temperature (inversely proportional to its mass, as found by Hawking) at its surface, and an entropy (proportional to the surface area of the event horizon which is in turn proportional to the mass). It may also have charge. As the black hole slowly shrinks by evaporation over a very long time, its temperature (at the event horizon) rises and its entropy falls.

The event horizon of a black hole is where fields and the space-time interact in ways not yet fully understood but it is certainly the point somewhere beyond which uncertainty in location ceases. Quite what this does to momentum on the basis of Heisenberg's uncertainty relation, no one knows; since time does not pass, energy and momentum have no traditional meanings.

We mentioned earlier this way of looking at randomness may not deliver multiverses in a spread of spaces at the same point in time as in the de Witt picture, but produce multiverses in a spread of times at the same point in space. We may call this effect *inertia* because time cannot be divided indefinitely – the Zeno paradox – otherwise it doesn't work. Thus, at the event horizon of a black hole time slows right down, space deforms, uncertainty disappears and every state takes on a single value expressed over a long period of time. In other words, space and momentum are translated by gravity into total inertia and where all frames of reference merge into one. Black Holes may radiate not just Hawking radiation but also tychons, concentrated along space-time contours which might explain the superfast ejections of matter and heating in the surrounding material. Thinking about Black Holes in this way suggests that the origin of the universe requires something else than a gravitational singularity to begin expansion.

A note on the beginnings of history

Ahead in the book I describe how the influence from the future works on mind and the human-scale narrative. At this point, though, I am going to re-examine the role of time in the origins of the universe in the light of this connection between gravity and entropy.

Before things began, the universe was everywhere (although this term

When we talk about the horizon or frontier of this region, we are really talking about limitations to a given structure rather than about a specific geometrical artefact in the vacuum (e.g. a black hole's event horizon overlays the space-time structure that is so distorted that where pairs of virtual particles are created at points in the vacuum the stretched geodesic separates them quickly from each other before they can recombine: one goes over the event horizon while the other can escape, giving rise to Hawking radiation).

has no real meaning in this context) extremely dense and hot. After a random fluctuation in the value of the energy density, everywhere began to expand. At a later stage quantum fluctuations create regions of which our universe is one. By going back in time all the galaxies we see come closer together into a tiny region which is what we call the start of the Big Bang, the moment when the energy of expansion reduced sufficiently for the residual energy to turn into a subatomic particle soup which in turn cooled enough for matter as we know it to clump into the seeds of galaxies and eventually stars.

In the chaotic inflationary universe scenario, however, different regions of the expansion fall to a lower value of the vacuum state at different times, creating separately expanding regions or 'bubbles'. Each of these bubbles would have a 'frontier' with the neighbouring differential rates of expansion. The origin of universes in this scenario would have a structure.

What's interesting about the big bang is that although it happened in our past, the fields that make up the universe we observe play themselves out over a time-dependent state in space-time that we are 'following'. Energy has some connection with the space that has been created by this expansion and adheres to it in some way we don't yet understand because without such a mechanism, consistency in the behaviours of matter throughout space-time are hard to explain. Early thinkers like Boltzmann thought in terms of rare random departures from a thermal equilibrium, but we know nothing about the state of the universe prior to the expansionary episode so quite what the presumed fluctuation in the false vacuum fluctuated away from cannot really be examined. While early inflation explains the lack of great *contrasts* in the distribution of matter and energy in the observable universe and also the lack of wild probability and evolutionary divergence in it, it is not in an equilibrium with itself since it's expansion is now accelerating; the universe is not a balanced entity.

Our existence creates a place for us in the omnium so we cannot help but think of our place as defining a centre to it. However, this is also true for any observer anywhere. The fact that our existence gives a centre to our observational horizon does not mean that we are at the centre of the universe. Observers as we know them evolve over time, so there were no observer components in the beginning. The question, however, of whether there was any kind of structure in the highly dense universe before it expanded or even whether it was itself infinite in extent, is something we cannot give an answer to. If there are other observers in other universes we do not know if they are all a part of the same original feature.

The question of whether structure existed before the expansion matters because it decides the amount of entropy the universe started with and therefore the eventual size that the universe will be, since the less entropy the more there is energy available for expansion.

The Universe is thought to have begun life with a very low figure for its energetic entropy since the few energy micro-states must have been fairly evenly distributed in the very hot and dense space-time volume. Dispersal was very limited where the points of space had little distance between them. As we shall see, however, there has to be more to this picture of entropy at the universe's origins. For example, Richard Gauthier proposes a single quantum particle origin as accounting for this since a single particle has very much lower entropy than several adding up to the same energy (the lower the entropy, the more free energy of the system there is to be used). Since entropy is about the distribution of energy states, a single particle, while it may fluctuate in a quantum fashion, implies either zero entropy or infinite entropy. Since neither of these seem to be reflected in the universe we see we can expect that in the moment of origin, some other value of entropy was acquired as states were created.

The expected fluctuation in the inflaton field to produce the orthodox chaotic inflationary scenario seems to need to be somewhat more complex than just a rise in the scalar value at a region in the field. Consider that, since all the energy of the universe was highly compressed, it's gravitational attraction should overwhelm the negative gravitational pressure in the fabric of space (as revealed in the Cosmological Constant). The two appear to differ by a range often given as between $10^{60} - 10^{121}$ in magnitude. If, however, time fluctuated into negative territory then the compressed energy can indeed begin its dispersal regardless of the energetic mass excess holding it in. Indeed, one might expect that the false vacuum is not simply a sum of all the fields but the sum of all evolutions in time both forwards and backwards. Thus any fluctuation in the value of the false vacuum should be considered as a providing a 'parallel' probability fluctuation in an opposite direction of time since there was yet no *space* to accommodate the fluctuation. Space is created by a fluctuation into the future which gave the false vacuum energy the means to escape its confines, and thus the expansion is seeded (There must always be a tension between states that are possible and the space needed to allow them to happen.) In such a high density the quantum probability of the energy actually having a presence 'outside' the region it occupied was so high (i.e.negative) that it 'forced' space to expand just to accommodate that probability. (The universe is obliged to put out more tables for dice throws if it wants to maintain consistency).

As we noted earlier in our discussion about location, probability can only occur when there is space for it. Thus the 'handshake' of the state of energy or matter with its future options of position creates the spatial expansion which then allows for more options within the confinement to present themselves.

Maybe the information about how to begin a universe never entered the Universe for us to get to know? (See **Thinking it up**.) Maybe the crucial knowledge is forever excluded from the Universe we inhabit. It doesn't matter how much information we extract from the Universe, we will never get to know how and why it began. After all when a bomb goes off, is there any information in the fireball that tells you about who set it off and why. It will tell you about the ingredients of the explosive say, but not much about the political dynamic which led to its setting off. Even if the bomb belonged to a cycle of bombs, we could come to an understanding of the cycle but not of how the cycle began or why.

The state of location of a wave, initially at moment a projected beyond the confinement into a spread of values in which lies point b, feeds back into the location state at moment a to make the confinement expand to point b where the state is no longer outside the confinement, and with a future option again projected outside it. As more dispersion occurred, the number of energy states with location options existing beyond any moment in time of the confinement's limit increased thus driving yet more expansion into the future. Since this process includes negative time, it is outside of the speed of light limit for energy. However, as the confinement expanded, particle and energy density fell leaving proportionally fewer states to contribute to the 'future' location of that boundary, slowing the expansion and cooling matter. (The opposite occurs in a black hole where such a temporal connection is 'squeezed' out of space-time beyond the event horizon to create the singularity.)

This scenario also supplies an explanation for the apparent lack of anti-matter which should be on relatively equal footing with matter in most origin theories. Anti-matter has different probability requirements; it disperses more quickly. When the heat bath at the end of the inflationary expansion produced particles anti-matter traversed the space-time more quickly than matter effectively attracted by the horizon confinement of the expansion. Anti-matter thus accumulated up ahead of us. There is some evidence that anti-matter travels more quickly than ordinary matter in what we see of it in cosmic rays. Anti-protons appear to travel many times faster than ordinary protons which appears to suggest some form of symmetry breaking with charge, or that it reacts differently to gravity or to dark matter. (There are experiments afoot to try and see if this is so and so, and also to examine mixtures of matter and anti-matter atoms more closely.) It may be the case, therefore, that all inflationary bubble universes are enclosed by zones of highly dispersed anti-matter and which may act as a barrier to escape.

One of the ways time itself can be defined then, is by the expanding probability volume per unit of gravity. (There is more to say about time and probability but later on.) One solution to the problem of the informational content of the universe retaining its consistency while its volume increases is to supply more probability. And further, the puzzle of quantum uncertainties has a more natural explanation. Any measurement with dispersal information in it will be more unlikely than those with say, only charge or spin information

The 'randomness' of the quantum fluctuations would then exist within a probability volume larger in time than their present moment. The fluctuations that occur at each 'now' must in some sense be altered by what actually occurred, in the future. Since these fluctuations are slightly 'more' random than we might expect, the direction of conformity is with the future not the past, and since they are also to do with momentum rather than spin or charge it is not surprising the experiments testing the Bell inequality using say, spin polarised photons (which do not obey the Pauli exclusion principle), have not shown any non-random correlations.

Right after the Big Bang the single field began to separate into the four fields and forces that we see now. Gravity separated first leaving the three other forces in a unified state before each separated from the rest. Each of those fields will have a different 'version' of probability volume associated with them, and the particles of those fields will interact with the information in the advanced waves (or has a relationship with the future), which we can think of as the tychon field, in a slightly different way.

As we examined in our discussion of Schroedinger's Cat, the weak force of any nuclei may reach perhaps as far into the future as its radioactive half-life. The strong force, on the other hand, may repel the interactions, while the electromagnetic force perhaps interacts more horizontally across the present moment, and gravity interacts with Cramer's advanced waves as entropy.

There is the question of mirror matter, the mirror image of ordinary matter, a proposal put forward to complete the symmetry of the universe, and which is now being suggested as supplying the Dark Matter in the universe. It reacts to gravity but little else in our world, yet it is abundant in the galaxies, forming its own stars and planets that we can't see. However, Dark Matter supplies about 80% of the gravitational matter in the universe but only about 25% of its energy density. Which is hard to account for since there should be mirror background radiation and mirror black holes and so on. The supposed symmetry supplied by mirror matter does not itself appear complete.

These effects on particles become effective over the large scales, so how do they work in the selection of observations or measurements in the human narrative? Just as quantum effects have been selected by evolution in processes like photosynthesis or bird navigation, we can show similar selection effects in consciousness itself where memories and predictions are laid down together.

Conformity

Our argument has led us to the expectation that there is a correlation of randomness going back in time as a counter current to future probabilities, for this is how the universe can conform to itself as it enlarges, and where a wider meaning of entropy describes the *conformity* of each event to a future-past connection including the information of the origin still existing on the expanding horizon of our spatially non-infinite universe. Each successive future state of the expansion has consequences in the universe's past. (Using a similar argument to the one Hawking used for black holes it is thought that this 'horizon' may be 'active': it will radiate at least heat energy which we know carries entropy.)

So let us hypothesise for the moment that everything that happens within the universe must continue to conform to the pre-loading embodied in this expanding 'horizon', an idea close to a number of suggestions for a holographic universe (the term originally coming from string theory) whereby everything that happens within the 3-D volume of space is determined by what is happening on the 2-D surface enclosing it. For example, Hawking showed that the entropy of a Black Hole is proportional to the area of the event horizon rather than to its volume (although it is still not yet decided what the volume of a Black Hole actually means, and the standard speculation has it at zero). But the details of how the boundary information gets into the interior quickly enough to prevent inconsistent local behaviour is not explained – it is the consistency problem again.

The basic mystery of the 'now' is how things keep their identities while at the same time supplying endless fresh possibilities for change as probability disappears into the multiverse. What does conformity mean when it comes to information and uncertainty? A simpler way, perhaps, to ensure consistency that avoids the problem of spatial separation and speed of light limitations would be a continual 'handshaking' with the future as suggested by the transactional interpretation.

As our thinking suggests, the horizon of the universe expands such that it's surface area can still connect with the increasing numbers of states that entropy is creating within or 'behind' it. Indeed, the surface area of the zone must be connected to the number of states it encloses as well as to the number of states that can project a location probability beyond it. But how? How can a *local* change of state be related to a boundary so far away? Mach (and later Sciama) had an idea where local inertia was connected to the gravitational energy of the whole universe. It can be done with a connection in time rather than in space.

With this perspective, the horizon, way beyond the distance that information can travel at the speed of light, can continue to exert its influence on events in our epoch using a dimension of time. Consistency in physical law is maintained by an influence from the origins of our

Physicists have had their suspicion about pre-existing correlations for a while. One criticism of experiments designed to test violations of the Bell inequality is that perhaps the states of the particles are not truly random but are in some way arranged back in time at the Big Bang, so that apparent disagreements with the existence of local variables is just an artefact of pre-history. The violations of the Bell inequality seem to suggest that there is an actual irreducible randomness in a value that does not come from say a simple sum of the randomness of two states. A big test of this was conducted on 30 November 2016 where three kinds of random numbers where used to test the violations: one set came from a random number generator, another from numbers created by the general public at the very moment of the test and one from numbers created by the general public in the days before the test. The results using these groups of 'random' numbers were examined for patterns and none were found.

universe passing backwards in time such that, whatever the epoch, the horizon remains in moment-by-moment contact with the depths of the universe. If 'consistency' *is* a requirement (like the conservation of energy) then events of today are, in a very real way, connected to the events occurring on this two-dimensional surface way ahead of us in time and space. Remember our two-dimensional creature experiencing probability as a measure of time and space? This kind of connection provides for an order over and above the separation of points of space into a 'foam' of virtual particles of random energies and for the absolute randomness in quantum states that still permits for sensible observations.

The notion of the holographic universe (Hooft, Susskind) does embody something of this idea. The multiverse scenario suggests that there is no limit to the amount of information that can be stored in a given volume of our universe. As randomness increases with growing entropy, it is hard to see what will limit the numbers of multiverses arising in random configurations of states. The holograph scenario states that events enclosed by a surface are described simply by the information on its surface, i.e. proportional to an area rather than the volume, which we have been arguing characterises probability. This area has been calculated for our universe to be around 10^{121} Planck areas which is a lot more than the $\sim 10^{90}$ quantum states estimated to be present in the visible universe right now (and weirdly fits with the maximum difference between the Cosmological Constant and the total gravitational mass of the matter and energy in the universe). This informational energy contributes to the total heat energy within the Universe, and the heat energy is a measure of the distribution of states.

If this number seems huge, then consider that the number of stars in the observable universe ($\sim 10^{23}$) is nothing like the number of atoms or molecules in a modest amount of material here on earth. There are the same number of molecules in just 12 gm of pure carbon (or one gram of hydrogen) as there are stars, and each of those atoms embody a number of quantum states. If what we understand about the consistency across the universe is correct, the disparity between the observational horizon enclosing the expanding universe and the number of states within suggests a very much more complex universe than we can so far imagine.

The paradox of the inflationary beginnings to our universe is that it was proposed to account for consistency in the universe. At the scale of around 100 million light years any volume of the universe looks like any other. At around 300,000 years after the appearance of energy and matter, the universe was so uniform that the light between matter which we now observe as the microwave background radiation varies in temperature by about one part in 100,000. The variations of probability which supply disorder, entropy, should have another source, since the inflationary scenario is expressly designed to eliminate uncertainty from the past or at least the influence of the fluctuations in the high density field which began it. So where, then do all the variations in state that we call probability or randomness come from? For any very large expanding space to include probability it must remain coherent throughout time and space since probability makes no sense in discontinuous processes. Since coherency is not possible with the limitations in space-time due to the speed of light, it can only be present by communication through time. The source of our probability comes from the future. So time and consistency are connected, and objects of the present connect forward (in time) to the influence of something we will call, whimsically perhaps, *permission to be* (like they are).

Let us call this influence a confirmation wave (after Cramer). As confirmation waves march backwards in time from the horizon, they would be marking time as it were in our present.

At least a single wave function for the Universe overcomes one curiosity about time: how is it that physical laws (i.e. the effects of the past) are actually sufficient to direct particles *forward* in time?

Let us freeze a moment. In each moment we would find a 'forward-moving' wave of probability matched with a signal going backwards in time and occupying the same space. Standing waves in vibrating ropes or in tanks of water occur like this. Waves moving in opposite directions sum to a stable pattern.

So what could this backwards acting wave actually consist of? What is the negative of probability? It is what we observe.

If negative probability is the removal of options in every random selection of states, then it could provide the cause for decoherence or collapse of the wave function. It would also mean a relative loss of space or probability volume. In terms of spatial dispersion gravity acts somewhat like negative probability. So let us take a look at some ideas of how this can work.

Expansion

By taking the transactional interpretation at its face value then the advanced and retarded waves have been with us from the beginning, and as space-time expanded so the connection of the most remote expansion points with the beginning remains in time. The energies that emerged in between the spaces of that initial expansion travelled out with the frontier wave and remain continuously connected with the beginnings through the transactional process. The offer waves going forward with the frontier have greater dispersion or uncertainty attached to them, while the advanced waves begin with this same dispersion but then drop some off at every observation as they go back in time. Since they cannot be precisely equal, the minimum difference between the variants of the future state is the origin of probability, our tychons. This is how space expands. In fact, by following this process back to the moment before the Big Bang, we can interpret the initiation of expansion as a moment in time where a tiny region was suddenly flooded with spatial probability coming from its more remote future requiring that the region expand.

So what is this informational energy that we call 'probability' actually? Advanced waves going back in time deliver up information about future spatial options in each observation they are involved in. They 'tweak' an uncertainty in spatial occupation that allows for expansion.

We will discuss this more in the next section. But let us end this section by considering how energy conservation works in an *expanding* universe. Waves are in motion. So energy conservation either must occur in the same moment as redistribution (or in particle creation) or it needs to be relativity adjusted. Since energy cannot be used without dispersal it seems conservation requires some negative time input − in effect a backwards flow in time.

In fact we might be observing this very thing.

When life converts entropy into order, it raises the heat level. So the local future of life will be hotter than its past. The genuinely final panic will be when life reaches the point where the density of low entropy processes is insufficient to make an incremental increase in local order. That really will be the end of growth for that universe.

On cosmological scales there two phenomena which are still puzzling, namely dark energy and dark matter. Currently, dark energy and dark matter are quite different. Dark energy is the name of the unknown source of expansion of the universe that appears to have arisen about 4-5 billion years ago and which appears to be accelerating. Dark matter, as distinct to dark energy, is observed through the additional kinetic energy of stars in the halos of galaxies and by the additional mass galaxies appear to have that holds them together under the faster rotations. Dark energy is a component of the energy in the fabric of the universe, contributing to the Cosmological Constant, while dark matter is a gravitational mass which does not or just barely, interact with ordinary matter and its fields. Just 5% of the mass energy of the universe is ordinary matter, the rest is dark.

In inflationary theory, the period of rapid inflation stopped when the energy density of the vacuum fell to a low level and the remaining energy was released in a hot 'bath' of particles, while the universe's expansion continued to coast on, like a car that has run out of petrol, gradually slowing down. However, dark energy is expanding space faster and faster once again, having begun this acceleration around the lifetime of the Solar system ago. So, did the repulsive action of dark energy also begin separately in the Big Bang, and was superposed upon the inflationary expansion and is only now revealing itself after the inflationary expansion slowed, or does it have a separate genesis?

So far we have argued that there is link between future activity and probabilities in the present. As general activity rises through the actions of life, one of the effects of the advanced probability waves passing through the past will be to increase the spatial dispersion. The secondary expansion that we observe and which began around 4 billion years ago can now be interpreted as the result of the 'pressure' formed by a degeneration in probability passing back in time requiring more space *in the past*. The accelerating expansion of the universe is happening in the past since that is where we look when we analyse stars and galaxies. We can consider how the Universe's future is confirming its

There is yet another idea to explain the dark energy fuelling the acceleration, namely an additional field to the cosmological constant called quintessence (Caldwell, Dave and Steinhardt). Quintessence overlies the pressure described by the Cosmological Constant but whose energy density and therefore repulsive action changes over time depending on the levels of the potential and kinetic energies in the field. Currently, Quintessence is more like a mathematical fix but there yet maybe some more compelling reason for it to be.

past evolution, and reflects the ever-rising interactions of activity and in particular the Coincidence Number of our present.

We return again to the accumulation of tychons, the signature of which, at the level of matter accumulations and the human narrative, is coincidence. Since coincidence is costless information which does not appear in probability distributions of states (except after the fact), it acts as a 'binder' to events and to allow entropy to generate greater numbers of causal intersections without being 'consumed' in the process. The effect of this at the atomic level would be to interact with gravity and increase energy 'utilisation'. This may be an explanation for Dark matter. It is, in effect, a super-entropy principle, observable over similar distances that the Hubble expansion is observed. As such it would interact with the energy of space-time. Dark Energy, while apparently different to Dark Matter may be in fact be two manifestations of the same cosmological principle.

As we have been arguing, however, a process of energetic addition to the fabric of the universe is increasingly moderated by the rising levels of coincidence and negentropy which such an addition releases, even as the temperature also falls. We can perhaps point to the beginnings of the expansion increases as resulting from some *event* in the universe and which may have altered the Coincidence Number sufficiently to allow life to develop. That is to say, a secondary universe-wide event occurred, long after the Big Bang, which was vital to the emergence of life. Based on our argument, the origin of the event of which this acceleration is the echo might be some activity symmetrically far in the future like the collapse of the false vacuum in which our universe is 'constructed' to another minimum or to the ground state itself, throwing up a cataclysm of fresh activity, or to the invasion of the expanding frontier of a neighbouring bubble universe or even to the collisions of two surfaces (branes) each holding a universe. Curiously, the time symmetry brings us pretty much to the death of our Sun (where it expands to a red giant in about 5 billion years making a total lifetime about 10 billion years) anyway. We will have to see. After a further 5 billion years our universe might be almost twice the size, which implies that the universe is going to get a whole lot more active than it is now (on the basis of more tables being put out for dice throws), and consequently a whole lot more threatening to emerging life. Humans might have emerged at precisely the key coincidence number evolution between too low (quiescent universe) and too high (overly active universe).

As we come to discuss information and Shannon entropy we can take the idea of the 'pre-loading' of information occurring from the expanding frontier affecting the evolution of states within and the probability values of our present moment and the outcomes of measurements. We have arrived at the point where we can see how the flow of backwards running information ties the future and the past together and creates the present. There are interesting implications for quantum computing, especially about how to make the choice of when to stop the calculation to extract an answer. So far it seems that answers in the quantum computer are only ever going to be to some arbitrary level of confidence. To improve that level of confidence we are probably going to have to recognise where coincidences among qubits will be a key manifestation of the answer, but that is another story.

So now that we have examined how the notion of coincidence leads to an idea about time and to the role it plays in the universe's origins, let us return no to the human narrative level and ponder further the significances to the human mind of how coincidental events occur in the flow of time.

A bit about time and human narratives

In which we consider the differences between going forwards and backwards in time.

In which we consider a teenager's room and the connection between time, order and entropy and how events are linked in chains of cause.

In which we ponder we ponder the meaning of the historical record, the present completeness hypothesis, and whether retro-causation might be observable, and where the past and future might be if they exist.

In which we examine the meaning of a message, the limits to information and the resistance to change.

In which we look at biological reproduction and the origins of life and the crucial role of prediction in evolution.

All the while, we mistake chance for choice, our labels and models of things for the things themselves, our records for our history. History is not what happened, but what survives the shipwrecks of judgment and chance. - Maria Popova

Before we consider time travel narratives, and while we have the multiverse scenarios of quantum realities fresh in our minds, let us muse a little further on ideas about the three zones of time, namely the past, the present and the future.

Where is the Past?

If the universe is just an ever-moving, ever-evolving 'sea' of fields and wave functions then it is hard to imagine that a past exists, since this sea of movement and probability is always changing, never fixed, and holds no memory of the shape of the surface of the sea just an instant ago. The present moment just disappears into the next with no residue of what it once was remaining somewhere for reference. The analogy, however, is not very accurate in almost every aspect. It does not consider the geometry of space-time over which the fields act, nor the stability of the states of some wave functions, nor the distances the wave functions describing a particle stretch over time. While the picture may convey a helpful image of the idea of random fluctuations in the fields, much of what goes in the universe is far from random, and, thanks to relativity, we know the present moment is not found in a change sequence coordinated everywhere but is composed of many pasts and futures, some of which cannot exist except in the eye of an observer.

Wheeler, commenting on the pragmatic Copenhagen interpretation of quantum measurement, agreed with the notion that the past had no existence except as that recorded in the present. That is to say, a memory of the past is not in any way *the* past (we will test this notion further on). There is only a single state at any one moment undistinguished by other values of time. Certainly around the time of the Big Bang there could have been little if any differences between states of different times, but that is no longer true as Relativity shows us. The universe on the large scale may be homogenous in matter/energy but not in time. It now has a past just like our minds, the new component of the mix, and which have a self-referential nature which requires a past to make them work. So also, as we noted, the universe may require a past to make probability work.

We also noted in our analysis of Schroedinger's Cat, that a superposition similarly relies upon a mix of times like memory and a prediction of states in order to work. So the multiverse notion implies at least a certain continuum of time between

> Wheeler and deWitt tried to write a function for the entire universe, but it was not a success. One of the reasons being that the equations only make sense by including an observer, which immediately undercuts the 'universality' of the equation. But can this picture be true of consciousness even if it is a brain in a vat or a simulation in a quantum computer?

past and future in any measurement. How long is that continuum? The 'volume' of time and space in which measurements may be made must include a continuum of time information sufficiently far in the future and the past to make sense of a probable occurrence, so it will be different for each class of measurement. Similarly, the transactional interpretation of quantum measurements requires a continuous connection between future and past times. This difference between describing the universe's behaviour as an ever-changing present and a continuous time surface is the key to solving some classical time paradoxes of human narrative time, and which we will deal with in the next chapter (see **Imagining time travel narratives**) by virtue of two time trips that one can make into the past.

If the past exists only in mind as a result of our self-awareness, and perceptions, thinking, being aware, all rely on delay, on stuff already laid down and stored in some way for future access, then consciousness is the very mechanism by which the past and the future begin to make their home within this universe. The present moment is essentially a ratio made with the past, while the future becomes the place to move into.

Since some memories (at least those to do with intention) are derived from a moment, they have to be held somewhere and they have to brought into play when needed – a form of pre-loading, in fact, on which we rely to think. This necessitates at least a *relative* past and future to be written into the fabric of the universe. The question is, do these relative times, now that they have come into being, also exist independently of consciousness and the brain? Does the human mind give to the whole universe the past and future that once wasn't defined for it? As if it created fixed points or standing waves in the universe's whole wave function, constantly maintained in its restless moment of change and movement.

We will come back to this question since it bears directly on our answer to Fermi's famous question (where are the other intelligences in our universe?), namely the monasmotic universe. But for the moment let us follow a slightly different train of thought. What *the past* might mean for humans at their operational level, in ENT.

Arrow of time

The removal of probabilities in our world through universe multiplication and decoherence may explain time's arrow but it doesn't really fit with an increase in entropy in our universe. In the multiverse view, probabilities of states, for example, of where a particle is, get removed by observation, so general uncertainty, as expressed in the superposition, is *reduced* in the observation. If we were to pass backwards through time at this level of magnification rather than have time go backwards, we would be continually moving from the more ordered state that came with the removal of probabilities of the observation to the uncertainty of the superposed state of the spread of probabilities before the observation had been made. In other words, viewing the succession of observations of the timeline in reverse order looks suspiciously similar to local disorder *rising*, in contrast to the world of the present where disorder rises going forward in time. The situation recalls the same illusion given to us by the film experience, and also describes an essential divide between events at the atomic level and events at the human narrative level, in ENT.

As we saw with the many worlds scenario the present retains its right to a mixture of certainties and probabilities, but where, through the reduction of uncertainty in the observation, unused probabilities are allowed to pass down the chains of causality from an option in the future *into the past*.

Thus, what appears to be an ordered and even static past will also become increasingly disordered. Just as each *present* involves increasing disorder, so does the past. Using entropy to create local order may cause disorder to occur further in the past and alter the reasons for being. Now we can see a new reason for the multiverse. It enables the future to communicate with the past.

So let us consider the case of the multiverse where probabilities or options are removed from the past by a measurement or observation. Does this consolidate the forward arrow of time?

Ministry of Truth

The role of memory doesn't always do the job of creating order in our narrative time (without pre-loading it does not provide negentropy). For individuals, memories are mutable – we know this to be true, but they are, too, for the human collective, for culture. We are endlessly revising our histories. A new understanding – it doesn't have to be a new fact (a telling point that we will examine later) – may well cause a re-interpretation of events or bring something else into view that had been ignored or unperceived but which can also demolish an understanding and undermine established facts as well as solutions to puzzles.

We can give historiography a new angle: to analyse the rate of change of historical understanding when there are no new factual discoveries to precipitate them. We might expect a trend in this understanding to move towards analysing why we choose the interpretations we do rather than the interpretations themselves. The counterfactual historian gives way to the non-linear historian who views history in terms of phase changes (catastrophy lesions in the surface of action) and superpositions of trends.

What we know about the past is in continuous revision. This is puzzling, given that we were there after all. How is it that we don't know what went on then? Some of it is missing. We make decisions about casualty in the present but these are always intermediary notions. The present we live is as provisional as anything else. Our consciousness swims in uncertainty and doesn't only live in an arid past of decisions made. (We proposed a reason why earlier on.)

But what happens when, instead of probabilities being altered in the past, we actually remove factual items entirely from the past? What happens to the collective experience when the mind returns to a moment in the past only to find the contents of that moment are gone? Can it even find that moment if there is nothing there?

It was Winston Smith's job in Orwell's *1984*, to revise the record and destroy anything that contradicts the 'truths' of the moment. This process embodied the following paradox that Orwell, the writer, either didn't fully see or just ignored (writing only that, as facts were eliminated, the past existed in memory). Smith worked in the Ministry of Truth (Minitrue) to re-fashion the historical record so that past predictions agreed with current facts, or say, to remove a once approved person, who had fallen out of favour, from all records that mentioned him or which showed him in photographs or newsreels. The old 'facts' were destroyed by document destructors called 'memory holes'. Later on such *unpersons* may be rehabilitated and their presence written back into history. But once the facts supporting one history had been entirely removed, how could the process be reversed? There would be no material to create the same history again; nothing in the archives to describe this person, where he was born, what he did. It would all have to be invented anew. Smith, indeed, on occasions, elaborated narratives, writing 'forgeries' as he called it to replace a deleted set of facts. (Once you start down this road, then History becomes complete fiction and there is no need for the historical narrative to contain any truth or fact or real people whatsoever, and by merely replacing one name with another one a single history could serve for all. Rather like reducing the pre-loading to a one-dimensional genetic strand.)

Smith's work reaches back into the past in order to fix the arrow of time and to destroy alternatives. In fact, the consequence of his interference makes for an endless present; it stops the arrow of time. Nothing develops, life marks time, running 'up' Escher's steps forever, not even cyclical. Although events occur, the re-writing of history ensures consciousness does not go forward.

Smith's 'nation', Oceania, was constantly at war with one or other of the remaining nations of the world, Eurasia and Eastasia. Whenever the opponent changed, all references to the time when they were at war with the other were removed from history, so that history showed only that they had always been at war with the current opponent. This revision is much easier to do, of course, than rewrite the complexity of a life since whatever statement one finds with a war combatants reference in it, only two words need changing.

But it is a mistake to think that by simply removing any given item from history that is all there is to eliminating it. Winston Smith's job is a particularly *literary* job, an editorial job arising in the imagination of a struggling writer alone on a Scottish isle, and is not one that could work in the real world.

> I am writing this on a Macbook. My wife has a few years younger model. There are differences between them obviously but there are stranger subtleties also. News stories differ, and perhaps one might expect this, but the same weather app, presenting a superficially identical appearance does not show the same weather details on both. Daily temperature ranges differ. The forecast differs. What hidden variables govern this?

People have memories and experiences which interlink with an object or a person but which exist apart. Objects leave all sorts of metadata around, like the ideas that brought it into being, the design and manufacturing systems and the shaping in the resource material from which they were were made. People have parents, siblings, offspring, jobs and creations. The past is not a rigid collection of words like a sentence on a page. Items do not have clear boundaries; they cross-refer in complicated ways. To have an item removed from the past and from the influence it traces through time, it needs to be not just removed but more subtlety altered or overwritten with something else that has just as real reasons to be. A narrative may be fragile in the sense of a lack of inertia or invariance but existing narratives are still hard to get rid of. Douglas Adams' hero Dirk Gently's rubric, 'the fundamental interconnectedness of all things' is an understatement. *Potential* existences of things are also interconnected and they change over time. Winston Smith, as he works, would have to have observed at least some kinds of undetermined alterations to non-material parts of history he hadn't touched. But the task of re-writing human narratives is even more difficult than Smith believes because it involves consciousness and requires intervening in psychological states to make narrative changes, even fictional ones, convincing, believable, true. (This is very nicely illustrated by Trevanian's two stories that I mentioned in the introduction, although it is a truth roundly ignored in discussions of time travel and by most writers of time travel tales. See **Appendix 2**)

While Orwell may not have deeply explored the paradox of re-writing history without memories, he did see how meddling with history changes minds. He was focussed on how the Party tried to edit language to enable complete obedience. It's a popular idea even nowadays that language determines what ideas you can have, although as we have been arguing memory is a greater determinant of what narratives are possible. Winston Smith could end up loving Big Brother only because he had no longer any memory of hating him or loving Julia.

> The objective of the party in Orwell's 1984 was not just to eliminate certain facts but to eliminate change in the mind, to eliminate coincidence, to create a constant state of awareness of love of the Party. Orwell was effectively describing the contemplation of the godhead in Heaven, or the extinguishing of self at Nirvana.

We can find a good illustration of the basic fragility of any narrative by looking at what happens when we browse the web. The future steps of our web browsing are (hardly surprisingly) mostly determined by the page one is looking at. But, since cookies residing on your system and, more importantly, your IP address, all of which stay with you as you browse, often determine what is shown you on each web page, a Markov chain analysis only goes so far.

Some browsers store your most commonly visited pages in order to serve them up to you quicker next time. So your web experience is shaped by habits, by stored information about choices you have made and your point of entry into the system in the first place. Nothing much about a Markov chain in that. Even without a browsing history or reference to cookies as a guide, at the least, web site software reads your computer type (or whether it is a mobile phone), operating system, browser type, language system, country information and location. Your web browsing experience is shaped by underlying narratives with entirely different purposes to your own. (When you block this, you will see a somewhat different web world than before the block.)

The browser analogy is interesting but what about the real world narrative? We saw how the sensation of the arrow of time can be fragile if the past remains in flux. All we have to ask is ourselves, Does our history really change without our realising it or does just the data we continue to hold in the present about the history change?

Certainly it appears that information we hold about the past readily changes into new ideas or opinions without necessarily any new factual discoveries to provoke them (at the very least due to interpretive drift in language). But for this to be possible, the records or memories we keep must continue to contain uncertainties which permit such mutations of thought.

We can 'mine' the records we keep with us in the present for stuff we missed and didn't observe. But that again suggests that what we keep must be a record of uncertainties as well as certainties existing in observations or measurements during the moments that the past was our present.

So there is nothing ineluctable about the past and that our proposition that the past is influenced by the future is compatible with what we experience of the past in our consciousness. We are getting closer to where observation in the quantum world and the outcome of events in the human narrative are more logically connected.

While the quantum world, however, appears to be eternally giving up probable values to observed values, outcomes in the human world are determined by a whole complex of narratives whose events exist only by chance. Chance plays an enormous role in the outcomes in our consciousness, and far from being removed by events in the mind, chance enables us to anticipate.

> Questions: Is the future consistent across the wave front of change? In remote places as well as near ones? Could there be local futures developing unobservable from any point other than in their own time lines? Not necessarily full universes, but smaller regions with their own time, little pockets of specialised future unconnected with others around them? They may or may not connect later. When we talk of an homogenous universe we are really talking only about the average mix of matter types and average energy over large volumes of space. The universe has distribution variations even at large distances, so, too, the microwave background radiation which has had nearly 14 billion years to even out. So how uniform is the time line in the Universe? Does inflation eliminate every kind of difference? How could we move from one part to another? What would the borders or horizons between these 'zones' of future look like? How would they react? Could they be penetrable and what would a particle have to give up in order to cross a 'membrane'?

We all have practical experience of this. Sometimes just a word, a glance, a smell, can shift us into a new path, a reorganisation of matters reaching down into the quantum realm. Connecting our knowledge of entanglement and superimpositions in the 'sea' of quantum states with our day-to-day narratives seems to be getting harder, not easier.

We can describe particles and the situations between a few of them, but we have no way yet of superimposing or layering the successively higher level states and descriptions in which observers sit to make up the reality we experience in quantum terms.

What happens when we zoom out, as in the *Powers of Ten* animation, from our subatomic particles, widening our frame of observation. We can describe the entangled behaviours of a few molecules in a chemical bond or in a crystal, but that's it. The great gulf between the causes in the subatomic world and the macroscopic world of ENT that humans occupy has yet to be filled with anything we might call understanding.

So let me repeat the suggestion I made earlier. Let us look at the entry point of probability as being somewhere in the middle ground, located in and about future states of consciousness and sending influence back into the past and down into the quantum realm as well as up into the human narrative, and in the light of what we have been discussing about multiverses, let us think about the present.

Present Completeness Paradoxes - Problems with the now

Time is an illusion. Lunch time doubly so. — Ford Prefect

Past, present, and future. Do they exist independent of our minds?

Some will say that past and futures are only beliefs since we have no contact with them. We can only sense the present. Others say that these pasts, presents and futures existed before we came on the scene and are independent of what we think about them, and have some objective reality.

But even if we experience only the present we seem to live in, there's still a problem. Physics is starting to hint at the notion that our consciousness and the physical world are inextricably linked way down at the causal atomic level. In which case our *beliefs* about the past and the future may still interfere with our interactions with the physical world as if they actually existed.

Thanks to entanglement we have non-locality to think about.

As we earlier described, the Bell Inequality suggests that causes are much more spread about than any local theory of causes can explain. This also means that what we think of the present, the now, is really a spread of time, and weakening what is known as the Markov condition. The Markov condition considers that the stages of causes should not remain dependent on the stages of the past. If we knew everything about a middle stage in the causes of an event, we would not need to know anything about the earlier stages. This is to say that causes do their work and then finish. Like Schopenhauer, who thought that anything produced has been produced by sufficient cause for its production and nothing more, Markov refused to accept hidden influences on the progress of events.

Interestingly, Markov excluded circularity from his condition, recognising that repetition occurs because of a continuing underlying dependence on the past. It is impossible to separate out causal influences in systems that keep repeating their states or which self-refer.

I will re-state the Markov Condition as the Present Completeness Hypothesis (PCH), which hypothesises that all the causes you need for a future event are to be found in the present – i.e there are no hidden influences from other times. But if we take Markov strictly then we are faced with a process occurring only by virtue of a prediction about its future and no past at all, no memory. How likely is this?

If the PCH is correct then time travel is unlikely. Unless connections with other time zones are *already built in* to the present, which could explain at a stroke why locality, freedom of choice of parameters to test realism fail the Bell test and confirm quantum theory.

Present Completeness Paradoxes – problems with the Future

Gödel, a logician who also found solutions to general Relativity that hint at time travel (by large closed time-like loops), proved that in any self-consistent logical system built on axioms, there is going to be at least one logical statement that is true but which cannot be proved to be true with all or any of the proved statements in the system. They are real but undecidable. Can these non-provable statements, called Gödel sentences, have functions in causality?

Certainly the physicist Roger Penrose thinks that such things belong in consciousness but others are not so sure. Gödel's result is derived from a set of limited logical conditions that may not be very useful in the Universe at large because his result revolves around a self-referential statement.

> Traditionally in logic a contradiction implies any statement. Gödel showed how a contradiction can imply a true statement which cannot be qualified as true. Sounds odd, but you find this notion in practical use all the time - by anyone who wants to get their way regardless. But you also see proto forms of it in heavy use by children, e.g. 'You don't make my brother go to bed at 5 therefore I deserve that toy.'

Here's what worried Gödel.

If you find a sentence G that's not provable in a system of axioms T then you have proved it. But if you have proved it then surely it's *not* provable in T after all and you can't know if it is true. There's a deadlock here. Gödel managed to break the deadlock by constructing, in a formal system of his own making, another kind of meaning to statements whereby such a sentence G can be true even though it's not resolvable in proof – a Gödel sentence.

There is something about this finding that connects with what makes a set and what makes the elements of a set that we talked about earlier and will come to again. But I like to think of Gödel's analysis like this. Proving that you *can't* prove all statements in your system are provable is a contradiction to the process of proving things, because proving you can't prove all statements requires all statements to be proved (one way or the other) including this one, before any proof that you *can't* prove all of them can be true, which is a contradiction, and thus, mysteriously, proves it, in a way.

So, as long as you can fairly compose the equivalent of that statement in your system, the proof that you cannot prove everything in your system will be valid. Some statements (at least one) cannot be proved and that you don't know which one it is (or they are) because if you did you could exclude it (them).

Now, it is generally thought that this result has limited meaning, and only within a branch of logic, but maybe there is more to it. Think about our brain-in-a-vat. In order for the vat brain to establish that the world is false (or that there is an external reality that is true and his version is a fake) it needs to know that every state it observes can be proved to represent the fake reality. Since this is a contradiction it can only at best know that some things in its reality are not real but others it will never know which. Similarly, it will be unable to show that everything in its head belongs to another reality. Some things it will have generated but it won't know which. Similarly, we, here and now, will never be able to show that our reality is not a simulation or does *not* have things from another reality in it (at the end of the book there is a proposal about how we might be able to use this very fact against the reality we perceive).

Penrose comes to a rather different conclusion about the significance of Gödel's propositions and thinks that it proves that consciousness cannot operate like a logical system built on axioms – like a computational machine – but on quantum uncertainties instead. Rather than unprovability being the causes of uncertainty (at some point the computational machine comes to a stop because it hits an unresolvable contradiction), it is the essential component of an awareness that can grasp contradictions and move forward in spite of them.

It seems, however, that this state of affairs depends upon how events are embedded in each other. Gödel sentences – let's call them gödels – exist within a consistent logical system. Simple arithmetical gödels may be proved by other kinds of mathematics like sets. The system of sets will produce it's own gödels which may only be proved by yet higher levels of mathematics. In any reasonably complex world some theoretically usable information about a system (i.e. about whether two states that contradict each other can actually exist in a state of contradiction) within it remains uncertain – its description is incomplete. So there are wider implications to gödels than just for logic. (Entanglement is about alternative states existing at the same time and not about contradictory states)

If the Universe embodies all logical systems that there are, and if one event evolves into another in a Markov fashion, i.e. in a set of logical allocations between the past and the future, then Gödel's analysis suggests that some anomalies in the physical Universe are going to be unresolved; are going to be gödels. In fact, you can turn the whole thing around and say that for the Universe to be consistent, gödels have to be in play! A universe without gödels probably couldn't exist.

Since gödels are states whose condition of validity cannot be known, even though they exist, or be placed in a logically consistent way anywhere, even knowing everything about the rest, then our present is necessarily going to be an inadequate explanation of itself. (*Get down, Hume!*) Even though the Present Completeness Theorem is useful, gödels imply that the present contains its own unprovables (by reference to other proved statements) – something about it is provisional. Where we find these gödels, of course, may not be predictable; their location and when they are going to crop up is also uncertain, but let us simply say that space may be shot through with moments – and the simplest moment is the creation of a quantum state – that cannot be resolved into a state in our space-time. (As if dice are being thrown on tables behind curtains and the results just called out.)

> Could gödels provide the beginnings to universes, seeing as they may contain unlimited information?

Even more curiously, in Shannon entropy terms, an unresolved quantum state or gödel seems to have unlimited entropy and thus unlimited information content, which seems paradoxical, though fitting, if true. Was it a gödel that caused the Big Bang, or at least initiated the expansion of our universe? We will get back to this.

But the most interesting things about gödels is that they appear to have no energy, no mass. They exist as information (or a lack of it) and can appear and disappear without adding or subtracting mass in a system. Gödels may be all we need to show information goes back and forth in time.

Gödel shows us that gödels, and incompleteness, are a necessary effect of consistency (i.e. continuity), (suggesting that The Present Completeness Theorem may not be correct after all). Even the ancient Greeks struggled with ideas of incompleteness in time and space, especially Zeno.

Zeno wondered about the present and how complete it was. He believed in a static universe where movement was an illusion (although why this was thought of as a good idea remains mysterious), and used the notion of infinity to create apparently inescapable paradoxes.

What Zeno really showed is the inescapable relation between time and space (Einstein is his inheritor). Locality is all about time when we refer to the 'stuff' of the world, and time is only continuous and capable of change where space is also.

Two Paradoxes

Zeno looked at the race between Achilles (a famous Greek warrior and half god) and a tortoise. The tortoise was given a head start of 10 metres. Zeno bet that Achilles wouldn't win the race. He said that by the time Achilles got to where the tortoise had been at the start, the tortoise would have crept on another metre, and that by the time Achilles got that metre further on, the tortoise would have crept on another 10 cm, and by the time Achilles got to *that* point, the Tortoise would have crept on a little bit further, and so on and so on, so that no matter when you stop the race, Achilles has never quite caught the tortoise up.

The infinite series of divisions that Zeno describes, while logical, doesn't seem to be *true enough* to apply to the real-life situation. Achilles wins in a little over 10 meters and Zeno loses his bet. How can the logic of the series and Achilles' win both be true?

Zeno wisely ignored the fact that the two frames of reference are not independent. The relationship exists by virtue of the clock they share. Achilles' motion defines the time intervals in which the Tortoise can move. As Achilles crosses the successively smaller distances at constant speed, the tortoise has less and less *time* to move on, leaving Achilles even shorter distances to cross to reach him. The gulf between the two reduces increasingly rapidly – it's an example of exponential feedback. At some point, half the infinitely small is the same as the infinitely small. The remaining interval is no longer distinct from the previous interval and Achilles can thus cross it in zero time.

Actually, Zeno probably understood the mutual dependence between the frames of reference very well because he was a member of a religion that did not like motion and believed that existence was an independent quality without divisions and was always at rest. He deliberately set out to prove that motion was impossible. He delighted in his invention of infinite infinitesimal distances taking infinite time to cross, and used it wherever he could to poke fun at those who believed mathematics said something real about the world; a question which puzzles mathematicians even today.

So, when Achilles' frame of reference meets the tortoise, it is moving at a relatively high instantaneous velocity while the tortoise is moving at zero instantaneous velocity.

This is how he overtakes the tortoise.

Because of the interaction between time (the time the tortoise has to move) and space (the distance Achilles crosses) in this paradox, the final interval of the ever-decreasing series is the limit both to space and to time.

A smallest unit of space may be easy to understand, but the smallest unit of time (a chronon) has its problems.

Here's The Encyclopaedia Britannica on the subject.

'The relation between time uncertainty and energy uncertainty, in which their product is equal to or greater than $h/4$, has led to estimates of the theoretical minimum measurable span of time ... and hence to speculations that time may be made up of discrete intervals (Chronons). These suggestions are open to a very serious objection, viz., that the mathematics of quantum mechanics makes use of continuous space and time (for example, it contains differential equations). It is not easy to see how it could possibly be recast so as to postulate only a discrete space-time ...' [– EB CD 1997 edition]

Zeno pointed this out with yet another paradox, the paradox of the arrow, and with which we will take more than a passing interest. At first sight this 'paradox' assumes what it sets out to prove. Choose a moment in the flight of an arrow. The arrow rests whole within its dimensions. Choose a later moment. Again, the arrow rests whole within its dimensions. It's not going forward and it's not going back. How did the arrow get from the first moment to the next when it is not moving beyond the dimensions of itself in each moment?

When the arrow shoots forward, Zeno's analysis relies on there being such a thing as a moment, a snapshot, of motion, that itself has no motion. No wonder there is a paradox. Moment by moment, the arrow has no weight or momentum either because these rely on time to calculate or make sense of. However, Zeno's paradox does hint at a decent question. How does the arrow actually make its path of flight? The track to the next point along its path must either be somehow stored up in the arrow, or the point that it sits in already contains the path information for it. Since a point must be capable of 'carrying' any track of any arrow, one must guess that the path information for that arrow is stored in the arrow. Einstein showed the opposite, or rather both. The point directs the arrow on a path according to its own energy while the space-time point is not neutral to the presence of an object at the point.

We know that we can use the geometry of space-time and the way it interacts with the momentum and vector of motion of the arrow to calculate the path the arrow will take, but Zeno suggested, at each moment, the arrow has no intrinsic information about where to go or how to get there. There is obviously a conflict between a moment, when nothing can be happening, and motion.

Zeno's paradoxes expose the logical conflict between tiny moments of movement and rest. If our mathematics uses infinite tiny moments of rest then what is the basis for causal continuity? How do causes get going if at zero time nothing has yet begun and everything is only at pure potential. We will come back to the problem of causes and chains of causes, because no one has yet established rigorously what is meant by a cause (Hume punches the air: see **The linking of events**).

In fact, as we now realise, it is in the tiniest divisions of space where everything happens. The tiniest division that seems to have any meaning in our physics is the Planck length, and it is pretty small, around 10^{-35} m. As it happens, string theory's smallest string is about the same length. Smaller than that cannot be described. Similarly the smallest unit of time is arbitrarily taken to be 10^{-44} seconds, which is the time light or information takes to cross the Planck length. If the Plank length has no meaning in terms of space, of course, then quite what the traverse time of light across it actually means is not clear. As we mentioned a while back, subdividing time as if it were like a spatial dimension or a real number makes no sense. There can never be infinite time in the gap between two moments in time, as there can be say between two neighbouring real numbers, since time wouldn't happen. Zeno assumed that by making space infinitely subdivided, time would also have to be infinitely subdivided. But this cannot be so.

How small is this unit of space? If we took a subatomic particle like a proton (say 10^{-15} m) and expanded it to the size of the observable universe (a diameter of around 8 x 10^{26} m), the Planck length would be of the order of the diameter of the Earth (13 x 10^6 m) on this scale. (Some physicists are imagining a granular spacetime filled with holes, but it is difficult to reconcile Relativity with this limitation because it would imply a limit to the shrinking of relative frames moving at speed.)

An intriguing is the idea of fractal surfaces surrounding black holes where elements of a surface subdivide into an exact but smaller copy of the parent surface and where that surface element subdivides again into an exact but smaller copy of its parent. Eventually the surface becomes infinite in extent, not in the way of a sphere which has no boundaries but in the way of a road that never ends. A particle trapped on such a surface should not be able to find its way out along it. Zeno would have liked that. Continuity is still a mystery.

Within this unimaginably tiny space and time where the energy density – the Planck density – is extremely high, everything happens.

Pairs of particles are being created out of the nothingness and then get absorbed. Quantum fluctuations contain so much energy that they can become entire universes. Our universe appeared out of one such extreme fluctuation, or so it is thought. The smooth peaceful continuity between one point and the next doesn't apply at this level of magnification. And yet, while these moments can produce universes they obviously don't in our universe (perhaps because the energy density has fallen so low), although how would we know? So what constitutes consistency between one moment and the next is an open question. If these tiny moments are filled with probability then consistency and continuity between moments do not seem to be possible since a probable connection between one point and the next is not smooth.

The continuity of a field is also challenged — indeed, reversed — in the larger volumes within atomic particles where their constituents, quarks, interact more weakly at high energies and more strongly at low energies, so preventing the infinite rise in the energy of interactions implied by any two things in a field getting ever closer together (like Achilles and the tortoise).

If we think of continuity as that property which enables information to pass from one place to another without being changed, then the idea of the foaming structure of space with points of enormous random fluctuating energies seems to be missing something. One wonders how probability can be the basis for consistency across space and also do its job of providing variations.

Thus, there seems to be a disconnect between the behaviour of discrete items that make up the whole and the behaviour of the whole. Any causal stream of connections necessarily becomes fuzzy at a certain level of logic, that is to say, with a rising number of gödels. The lying Cretan, Russell's Barber, Achilles and the tortoise all show us that the whole is something different to the sum of its parts, and when it comes to quantum states, dividing a wave into component parts is not as simple as it might seem.

The future in the present – the Relativity question

The Relativity paradox about the future can be described like this. Let us consider the case without the contractions of time and distance for a moment. The light from Alpha Centauri takes 4.3 years to reach us. We are observing the past of the star – how it was 4.3 years ago. And the same is for any inhabitants of planets around it looking at us.

They see our past – what we were doing 4.3 years ago. Yet if we started off towards it accelerating very close to the speed of light we would, theoretically at least, because of time dilation, see their future because they would be moving through their world-lines faster than we are. The travellers heading for Alpha Centauri would certainly see the life of the Alpha Centaurians (let us call them the ACs) speeding up with respect to them, but speeding up with respect to the past that they observed at the moment they took off – that is 4.3 years into the past of the ACs. When our travellers landed after ≈4.3 years of travel, the AC present moment would have advanced 4.3 years further on. The travellers, by the end of their journey, seem to have actually 'observed' 8.6 years of development in the AC worldline, compressed into their travel time. (Funnily enough this actually coincides with the Relativity time dilation calculation for a speed of around 9/10s the speed of light.) On this basis, they have not travelled into a 'future' that was already in existence when they set off, they have just 'hung around' for a while until the AC developed their future.

Since the observation made on Earth of events at AC must always be separated by 4.3 years, what should we expect our Earth astronomer to observe when she tries to watch the travellers on their jour-

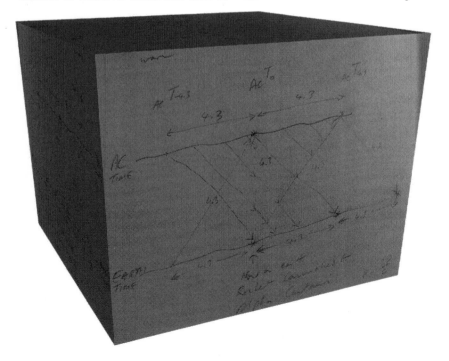

The Cosmology of People, and the Time Travel Solution

ney? She observes their journey for ≈4.3 years just as she observes AC advance from the past through 4.3 years of development to the point in their present when the rocket from Earth took off. But what does she actually see? The rocket ends up moving so quickly that she is unable to get a fix on it. As the rocket approaches AC, she observes it becoming so indistinct as to be unable to tell when or where it has finished its journey. Only after 8.6 years of observation would she be able to tell conclusively that the rocket had landed. I sketched this situation out in my notebook one day long ago.

Time dilation works like this, as Lorentz showed, and using the 3000 year-old discovery of the triangular relationship we call Pythagoras' theorem to explain it. If we set a mirror at a distance D from us and flash a light at it. The light takes the time of twice the distance divided by its speed (the speed of light, c) to return to us. We will call this time $t_0 = 2D/c$ (and $D = ct_0/2$). If we get in a rocket and set off at speed u parallel to the mirror and shine that beam of light, the light travels a longer distance L to the mirror and back. So its time $t = 2L/c$ (and $L = ct/2$). In that time the rocket has travelled a distance of ut. By Pythagoras, $L^2 = D^2 + (ut/2)^2$, so putting it all together we have, $(ct/2)^2 = (ct_0/2)^2 + (ut/2)^2$. Collecting terms we have $t^2(c^2 - u^2) = c^2 t_0^2$, and thus $t^2 = c^2 t_0^2 / (c^2 - u^2)$, dividing right hand terms by c^2 and taking the square root gives us $t = t_0 / \sqrt{(1 - u^2/c^2)}$, the time dilation formula. (What is so interesting about this is how time dilation exists with reference to a *local* variable, namely the local time in the frame you are using as your reference point.)

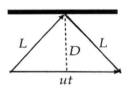

The AC, on the other hand, should observe something different. At the point of launch of the travellers of Earth, they see an Earth 4.3 years prior to that launch. As the travellers voyage on, the AC observe Earth's timeline progressing to the point of the launch of the rocket. Then the rocket, having been travelling all this time unknown to the AC, is set to arrive within a couple of months. What do they see? Since the rocket does not travel faster than the image of it, the views of its launch must precede it, but what of the rest of the rocket's voyage? Since the rocket cannot overtake its images the AC would see nothing until a view of the launch quickly obliterated by a burst of radiation and then the rocket lands.

In spite of our thought pictures, this is not what actually happens if we use elapsed time to define the distances. Einstein's brilliance was able to intuit that the speed of light was constant for every frame of reference and explained the Lorentz contractions due to velocity in relation to the speed of light. Because of time dilation, the voyagers in the rocket experience time speeded up outside their frame of reference by large amounts. At 99.999% the speed of light, one second passing in the rocket is around 223 seconds passing at Alpha Centauri (as well as on Earth).

According to this formula, the voyagers of a rocket moving at almost the speed of light will arrive at AC at a point where the AC time line has advanced ≈ 961 years, i.e. well into their future. (It seems difficult to accept that for a companion rocket launched at the same time but whose speed is only 99.9% the speed of light the two rockets arrive at moments separated by 865 years in the timeline of AC, but that is the implication of Relativity. The second rocket travels more slowly, with a time dilation factor of only 22 and so arrives at AC 96 years in the future but this is in the past relative to the AC time when the first rocket lands. In terms of how we think about history, the second slower rocket got to AC first (the AC inhabitants and their history books would agree)! But where does this future exist if it is not visible to those in the rocket when they begin their journey, or for those inhabitants of the AC who have yet to experience it? Furthermore, as futures need time in order to develop and if they already exist with respect to the voyagers, in what form of time or dimension did they develop before they were observed to exist in our time-line by virtue of the motion of the rocket?

The above is a thought experiment. In physical terms such a voyage could not be made even supposing some 'magical' propulsion had been invented since the mass of propellant required is too great. The rocket has to accelerate up to the near light speed, and in order to land at Alpha Centauri, the rocket has to negatively accelerate (decelerate) and return to the common frame of reference that they both will share. Taking accelerating frames of references (say your rocket is constantly accelerating at the same gravitational acceleration we experience on Earth) into account, clocks along the lines of flight run faster in front of you and slower behind you until a 'horizon' point where they run backwards. AC in the distance would appear to accelerate towards you as time went on. You would, however, still need ≈ 3.6 years of your time to get there.

The problem about accelerations are, however, more complicated because an acceleration is a force, which it is why it is difficult to distinguish from gravity, and since almost every motion in space-time involves an acceleration of one sort or another, matter and energy are constantly exchanging differences. We can calculate these interactions for particles (time delay is certainly true for atomic clocks travelling in Earth orbit or for particles like muons whose calculated decay into other particles appears to be delayed in cosmic rays or when they are contained within a particle accelerator), but what does Relativity imply about information? What does this means for consciousness?

Some philosophers believe that Relativity tells us that every future must already exist and that you can access any amount of future, including minds and histories, by moving along the dimension of time by virtue of your velocity through space-time (a trip at relativistic speed is rather like making Trip *A* forward from *T2 to T1* in our diagram of the timeline (see **Imagining time travel narratives**). On the other hand, we do not know what role mind may play in this mixing of times, because mind, as we will discuss later on, is rather different to matter and cannot be based simply on the mounting accumulations of particle behaviours. Furthermore, the implications of having minds moving at high velocity whose apparent processing speed is much slower to a relative non-moving observer have yet to be understood. We only have evidence from the time to decay of subatomic particle states that their internal 'clock' slows at relativistic speeds but we know nothing of the implications of this for a human mind. We think that the internal clocks of astronauts differ from terrestrial clocks by micro seconds but do not observe perceptible differences due to this alone (atomic clocks brought back from orbit are slower). If there are any, they are surely swamped by other physical changes. But travelling at significant fractions of the speed of light may be something else again. It is worth pondering whether consciousness presents some kind of information constant to the merging of frames of references with different times just as the speed of light is the limit to the rates of energy transfer between frames of reference. Again, the facts of the real world appear to suggest some kind of divide between the behaviours of particles and human consciousness at the level of narrative, of prediction and memory, and this may end up being significant to pilots of interstellar craft (see **The case for the human pilot**).

While Relativity has destroyed the idea of being able to establish the same moment across moving frames of reference (simultaneity), it has yet to demolish the idea of the now which *we* occupy.

In the human narrative, we can reflect upon this *now* with the help of media systems like Facebook and Twitter which have rolling content continually updated. While these systems have a memory it is not a memory of the entire picture at any one moment. Twitter memories are the histories of individual twitter users contained in each user's account. Make a search for thematic content (hashtags) or for subjects that are popular with tweeters at that moment and you will see a continuously updated public list of tweets (not necessarily from your followers) tailored for your geographical area that has no long term memory. (Historical information is retained in statistics about tweets and accounts which businesses make use of.) You cannot go back in this rolling feed to a given time in the past to see what anyone else was tweeting on a given topic. Imagine what it would be like to maintain the history of the rolling present for every point in time for every user. The rolling content has no purpose outside the now (which contributes to its addictive quality.)

With this analogy we can understand how local pasts and futures may combine in consciousness to give us the sensation of the rolling global present but also we can see how strictly *local* history could be changed without causing serious paradoxes in the wider world of the sum of all events. Local pasts may be stored even if the actual now is not. This idea, however, may well be contradicted by tests of the Bell Inequality and we will discuss this at length later on, but for the moment let us recall the transactional interpretation and consider that a measurement involves a past and a future, and that we can interpret how we make an observation, namely by the *arrival* of a wave function from the future confirming the observed probability. Until this confirmation arrives then no measurement happens. We can consider this wave as a wave of conformity coming from the future, a form of backwards 'adjustment'.

Why should we look at the situation this way round? We need to look again at notions of time's arrow and how information works.

The symmetry of time

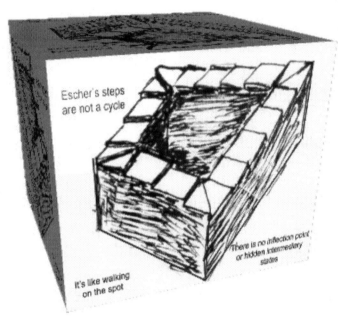

Pondering the final panic gets one thinking about the future. Not simply about how little time we have allowed us to defend ourselves from the mathematics of self-destruction, but whether time is a true barrier to access to other regions of narrative time.

Let us first look at what most people agree on: the ability of time to distinguish events between a before and an after. Though, since the advent of Einstein, we realise that time doesn't always separate events in the same way for every observer.

It used to be our sensation of external narrative time was just of what came round again. Early Human's awareness grew to the point where he wanted to remember and where he needed to predict. He started out satisfying those two concerns with the simplest method possible – a cycle. Static humankind sat, comforted by the repetition of sensations and the riddles of conformity. The idea of re-incarnation completed the life-cycle: death being the cause of life's beginning anew,

Cycles prepare us for what's to come and they 'remember' the past. The cycles of the sun and moon, of living creatures, of the repetitions of the seasons, of life and death, were able calendars of events and of man's activities within them, and it was from within them consciousness began for it is by remembering the past we can envision the future. This is how we think. Past and future are inextricably linked in a cyclical view of life. We cannot truly create the future in our mind's eye, we can only play with the tenses of what we can remember.

With more acute observations, humans saw that life's response to the natural cycles changes over time. A winter is never the quite the same as the one before; one winter's hunt never repeated the successes of the last one. Humans realised that the cycle of seasons must have a more interesting cause than mere repetition of a god-created situation.

A true cycle is a sequence that ends up with the same beginning. There are many of these in biology. The operation of muscle tissue, for example, or of nerves, reveal these kinds of cycle at work; so do the action of enzymes and catalysts. But where does the repetitious element come from?

A continuous wave attributed to a particle, however, is rather different, and its very construction may contribute to the arrow of time. Rather like Escher's steps, although there is no end, there is an interesting difference between going forward and backward in time which is that a ball say, bouncing along them 'ascends' in one direction and 'descends' in the other. We will come to this later on.

For a system to wholly and completely recapitulate itself, some component of it must be working backwards to provide the preparation for the repeated beginning. There must be some hidden states that allows for the progression to repeat. Since symmetry and cyclical components gives an atomic particle its spin, and maybe even its identity, this suggests there are differences in character between 'lifetime' or 'duration' and the time in which events occur. Two times that are relevant to our views about past and future. (something along these lines was suggested long ago by the astrophysicist Edward Milne in 1935, who devised a theory of gravity, where there were two times – one (t) for atomic phenomena, one (τ) for gravitational phenomena – linked by $\tau = \log(t/t0)$.)

If there were any symmetry in time it would certainly be more complicated than humans once imagined. Even Stephen Hawking thought that the Universe collapsing would be just like its expansion run backwards. He later called this his 'greatest mistake'.

The perceived direction of time is still the unsolved mystery both for the universe and for us at the human level of mixed cosmologies where time is different depending on our decisions. Why do we feel so sure that the action of time is non-cyclical, non-symmetrical, and which goes in an irreversible direction, at least in this universe?

Does sentient time go backwards?

Run a film backwards. It was a style very popular in commercials a few years back. It's thought to be a neat illustration of time going backwards. But really, the illustration is a bit of a red herring. We are outside the frame and watch the backward running film in forward running time.

We can run many processes backwards and not know the difference from our frame of reference. I once watched a very blown-up and very slowed down view of a sprinter doing the hundred metres sprint. When played backwards it was only the background and the end points which gave the game away. Repetitive actions like a pendulum swinging could be played forwards or backwards without knowing the difference.

To have the time go backwards in which we had previously gone forwards in, we would have to unpick our perceptions and our sensory apparatus as well reverse all movements. Nerve impulses would have to disentangle themselves from consciousness, retreat from the neural networks they stimulated, run back along the nerves and emit themselves from the skin or sensory organs. We would have to *un*experience the impact of the experience, de-construct the narrative, and this would be quite different from merely *doing the event backwards*.

Write the word 'time' backwards and you even get the word 'emit'. Hmm. I often wonder if, when we listened to someone who was going backwards in his time, we could hear a buzz in his ears of all the sounds he or she has heard coming back out.

The old-fashioned film process gives us a good illustration of the difficulties of establishing just what is the relationship between events occurring and the direction of time.

In a film projector, the film rolls off a reel above and is fed through the 'gate' so that the light can shine on each frame in turn and throw the image onto the screen. The film passes down through the gate and gets rolled up on another reel. Reverse the motion of the spools and the film runs up through the gate. People walk backwards; balls fall down from the sky and hit your foot.

We are accustomed to think that time runs in a set direction. Some film projectors also behave like the 'arrow of time' and are only geared to run the film in one direction (say top to bottom). How can you have time run backwards on the screen using such a projector?

To make the illustration below, I picked up a pencil and drew out the frames of a sequence that would show an arrow projected on the screen entering at left and falling down out of the screen at the bottom, as I would observe it in real life.

I began at the top of the filmstrip, and drew the successive frames going down the page, just as I expect to see the arrow behave on screen. But this means that when I come to project this event onto a screen, using the one-way projector, the film passes through the projector bottom frame first, and the arrow flies backwards: appearing on the screen at the bottom right, flying tail first, and exiting at the left.

No problem, I thought, I'll just turn the frames upside down.

But it can't be done this way.

The arrow flies upwards (see drawing).

Since I have a single-direction projector, there is no orientation of the film that will show the arrow behaving on the screen as I intended.

It cannot be done by any form of rotation of the *whole* filmstrip, backwards forwards or upside down, mirror-imaged or reversed.

The only way I can get my arrow to behave correctly is to have the film reverse *direction*. If I pull the strip I first drew *up* through the projector, on the screen, the arrow flies as it should. But if the projector reels cannot be made to reverse dir-

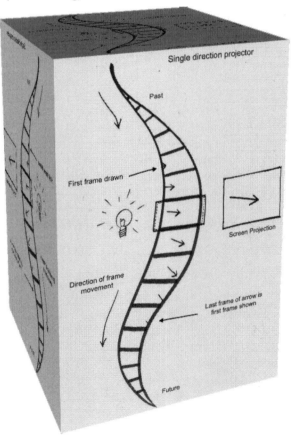

ection, how can I adjust the film to show the real-life movement of the arrow?

The only way I can alter this film to show the arrow on the screen coming in on the left and falling downwards to exit at the right is to *change the order of the frames*. I'll have to cut out each frame and splice them together in reverse order. Which is to say, in order to show the arrow behaving correctly, I would have had to have drawn the sequence in the opposite direction to the way the film must run through the projector.

This little thought experiment focuses our attention on the mystery of how the mechanics of the brain recovers the sensation of time passing in a recalled event.

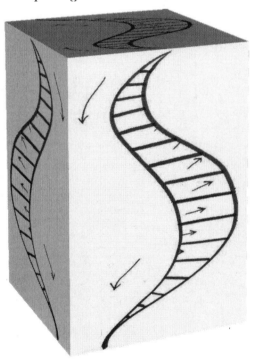

Imagine watching the arrow in real life. It exists within an environment you are familiar with, so old memories are also being delivered into your awareness. Suppose the brain 'writes' each successive moment of the arrow's movements into memory as it experiences it (drawing on other memories of and connections with the contextual stuff at the same time) while your consciousness follows the motion in parallel with the actuality, each frame superimposed on the previous one. There is a reversal here or at least a discontinuity.

In fact there are a number of discontinuities in the brain. Vision is upside down on the retina, the nerve bundles from the eye themselves split into two and go to each hemisphere, the cortex of the brain is in two halves which have to communicate with each other, and so on. We will meet a more revealing biological example if this kind of 'handedness' later on.

Memories appear to be stored features in cells as well as being involved in increased neuron connections, so something is set down in the direction of world Y senses Y brain network Y cell. What then does the brain do when you recall this sequential movement later and experience it in the 'mind's eye' repeating the scenes? Find the beginning frame address and follow it through like a computer does with a video file? Just how many proteins in a cell do you need to store a single frame of video? For the whole of a life? Where do we keep the list of addresses for every memory? The digital analogy does not work well anywhere in the brain, let alone in recall. (There is another radical proposal that you will find in the conclusions of the book.)

What can we make of people with eidetic memories who can actually view the arrow's movement in their mind's eye as if they were watching it for the first time. Hallucinations similarly, are scenes played out over the top of what one is looking at in the direction in which they happened in the world. How is this done by the brain? It isn't just in recalling things that the order of frames is puzzling. If we think of consciousness as some kind of projection of the brain's activity into our internal senses, then which way round are these projections?

If we think of perceptions as if they 'project' events on the brain and *if* these 'projections' enter the brain serially just like the film, then consciousness is likely to be 'experiencing' the sequential order with respect to time in

Modern digital video suggest other problems with going back in time. Since a video picture is made up of scans that are interleaved, running a video backwards should look very different to just a sequence of frames in reverse order because the scans would have to go into reverse as well.

the opposite way to how it occurred in the perceptions. Either we are looking at the whole world backwards (which might be true as we will see) or consciousness is not strictly constructed from sequences of data that enter our brains, in which case, the moments of sensory experience could 'arrive' or be expressed in any order; they wouldn't have to conform to any direction of time on entering the brain (see **Boltzmann Brains**). The arrow of time we do experience is merely a function of the way our consciousness is currently extracting information from our brains and not a rigid feature of the world we are perceiving.

We see the puzzle of self-awareness and the problem of the internal '(I)eye' (Hofstader and Dennett) when we consider recall. For the perception of reality, at least our awareness appears to be in the correction position in the temporal and physical sequence of events. Awareness is the end result of the process of perception, and we feel that we are within the experience because we are physically inside the perceptual apparatus and at the end of the sequences. However, how is that positional relationship repeated when we recall an event? How do we get 'inside' an experience which is, one expects, to be yet further within the mind, both in time and physically (by way of storage)? If awareness and our moment by moment perceptions are separate systems and it is the same awareness that 'reads' or 'frames' the projection of the brain's workings into our mind-space during recall, then that reading of memories should be in reverse order for us to recall a forward running sequence of events, the arrow of time.

So the question arises, is backwards running time the key to consciousness? Should we consider that, just like with the film strip, the future (i.e. what is about to happen) arrives at the 'gate of our awareness', is seen in the present and then passes out of the gate into the past. We don't arrive at the future; the future comes to us. (Perhaps this is the essential ingredient over and above quantum uncertainty that gives a collection of states and memories awareness in the first place.)

There is the argument that our modern age, dominated as it is by images and specially moving images, has forced the mind to think more strongly in terms of sequences (*the medium is the message*) and to strengthen its sensation of both the 'framing' of experience and of time passing.

So thinking about running a film backwards has brought us again to the real problem of what consciousness is actually doing. For some, consciousness is just a phenomenon arising from the actions of the brain, and our sensation of being in charge of what we think and intend is just false. The brain goes about its business in its deterministic way and consciousness just appears after the fact as it were. Plenty of studies show this to be at least partially true, for example, in making decisions, neurons get aroused *before* the mind is aware that it is making a decision. The brain may well be leading the mind in these cases, and yet we also know that having beliefs can suppress this arousal, i.e. where the mind is *leading* the neurons. Even when we are aware of what we are thinking, we tend to rationalise our reasons for doing things after the fact (probably evolved for social dynamics like convincing someone to follow your lead or explaining away mistakes). Still, the 'how' of sentience remains, and trying to decide when it occurs does not ease the problem of how it occurs. Some say that this is a matter of unconsciousness not consciousness, but, again, the experiments are really showing a discontinuity of timing, the order in which real world events and the mental events are occurring. Experimenters call this problem *post-diction* and they are able to show that within a sensory delay (that is small) the mind doesn't make up its mind about its decision until after the decision has been made. So unconscious processing seems to have made the decision which the conscious mind merely approves. The mind is its first spectator.

People use the idea of our snap judgements as proof of unconscious processes going on in the brain all the time. For example, in *Blink: The power of thinking without thinking* (2005), Malcolm Gladwell discussed many occasions when we seem to make up our minds about people and the outcomes of events within seconds. People seem to be able to guess the professions of those they observe better than chance. The main thrust of his argument is rather beside the point, however, since he is talking principally about the internalised knowledge of experts, or of anyone who has spent lots of time interpreting specific complex subject matter, for whom a quick response is effective. For the untutored general public, intuitive guesses are still just that – guesses. All the same, the interplay between the assumed unconscious and our conscious awareness can be just as profitably described in terms of different time directions in the brain. We will come back to this towards the end of the book when we look at language and dream. But for the moment let us run with the idea that the brain has a more fluid appreciation of the direction of time.

It's rather curious to consider that moving against a gravitational field is rather like having time going backwards. Time slows for you as you

> Curious audience member to struggling comedian, "So what's the secret of ...?"
> "Timing."
> "(sigh)...comedy?"

descend into the gravitational well and speeds up as you move away from it. This is different to the idea of time slowing at high velocities with respect to an observer, where clock time passes the same in each frame of reference but only changes in the observations between the two. Nevertheless the idea that the value of time only depends on the circumstance of its observation suggests that other things may happen to it, like reversals, or where the entirety of creation is connected by time going back and forth in great circular arcs of causes, or where any point in space could spontaneously launch universes with the opposite time.

We saw earlier that travelling back through time in the casino world seems to mean that there are no probabilities in that direction, every event would be thoroughly determined, which nicely conforms to what we expect from the past: that it is always there, and it never changes (we also know that the past has to conform with the future from the discussion of dice and tables). So we can see a direction to time simply from the physical removal of probabilities from our universe. The future has more of them (in the sense of entropic disorder), the past has fewer. This is why time is different to other dimensions.

How entropy got into the system is similar to a problem that the Greeks pondered— whether the laws of physics, like the law of entropy, exist on their own as some kind of eternal and non-physical reality, or that they somehow pre-date the existence of our Universe. Many thinkers puzzle over the fact that, when it comes to understanding the Universe, mathematics seems to work not just well but *too well*, and that some maths derived just from other maths turns out to explain physical processes that hadn't been thought of originally. (Some thinkers consider that mathematical structures are in fact the only reality.) In other words, maybe we have *inherited* an entropy as well as a maths that comes with this universe from somewhere else, like a previously 'tuned' universe, or that we really are in a simulation.

This casino experiment also suggests an important asymmetry in our consciousness of time, namely the creation of past and future, of the pairing of prediction and memory. None of these were needed before minds arose, and it is perhaps only by virtue of the observational modes of consciousness that these regions of time have a reality.

The casino reality, however, is not the way the quantum world actually works. The quantum multiverse is the opposite. Going back in time re-creates probabilities and uncertainty. It appears to be entropy in reverse. We have already broached the idea that entropy in reverse is connected to consciousness. So, before we can get into the heart of time and causality, however, let us look more closely at entropy and its traditional role as marking the forward-running one-way arrow of time.

The teenager's room and the arrow of time

Entropy is another reason why there appears to be a direction to the motion of time in our universe. Irreversible changes in entropy occur as time passes.

Entropy is often referred to as disorder (although it originally described heat), but it is really about the distribution of energy states within a system.

We can make use of an instructive analogy — a teenager's room, viewed from outside. A teenager's room left to itself will become more and more disordered (disregarding the energy source — the teenager — for the moment). Specifically, it's a certain kind of disorder where it becomes easier to maintain the objects in new positions, like on the floor, and is often described, strangely, as an increase in the degrees of freedom available to the contents of the room. (Once you allow disorder, things will be anywhere it is less work to deal with them. A mess of CDs out of their cases is less 'work' to deal with than filing them in their cases on a shelf, except when you want to play a CD you haven't played for a while then the work done in looking for it is rather higher than one wants.)

The teenager does have a comeback, however, to the demands that she tidy her room (inject more energy into the system) because she knows instinctively that just re-arranging things in a room does not necessarily decrease the disorder — that is, reverse the effects of dispersal. This fact is embodied unconsciously or not in every dispute about tidiness. Simply *tidying* a room does not necessarily

Although the entropy of very small systems can fluctuate back and forth over small time scales. If you partition a box in two with a small hole in the partition and fill one side with gas. Over time the gas will fill both sides. With millions of molecules, the chances of the gas all occupying one side of the box again is small. If you had say only ten molecules of gas, then chance of the ten all accumulating in one side are much higher. But it is a far cry from there to say that because the ten molecules have accumulated in one side of the box again that time has gone backwards. We can say that the boundary between human events and the behaviour of quantum systems occurs roundabout here where irreversibility becomes almost certain.

make things any less disordered. A free CD can be 'tidied away' on a shelf between two books but that CD has very much higher probability of being lost (or remain dispersed) than one where it is set on a shelf in plain view in its case placed alphabetically by title with all the others. Floors can be cleared by grabbing clothes at random and stuffing them into random drawers but that doesn't help you find any specific item of clothing later. The information about those items remains uncertain. (What has actually happened is that the room has lost its *preloading*, and we will come back to this in a moment.)

From the outside, with the door closed, however, the changes in entropy have little effect, regardless of where the objects are. But what parents know full well is that a messy room has the habit of leaking its messiness out into the house. The greater the internal disorder permitted, the more chance there is of the objects in the room finding themselves outside it. (This positional uncertainty is quite an important feature of the universe which we will discover.)

We can see at once that the order in the room (in the sense of availability of an object) is a matter of information about the positions relative to each other of the items in it. Randomness in this sense is the loss of positional information (how much information you need to describe where everything is – the microstates). In an untidy room, an object would not give you any clue to where another object is (*Mamá, where is my book?*).

There is a notion called 'entropy entanglement' which essentially describes the degree of dispersal shared between two systems. It is as if two sisters live in separate rooms but share the same collection of CDs, books, clothes and makeup. The dispersal of those items are correlated with the entropy of each room. The correlation searches for a maximum, which will be twice the entropy of each room, and this defines the arrow of time. This idea of time, however, is distinctly local.

This is a puzzle when it comes to black holes, since it is not clear what happens to information contained in the matter which goes into it. It is often said that if you tossed a book into a black hole you could in principle recover the information in it from the radiative loss, from the Hawking radiation, but this is incorrect for two reasons. Firstly, the total information in the book also depends upon ink molecules following each other in a pattern that is not encoded in the particles that make up the book but merely by their relative positions in space which would be destroyed in a Black Hole. Information is essentially structural and the question about where this structure comes from in the origins of the universe is an important one. Secondly, the information in a book doesn't reside solely in the physical facts, nor necessarily in a brain that is observing the pages of it. It resides in the *preloaded* brain reading it. A story we will get to.

The general state of energy of something has to include this inclination towards random distribution of the micro-states within it, going from low entropy to high entropy as it expresses it. The lower the entropy in a system the fewer but more distinguishable are the ways in which its parts can behave. (Only in a specifically-ordered room does the concept of 'out of place' have any meaning.) The higher the entropy of a system, the higher the number of tiny but less distinctive ways its parts can behave. This gain in states is like energy passing through a one-way gate or a trigger: once entropy has increased, there's no way back.

Imagine two paddocks side-by-side. One paddock contains mostly quiet horses, but there are a couple of very frisky horses racing around doing crazy stuff and taking up all the room. The other paddock is empty. Join the paddocks together – that is, allow some dispersion – and all the horses have more room to race around. The frisky horses start to excite the herd, and, as they do, become tired. You have a troupe of more excited horses; some reasonably excited and others less so, and, as a group, a little more difficult to control. Now, even if you were to pack the horses into the one paddock again, you could not create the original *distribution* of horse energy, and it would take a lot more work than it would have done before (frisky horses need more room). The earlier moment has been lost for good: it is in the past and cannot be gone back to.

So the arrow or 'direction' of time seems clear, but this kind of illustration, while informative, seriously begs the question. Time and entropy increase together but neither one explains the other.

Entropy is a number. The Universe began with it as close to zero as it could be, and it will amount to a big value when the Universe is at its lowest level of order – the so-called 'heat death' of the Universe. Is this entropy level the highest number of anything that can exist? In fact, the highest meaningful number?

All the same, entropy in our world certainly appears non-symmetrical in time. The randomness of any system will increase over time and cannot decrease left to itself. If time's arrow is related to entropy then time going forward does not sort out the unformed mass. It does the opposite. It generates disorder... in a closed system.

But we also know that it is possible to ride on the back of increasing entropy in order to lower the local entropy of a system. The increasing order of life is a pretty good example of this. In replication, the combining of sex cells in sexual reproduction is helped by the fact that the sex cells have first split their nucleus into two halves, with an entropic 'debt' that allows the recombining into one whole again at fertilisation. It's worth pointing out here that the female's egg and the male sperm are very unequal to each other in size, quantity, movement and lifetimes, and this actually helps the re-combination. Had they both been more similar to each other the entropic debt would have to have been larger to make recombination likely (e.g. sperm swims while the egg remains static. If both were swimmers meeting up would be more difficult).

Although entropy is defined as that part of the total energy of a system that cannot be used, it is an essential part of a process's identity. It is that part of a system that *allows* the work the system can do to take place, without itself doing the work. Further, the disposition of entropy to initiate work also depends upon the relative energy values between the beginning and the end of the process (beyond a certain value). So entropy seems to be, in the end, irreversible. Everything will run down. Local systems can use this running down to lower their entropy (become more ordered) for a while (that's how life is going along) but it's a losing battle. At the beginning, no-space (low entropy) expanded into the volume of space we live in, creating matter and radiation on the way, and long into the future all matter will radiate away into an even vaster and emptier space again (high entropy). This passage of order to disorder may be a useful measure of the resources used and of what remains to go before the end of time itself when the Universe is in equilibrium, but it is not in itself the ticking clock that can tell us how much time we have left. For one thing, the famous formula for calculating the likelihood of a state is based on a formula that counts the number of possible states. We have to know what is likely before we can tell what is likely. We will come back to this enigma when we talk about information.

But if the Universe is an isolated system (with a net fixed entropy value), how did it become so ordered in the first place? (Physicists have calculated a very low likelihood that such a low entropy value could occur by chance, although this calculation is

> The Earth is, globally, more disordered now that humans have been on the scene for a while, even though some things about human life, like cities, look to be very ordered. As energy uses itself, it gives up a part of its ethereal existence to its surroundings. It becomes more creative, more real, and a little less imaginary. So in this sense at least entropy appears to resist time travel.

a little like Littlewood's law, see **The law of large numbers**) How did it acquire such low levels of entropy (high order) in the first place that enables it now to be running down? What is this potential to disorder? From one perspective the universe began with high order (low entropy) because the energy spike that began it occupied a very tiny space – the dispersed value of its energy was low – so its potential to disperse was very high. Distinct energy states need partitioning in variants of space or partitioning in time in order to exist.

Maybe the existence of entropy is all the proof we need that the Universe is part of a much larger system of many universes. But perhaps it provides proof of something else. Could the existence of life, running as it does locally against the trend of disorder, actually be the cause of our universe's beginning.

This is an extreme form of the anthropic principle and is something we will come to later as we try to prove that some interactions between events can go backwards in time as well as forwards? Just like the film strip. In order for us to perceive forward-going time, in order for entropy to create the narrative of our universe, at least some kinds of events need to go backwards.

We can take another look at this through the lens of Shannon entropy and ideas about how information gets passed on and used.

Information and entropy

We began thinking about time and it didn't take us long to get to entropy, but entropy is leading us into information and information will give us another angle on the meaning of time.

In the following discussion I want to highlight the notion that all the resources we need to move forward into the future come from the components of the present. When we consider what information is and how it works, we can see that the concept is paradoxical and there is something missing from our understanding of the present (see **The present completeness paradox**).

The basic definition of meaningfulness for a system which can have a number of states with probabilities p becomes thus, $H = -\sum p.\log 1/p$, (analogous to the Boltzmann formula for entropy $S = k.\log W$, where W is the sum of the probability of all the states the system could get into). Shannon differed from the earlier equation of Hartley in that he saw that each symbol in a message had its own probability, whereas Hartley gave the same probability to each.

The concept of entropy was originally developed for chemical reactions (Boltzmann), that is, for events higher up the chains of causality than space and time, events at the human level in fact. Then Shannon created the concept of the entropy of *information*, and that idea has been applied in all directions. The paradoxes of this way of looking at information may throw some light on a *time-line* and how the universe is arranged.

Information is a vague concept but Shannon gave it a definition (he gave life to the 'bit', an idea of John Tukey, a statistician). The amount of information in a symbol (with a minimum meaning of 1 bit – either it is what you expected or it's not) is related to its 'trustworthiness', or probability of being something else. (But even that idea is not as simple as it sounds. Where does your expectation come from?) Shannon called this information, the entropy of the symbol, and he made it inversely proportional to its likelihood of occurring, which meant that the more unlikely or unpredictable the occurrence of a symbol is, the more 'meaningful' it is when it occurs. (One could argue, on the other hand, that a truly random process tells us nothing at all. We will get to this later on.) For a message of symbols, its entropy is the sum of all the probabilities of each symbol occurring where it does. As the number of possible versions in a system increases, so does the entropy, and so does the information needed to describe it until a maximum is reached.

So what is Shannon entropy really about? As a mathematical expression of the range of possibilities available with any symbol set, it had been anticipated by earlier workers in the field. However, it did open up the examination of the world in terms of information, in terms of bits, the binary yes/no of data. Everything can be evaluated in terms of a frequency pattern, and the information approach has been successful in diverse fields like discovering who wrote anonymous works, who has plagiarised their PhD thesis and the encoding of mobile phone messages.

Shannon's idea of information was not about the meaning of any message but merely about its integrity when transmitted; how you distinguish it from noise. For Shannon, a message was just a pattern — a pattern of noise, of ink on paper, of waves in the ether — that was more or less well reproduced as it moved from one place to another, and giving rise to confusion when the word 'entropy' is used in this context.

> This is part of a message emailed me years ago for amusement. '...dseno't mtaetr in waht oerdr the ltteres in a wrod are, the olny iproamtnt tihng is taht the frsit and lsat ltteer be in the rghit pclae. The rset can be a taotl mses and you can sitll raed it whotuit a pboerlm. Tihs is bcuseae the huamn mnid deos not raed ervey lteter by istlef, but the wrod as a wlohe. Azanmig huh? yaeh and I awlyas tghuhot slpeling was ipmorantt! if you can raed tihs forwrad it.' It beautifully illustrates how the observer is implicated in the meaning. (It also unwittingly illustrates how word length is an important error-correcting statistic. Play around with word lengths and you destroy comprehension easily.)

Shannon entropy does not describe any form of energy or work that a *message* can do. It describes a property of the state of the whole system in which the message occurs.

The faithfulness of the reproduction had a limit and the problem of *meaning* was not included, although he did relate his entropy of information to predictability. The circularity of which seemed to have eluded Shannon, or he just ignored it. To be able to predict a symbol you need to have already an understanding of the statistics of the symbol set used, and to come to an understanding of the faithfulness of the transmission you need a sense of the meaning of the message (or to redundantly compare sent and received messages, which, if possible, removes the need for the message).

Message systems like languages have rules (equivalent to symbol set statistics) that you need to know, and rules are rather like negative entropy; they help you *predict* what the next letter is going to be in a message written in a language. But languages also contain tricks to resist decay into unintelligibility. All languages can add more symbols to a message without altering its meaning (for example, by adding repetition, or just commas). Repeating symbols lowers the entropy of the message; it adds negative entropy (essentially a form of error correction). Of course, in language at least, repetition itself also has meaning – that's very very bad; a rose is a rose is a rose – so the issue can be more complicated. It is recognised that prose in English generally has twice as many words than is required to make sense of what is being said, and words themselves seem to have more letters than they need since they can often be uniquely described by fewer letters especially when context is also used (hence shthnd and spdwrtng systems, and the game of H__gm__).

English vocabulary estimates generally fall along the following lines (Jespersen, Seashore, Eckerson and Oldfield): 500-1100 for a 2-year old child; 2,700 for a 6-year old; 26,000 for Swedish peasants; 70,000 for English-speaking college students; 75,000 for English undergraduates.

Thanks to this negative entropy, an English message is fairly predictable (English is generally considered to have one bit of information/intelligibility for every 8 bits of message) and thus has a relatively lower Shannon entropy than, say, French, which has fewer words than English (perhaps 100,000 words) and some of whose end letters and other accent marks are changeable with context (try typing on a French computer keyboard and you will get a good idea of how much more you need to know); and a lot less than ideographic Chinese, where, with even fewer 'words' than French, if you get one stroke within an ideograph wrong you have a different meaning or no meaning at all.

English and French share a great deal so the differences between them are actually small. Chinese, however, is rather different. Chinese speakers need to be aware of the background to any utterance, the conceptual references. Their education, in general, makes them aware of how the same theme was dealt with by past writers and poets and which helps them decide the meaning of an unfamiliar utterance. Chinese speakers share more background, more pre-prep and accuracy. Even so, a Chinese vocabulary estimate for a Chinese graduate student (working in the US) is around 46,000 signs and combinations, while for an English undergraduate the vocabulary estimate is around 75,000 words. An educated Chinese speaker could know more than 50% of his language; an unfamiliar utterance would be rare, although, for everyday use, around 3,000 characters are required to cover almost every situation, or 3 – 4%. Even university educated English speakers know only around 15% of theirs and are often having to decode complex or inaccurate sentences he or she has not heard before, which means they need more redundancy and context. For everyday use, an English speaker probably needs less than 10,000 words (1.5%) and perhaps only 3,000 (of the most used), which is less than 0.5%. English speakers do not have to reject an utterance just because its syntax or the spelling of its words is wrong. In fact many English speakers trying to speak another language of lower redundancy are often puzzled by the failure to be understood when their sentences are inaccurate but seem 'close enough' to the meaning. Redundancy and context is, therefore, highly significant for English speakers and contributes to its attractiveness as a second language.

(There are around 80,000 Chinese signs/words, including rare and obsolete ones, which is a strong indication of how differently the messaging systems function. Two Chinese character sets used for the internet, Big5 (traditional) and GB2312 (simplified), have 13,051 and 6,763 characters respectively, whereas there are an unknown number of English words, though estimated to be at least 600,000 – some put it it at over a million – excluding technical terms. You couldn't write a message in Chinese even remotely like the one I received in my email.)

Looking at the situation another way, there are around 8,000 different syllables in English, while Chinese has at most 1,277, and including all the various tones a syllable can be pronounced. If raw permutation of speech sounds is a measure of information, then English surpasses Chinese in this measure. On this simple basis alone, if Chinese and English are to convey the same amounts of information in their respective languages, the Chinese speaker needs to be aware of more language, i.e. to have a greater amount of pre-preparation than the English speaker.

Predictions that the Chinese language will become the new *lingua franca* of the world are likely to be premature given this burden of pre-prep on Chinese speakers. What is more likely is that Chinese will mutate rapidly into a something more atomic and flexible under the burden of transmitting information to those who do not have sufficient learning. (Standardisation and the reduction of dialects has already begun, and Pinyin, the romanising of the Chinese language, is a significant addition to the momentum.)

Lack of redundancy (negative entropy) in language exposes a number of paradoxes of language comprehension.

Makers of web sites can take this pre-preparedness into account in order to defy robot programmes that try to register with web sites automatically, like asking those who register to read twisted numbers and letters and write them into a box. These 'captchas' are little tasks that only humans can readily perform because they have so much more 'pre-loaded' information that helps them understand images better than a computer program recognising the strict the bits and bytes of the *captcha* image (although this is changing rapidly with newer pattern recognition software akin to neural networks that can make sense of almost any pattern).

Messages composed in English and French, say, of the same bit length may not carry the same amount of intrinsic information. On average, the French one will require a better 'educated' receiver to make sense of it (to predict its meaning) than the English one because there is less redundancy (or error correction) in its lexicon to preserve meaning. French messages in the long run tend to rely more on pre-prep (education) to extract meaning (where it can be terser than English) or on more words. This is easily shown as French translations of English novels contain more words than the English original. (One might be tempted, too, to make some generalisations on French literary style which would tend to use relatively more qualifying phrases and/or adjectives. I am reminded here of the English writer Samuel Becket who preferred to write his wordy monologues in French.) On the other hand, English sentences can easily contain too much negative entropy, or waffle. Too much negative entropy is, paradoxically, an effect close to noise in a message but is less easily removed because its removal can raise the entropy of the message (taking away the negative entropy) again, and reduce the ability of the receiver to predict meaning. It is hard to see how language could evolve at all without the development of redundancy, which is where artificial languages fail (see **Informational inertia**). (Redundancy, i.e. adding copies of qubits, as a protection against decoherence will probably be a key component of quantum computers.) This redundancy features in our return to an old debate about the foundation to all language, which we will discuss further on (see **Lies and language, fallible narratives and dreams**).

There are two ways you can know the integrity of a message and the efficiency of the medium by which it is carried. You can compare before and after message states, or you can load up with enough information to be able to know what the message is supposed to look like or to mean. Since the first method is only possible when it is redundant, Shannon entropy is really about the latter. A language message has *intrinsic* information for an observer (hence the message), and the entropy measure is no longer relevant just to integrity. The observation or 'measurement' of a message will include the meaning for the observer such that information values no longer depend on individual symbol states (this is an example of how some kinds of cause can go both ways, namely down into the elements and up into meaning). Though, overall, as the message degrades, and both its intelligibility and intrinsic meaning reduces, since it is more difficult to predict what the message actually is, its entropy also increases. Thus, high unpredictability is analogous to the increased energy dispersion measure of ordinary entropy. (In fact, there is a teasing study done at the University of Lyon relating actual energy loss in talking in languages which appear to contain less information per syllable: they are spoken faster to create similar throughput levels of communication. The study is highly criticised however for some problems we have already described).

In some ways we can think of Shannon entropy of a message as a sum of intrinsic meaning spread over all the permutations of the message. Some permutations mean nothing, some mean something else and some may mean the same. Now we can begin to see what Shannon entropy actually tells us about information systems, and what it is actually measuring. (We will meet this 'paradox' again in the many worlds quantum scenario.)

To find a simple illustration of the huge range of information coding differences in languages, I pulled down a 1954 edition of the French pocket Larousse dictionary from my shelves. It has 32,000 entries and takes all the proper nouns and lists them separately at the back. Aside from this there are just 16 words under 'W' (and 10 of these are also English words. Incidentally, Portuguese has even fewer – just 5), 8 words under 'X', and 18 words under 'Y'. While a similar English dictionary has 37 pages of entries under 'W' alone.

It wouldn't matter how much negative Shannon entropy was in a message written in English if the recipient didn't know (was pre-loaded in) English. For messages written in a language (a set of symbols) to be understood completely, they need to be related not just to the number of possible words in the language but to conform to the rules that govern the meaning of *every possible* message. The true key to intrinsic meaning is that this preparedness is 'sent ahead of time', as it were, by learning the language used in the message before it arrives (at least $10 - 12$ man-years of language immersion, an enormous amount of work, has to be done by a receiver brain before *any* message phrased in his or her language is likely to be properly understood), and it is this preparedness that defines the core predictability of the message over and above the integrity of the message.

This pre-loaded requirement for language message comprehension seems rather like the presence of hidden variables in states of matter or like the old arguments for the presence of the aether in which light was thought to move. But maybe there is more to it.

Here is another example of how 'pre-loading' works. In the movie, *'twilight's Last Gleaming'* (1977, Aldrich) a team break into a nuclear missile silo to blackmail the US government into revealing military truths. At one point the leader (Burt Lancaster) has to read back a code over the telephone from a card. The card reads 'ROMEO AND'. He is quizzed, 'What's the third word?' But he continues to say, 'there is no third word'. Which was right, of course. Plato confronted this issue and concluded that we are all pre-loaded with understanding (hidden variables) and that our experiences just conform to that knowledge. Even Wittgenstein in his own inimitable way touched upon this problem when he suggested that 'truth' is a matter for context or 'culture', i.e. the preloading. A scientific 'message' only has meaning when interpreted by *science*.

In order to understand (to predict the meaning of) every possible message in English, you would, of course, need to know not only the meaning and usage of every single English word and all the rules that govern their use, but also how words are applied to the real world by the complete set of human minds who use English (even by those who do not necessarily follow all the rules). In other words, for a message to have zero entropy from the point of view of the recipient (that it contains no uncertainty whatsoever), the recipient needs to know a whole lot more than the dictionary of the language used. He or she must in some sense *already know* what the message is about (*pace* Socrates).

This pre-loading is what Shannon entropy really measures. The probability of each symbol is established beforehand by referring to its likely use, to information already established. As a receiver, you have to know the frequencies of the symbols in words, the likelihood of their being placed after any other symbol and all the rules that govern all such placings and the composition of all groups of symbols *before* you receive the message. The minimum possible message entropy value, therefore, is found, not in the message but in the necessary pre-loading required to understand it.

An undistorted message in English to a perfect English speaker doesn't actually contain any uncertainty: it's Shannon entropy (or uncertainty) is zero (well, perhaps not zero since the *fact* of the message also conveys meaning), and thus it appears to contain no information, and implies maximum pre-loading. Zero pre-loading, on the other hand, implies maximum unpredictability with zero comprehension.

For non-complete pre-loading there is both a greater tolerance for uncertainty, which is to say a message may appear to have more information than it has, and a greater capacity to make a mistake in interpretation. For full pre-loading, the opposite is true. For example, a message that reads, 'the cat sat on the atm', may contain an error in the last word or it may mean what it says: the cat is sitting on the cash machine... and clients can't get their money out. If you were not familiar with the sentence, 'the cat sat on the **mat**', known to (almost) every English child, then you are more likely to take the last word as given. This example highlights the paradox of Shannon's predictability (the more you know, the more capacity for ambiguity there is and thus the greater the need for context to resolve it).

> The paradox of pre-loading is that any distortion in a message is now *more* likely to produce ambiguity for a perfect English speaker. Because, out of the vast numbers of possible messages, any error could suggest another meaning and not an error. The more of the language a person knows, the *smaller* the error that produces uncertainty in meaning. Whereas a non-perfect English speaker may not always recognise an error and believe he has understood all the message or can safely ignore parts of it.

Error correction can be included in a message but its effect is limited. It can tell the receiver if there is doubt about the integrity of the message and therefore to the intrinsic meaning but not what the intrinsic meaning is meant to be. That is to say, the amount of work that the pattern of information in the message can actually perform always depends on the context of the information, that is, the combined system of message *and* observer. As each symbol arrives the accumulated information improves the likelihood of predicting correctly the next. (This has implications for the multiverse which we will examine later.)

If we imagine a message sent to say an alien, who has none of our pre-loading, the message has to contain within itself the means of giving it meaning. That is, it also has to be an algorithm and embody a formula for extracting or displaying what the alien would consider meaning. In general, the less pre-loading there is, the more the message is going to have to take up the role of explaining itself. In which case, we find that there is even less information or uncertainty carried in it (since algorithms must define the message, they embody little uncertainty). The more the message is algorithmic the less Shannon information it actually holds.

A simple error correction procedure might count the number of 1s in every group of say 4 bits (in a message coded in binary). This will tell you if any zeros have become 1s in the transmission or if any 1s have become zeros but it can't tell you if a both a 1 has become a zero and a zero become a 1. All error correction adds to the message length and is prone to transmission errors itself. At some point it is more economical merely to repeat the message any number of times and compare versions. This is what we do in conversation. It's better to say 'What?' and have the person repeat themselves than to try and guess what they are saying. It's a fine line. Saying 'What?' too many times does not endear listeners to talkers. A certain amount of effort (like searching around one's pre-loading for clues) from listeners to understand is expected and is more economical in temper as well as time.

The combination of pre-loading and intrinsic meaning brings with it another important realisation about information. Predictably is based on statistical analysis, which is to say that its measure depends upon some measure of consistency (that the statistics adhere to a persistent pattern, to internal structure), on *inertia*, or informational invariance.

Informational Inertia

As we have noted, for the symbol probabilities of Shannon entropy to have any significance (that is to say, the degree to which messages can be distinguished), a minimum number of message variants as well as the meaning of each symbol with which it is written must have been transmitted in advance. This is especially true for older cryptography and the source of all its weaknesses (except the one-time pad which is preloading to the max).

For Shannon, the number of ways of arranging the symbols of the message is a measure of its information content, but as far as the intrinsic information of a composed message is concerned one or at best a few arrangements can carry an equivalent intrinsic meaning. Most predicted arrangements of symbols will not convey the meaning of that message. So that the relationship of symbols to intrinsic meaning in a message cannot be down to statistical frequencies alone. For example, the 'garbled' message I received in my email, is still understood so its entropy is the same even though the statistics of its letter appearance are different to the properly spelt message (e.g. the statistical probabilities for the letters forming the word 'them' are different to the formation of the word 'tehm' as are the statistics of its placing in a sentence and yet its invariance is high. Out of 26^2 letter variations of the four letter word 't - - m', only 7 make proper English words and only one contains an 'h'. An English speaker would have no trouble knowing it is 100% likely that a garbled 4-letter word t - - m where one mixed up letter is an 'h' will mean 'them'. That is, virtually all those variations mean the same thing, and its placing in the sentence also adds to the invariance.)

> Surprise is a challenge to our preloading. Evolution has given us more pleasure out of things that defy our expectations. This is why the ladder of life exists at all. Surprises becoming known creates negentropy.

Shannon seems to recognise this almost in passing. He noted the very structure of a language gave it redundancy. He proposed a measure, *D*, of the amount you could *take out* of a message and still retain the information. However, taking symbols out is only one of the things you can do to a message. As we have seen, symbols can be in the wrong order, or added, and words can be combined or split up in various ways. Another measure of the capability of a message to retain information after transformations is needed. So we are going to say that the lowest number of alterations to a message needed to create an alternative meaning or a paradox indicates its *inertia*. The higher the inertia (invariance) the more predictable the intrinsic meaning and the more intelligible. So inertia in our sense is somewhat the inverse of Shannon entropy and somewhat proportional to the pre-loading.

SETI has this problem of meaning to overcome. It tries to consider what preloaded languages might be common to intelligences in the Galaxy and which of them might be used to send messages between them.

From our cat-on-the-atm sentence, we can see that certain word categories like nouns have less inertia than say, verbs, and we might speculate about languages in general and say that the separations between languages involves inertial differences. One can make the case that one of the reasons that English is so successful is not only because of its connection with colonising trade (after all, recent world-wide Portuguese, Spanish and Dutch trading empires existed before it and when England scarcely had an international presence; India and China covered enormous ground in medieval history, yet none of this led to any of their languages acting as a *lingua franca* outside of their cultures), but because its high inertia, and its quirky spelling contributes significantly to this (because extra or odd combinations of letters are negative entropy). For example, *q* followed by *u* in a word means that any letter could take the place of the *u* and the word would still be read (Shannon only observed that the *u* was redundant and could be removed; he didn't considered that it could be replaced). Add an *e* to the end of most English verbs, however, and it is most likely to be interpreted as a mistake. Not so easily decided in gender-biased French: French has less inertia.

Inertia gives us the idea of quanta of message entropy, the amount of entropy needed to irreducibly alter the meaning of a message. We can see inertia in practice with the distorted email message a few pages back. Clearly, for a beginner English speaker, the words in that email message would be almost entirely unpredictable and have no inertia, no re-constructable meaning. Inertia is related to redundancy and thus inversely to entropy, but requires the pre-loaded information necessary to resist chaos and the random. In fact, inertia and pre-loading are proportional to one another. Another way of thinking about it might be pattern persistence and we will talk about this we we talk about the brain and neuronal networks.

It is instructive to read about the creation of artificial languages and observe in many cases an opposite tendency at work. One of the latest of these efforts, John Quijada's *Ithkuil* (over which Quijada laboured for thirty years or more), tries to eliminate all uncertainty in language expression, by making people precisely mean what they say. (An example: *I offer you my gratitude in exchange for your kindness.* Uhispal ük^{h}u tô myal kô ednaul. Quijada also designed his own 'script' for the language.) Not surprisingly, given our arguments above, it is impossible to converse in because it is too algorithmic. It was created to avoid misunderstandings but fails to take into account that languages are spoken and interpreted as well as read. There is what might be called interpretive drift in the consciousness of a listener over which a speaker has no control, and never will. Redundancy and inertia was not taken into account which meant that no one was able to guess in *Ithkuil* where a message received was garbled or incomplete or to intuit the full meaning when it was. For SETI explorers, it is worth noting that *Ithkuil* was more algorithm than message, and ended up being much less informative than its creator expected (who has now embarked on a new language). In another example, Frank Drake's original 'Arecibo Message', transmitted to star cluster M13 was designed as a sequence of 1679 binary digits, uniquely factored into two prime numbers 23 x 73, and which when organised into a rectangle of these dimensions formed a simplified picture of some atoms, DNA, a human and the solar system, with numbers describing the sizes of these items, lengths or molecular formulas. Drop a single digit from the message and the result is nonsense. A single change in any of the formulae would create paradoxes impossible to resolve.

(Inertia in the multiverse scenario might be responsible for the stability of some states to maintain themselves regardless of measurements. We can certainly apply the idea to a reduction of the multiplication of them and explain how a measurement occurs at the first differentiation of value in the superposition.)

Ithkuil can be found on-line at www.ithkuil.net. John Quijada writes, "...For me, the greater goal is to attempt the creation of what human beings, left to their own devices, would never create naturally, but rather only by conscious effort – an idealized language whose aim is the highest possible degree of logic, efficiency, detail, and accuracy in cognitive expression via spoken human language, while minimizing the ambiguity, vagueness, illogic, redundancy, polysemy (multiple meanings) and overall arbitrariness that is seemingly ubiquitous in natural human language."

Biological language and the origins of life

We can also apply these simple ideas of pre-loading and inertia to the origins of life. As far as life here on Earth goes, an important fact about DNA is often disregarded. While a strand of DNA is in effect, a one dimensional string of data, of information, it appears to code for all those complicated three-dimensional proteins in the cell. How? It does it because of pre-loading. There is already in the cell additional information by way of the ingredients of proteins, all with their separate 'statistics' and behaviours. The proteins are constructed, or translated, using RNA which in turn is derived from the DNA. The DNA does not have to program every parameter of every atom to make the molecule, the ingredients and the environment in which they exist, the cell, supply what is missing to create the full complex protein shape. In fact, DNA is so one-dimensional it is not strictly speaking an algorithm either. It is the translational system that is the algorithm.

We shouldn't ignore the argument for the 'seeding' of life here on Earth from elsewhere, from space-roaming biological components (here we tip our hat to Fred Hoyle again who developed such notions, and to Timothy Leary who had wilder ideas of humans being purposefully seeded by aliens), and which also makes the origins of life even more unlikely (seeing that life depends even less on the site but on the presence of certain molecules made in dust clouds, say, arriving at the site).

The information held in the DNA is a very reduced level of information compared with its products, and it is hard to see in what way there could be a direction of simplification and abstraction from the complex molecule assemblies in the primordial soup or even in a primitive cell structure to the creation of the genome. (It is because the DNA is a reduced strand of information that a gene can be used in the making of more than a single protein. One wonders what process simplifies *backwards* from a soup of several proteins to a single manufacturing instruction?) Additionally, DNA contains a lot of redundancy where many strings of bases do not appear to do any work, which is puzzling if DNA either coevolved with its reproductive and translational systems or was somehow an abstraction of existing material (although some of what was thought to be 'junk' has now been discovered to have purpose). Given what we have understood about information in a message system, this material could act like the negative entropy in a language, or the inertia in a message;

it helps the DNA convey its messages even through 'noise' in the cell. It also suggests a way in which the whole system could have come into being in the first place. More in a moment.

It is now becoming clear that the idea of a gene being demarcated by the stop and start codons is no longer a sufficient definition and that the translational system is far more complex than previously understood. It is known, for example, that sometimes smaller strands of RNA are constructed from sequences within the gene which create proteins which then act back on the whole gene. A self-referential system unimagined only a few years ago. RNA from widely separated 'genes' on the double helix get spliced together to generate proteins while some RNA is translated but which make nothing.

Given our earlier arguments about pre-loading, the complex translational system is suggestive of another creative route implicated in the establishment of DNA in addition to the one we can trace in cells through the earliest organisms to those of today. DNA seems to have come late to the scene, and that the cell and at least some of its ingredients and systems were functioning before DNA arose as a repository of the information. (There is the co-evolutionary argument: that all components of the reproductive system, including DNA, evolved together. But this is just a matter of setting arbitrary boundaries to the system. Coevolution reduces to evolution, since it is always unlikely that coevolving systems can maintain their relationship in evolution over time because of chance changes in dependency of systems on one another. Instead of the idea of coevolution let us consider pre-loading where what ends up as contextual information is transmitted first.)

Mutations in DNA only have meaning and be reproduced with reference to its informational inertia and the pre-loading. Imagine replacing a letter in a

> The precursors of DNA were probably involved in some kind of cyclical self-referential activity because evolution doesn't really work on single molecular examples.

language with a new symbol. It will produce noise in proportion to the inertia: some words with the symbol will be read as the same. Imagine adding an extra symbol to a language to increase the number of meanings that can be expressed? With regards to DNA, alleles are accommodated because of the informational inertia in the system, but to add or to use new genes requires much more than the existence of inertia; it requires the pre-loading of translation mechanisms. Such may seem obvious, but when we start to consider chicken and egg questions, the mystery deepens because if a gene is added, the inertia of the DNA needs to also rise, otherwise DNA would work less well and become more fragile. Which is to say, responses within the cell system need to acquire more scope or to adjust their actions to permit the evolutionary significance of the changes to take hold, and to occur simultaneously with the gene addition. (This is referred to in evolutionary biology circles as 'phenotypic plasticity', though usually in the sense of an organism being able to survive in new circumstances sufficiently well for mutations to then take hold, and in distinction to the standard Darwinian slow accumulation of alleles before environmental changes. Others refer to 'Evolvability' which we will discuss later on.) It is hard to imagine how this informational leap might have occurred without it being through the *merging* of several already evolved simpler systems into one cell (e.g. as in the adding of mitochondria, or indeed in sexual reproduction itself). This simple fact may well be the reason why it took perhaps 1.5 billion years for primitive single cell bacteria to lead to the complex cells called eukaryotes. Yet another example of coincidence playing its role in the origins of life and pointing to an even more reduced distribution likelihood in the universe at large.

This might be a case for thinking that the logic by which events take place exists independently of the physical facts. In fact there are physicists around who think that each logical or mathematical construction entails an entire parallel universe of its own in which to 'house' it, but we'll not think about that now.

We can probably dismiss the concept of panspermia seeding the primitive precursors to life on this alone. If panspermia is to work then it will probably involve the distribution of more complex systems than simple molecular precursors, which would require either a higher level of chemical activity in space that we have so far observed or a coincidental arrival of several types of molecular systems from space onto the surface of our planet. If we can established the underlying levels of coincidence – the coincidence number of our universe during the epoch in Earth's history where life began – to estimate this happening we will know if SETI is wasting its time. As an aside, one can imagine that the new fertility techniques of using 3-person donors of DNA (female cell containing its mitochondria with the nucleus of another and the sperm of a male) might provide occasions for a new round of significant evolutionary change through the surprise merging of systems.

The merging of systems can also increase the information inherent in the DNA system beyond the requirements of a new gene, which is why we see metagenetic effects, the plasticity of response, or how genes themselves can alter details of their expression according to circumstances.

The merging of systems, however, means that sets of independent probabilities now have to interact and become interdependent. Merging of systems must require more than just trial and error to test its success since, as we argue, such merging is produced irregularly by coincidences that do not have predictable rates of happening, unlike ordinary gene mutations that occur at a steady background rate. Merging would explain the large quantities of non-coding DNA in the human genome and also imply that switching off genes is just as important as switching them on, and that there are likely to be genes that suppress the action of large swathes of the DNA strand that were acquired from merging with other strands. Merging is often negentropic in biological systems since it is more favourable for a network to merge with another network (with many degrees of freedom) than to add nodes sequentially. On this basis alone we might expect that viruses which jump species will tend to gain in virulence.

It turns out that at least one example of merging has been traced in the genome. Many creatures, even small ones and plants, have more chromosomes than humans, and some have way more chromosomes than they appear to need for their functions, so clearly chromosome increase and reduction is part and parcel of family trees, and indeed, the reduction of chromosome through merging can be traced in some species. (Wikipedia has a list of species by chromosome count.) Homo Sapiens has one pair fewer (46) than the Apes hominid branch from which it split, occurring through the merging of two chromosomes (5 & 6 in the apes set) to form chromosome 2 in the human set. Since the merging of chromosomes occurs with sets of genes that already have been 'proved' in evolution, it can provide a short-cut to change with less risk of damage from a mutation in a gene.

An explosion of forms from as yet unclear reasons (often thought of as due to extra energy available from an increase in oxygen levels), known as the Cambrian explosion might provide us with an example of this coincidental merging process. (Curiously, this explosion of forms and the creation of true animals was followed by an extinction episode. If the Cambrian explosion had not happened, primitive life as it was living in the ooze may easily have been extinguished in that episode.)

The merging of systems in cells provide evolutionary jumps in the size and scope of DNA precisely because the probabilities within each merging system have altered directly in the future beneficial direction. The merging occurs in the first place because there is a probability 'fit' for both systems. Which is to say, without some form of subtle channelling of activity, sudden multiplications of forms at rates beyond the molecular mutation rate within the cells seem unlikely to happen. The merging of mitochondrial DNA into human cells was one significant jump in evolutionary development — and the role of mitochondria in neuronal cells especially during sleep is an extraordinary story which we will come to — but we are only now beginning to understand the evolutionary significance of the merging of genes and DNA strands in other life forms, and see how coincidence lies in the foundation of the trajectory of life. It is beginning to look as if punctuated equilibrium and the steady state evolutionary paradigms are both correct.

These effects go beyond simple coevolution. Extended evolutionary theory is grappling with what appear to be great shifts in informational activity within cells which produce an increase in forms and species. In order to explain these informational shifts some biologists are now considering the idea that it is the trait (through subtle modifications of proteins say) which appears first as extraordinary behaviour within existing systems and which only later gets fixed at the genome level. What the cause of this beneficial channelling might be and the broader consequences of this cause fill the pages of this book.

Curiously it turns out there is evidence of at least one form of informational shift in the geometry of the DNA strand itself.

Let's recall the difficulties of ordering the frames of the film strip, and the observation that the brain may need to project event memories into consciousness by running the frames of the memory backwards. We might consider sequence reversals of primary significance as a type of event-time or information chirality about remembering.

Consider then, the discontinuity of informational chirality inside the cell. Any helical twist is necessarily chiral (for DNA the helix twist is right-handed), but the strand of DNA is made up of amino acids in the cell all of which are geometrically left-handed (glycine is the single amino acid that does not have handedness; it is its own mirror image, and, interestingly, does *not* appear in the DNA strand), whereas the sugars that make up the proteins within the cells are all right-handed. This begs the question of how chirality evolved from naturally occurring right and left handed molecules, since mixtures of left and right sugars or amino acids would inhibit protein formation, cell reproduction and the lengthening of DNA through a reduced reproducibility, though it does suggest that chirality of some type is necessary.

Did chirality come first before life could develop and, if so, what sorts of probability determined it? Perhaps we are looking at one of the fundamental processes of self-organisation that compose our time-line and our present, where flows of information go in both directions in time, and there is a 'time chirality' in the products in the three dimensions of the cell.

Time chirality is different to the differences that may appear between running frames of time one way and then simply reversing the flow (without changing the order of the frames). Reversing the frames sometimes makes no difference at all, while time chirality would make a difference. A pendulum swings the same when you run a film of it backwards. If time actually went backwards in a pendulum we would probably see it swing up overhead moving slowly over the top and most quickly at right angles. Not like a backwards running film at all. There is little symmetry in actual backwards going time.

Evolvability

Some biologists discuss similar ideas under the notion of 'evolvability' (e.g. Nowak, 2008) whereby precursors to life also select themselves preferentially through adaption. It is really about the ability of an organism to reproduce a mutated version of itself. Evolvability is about the arrow of evolution where the differential reproduction rates between mutations that are positive or neutral and survive to live on and mutations that are negative and die out. This idea has been extended. It is thought that organism's self replication becomes organised in a way to stimulate further helpful mutations. For example the complicated 3-D structure of proteins may be constructed in such a way as to favour the positive effects of mutations (by providing a higher number of self referencing sites where a mutation would do at least no harm. (And, too, the ability of folded proteins to carry a range of different molecular 'add-ons' may be one of the coincidental results that enables the brain to store the memories of a life.) That is to say mutations do not have purely random effects on an organism. This 'pre-preparation' for mutations is a form of pre-loading similar to what we discussed about language which allows the absorption, accommodation and response to a confused or incomplete stimulus as to an expected complete one. Evolvability is to some degree similar to our notion of interpretive inertia or meaning invariance. Chromosome recombination in sexual reproduction is certainly one example of organisms achieving this where alleles can be shuffled without causing damage to the organism to probe the fitness landscape. It is our contention that coincidences act in a similar way to the crossovers of chromosomes and allow some probability of future fitness, or at least resistance to damage, to communicate itself to the organism.

We contend that this notion of evolvability, however, leaves out the additional benefits from coincidence between chains of polymers and between behaviours that the organism is directed to by the reproductive system. We would expect evolvability to evolve in response to the coincidences in higher level narratives since the rates of organic evolution depend upon an evolving coincidence number. However modern human life overwhelms the effect of coincidences operating in the organic world with the use of manufactured coincidences. The result is derived not out of the future influence but out of the past. It is in conflict with with the natural evolutionary pressures that have taken place up to now and give us ersatz evolutionary outcomes. What impact this will have on the future of the human narrative is unknown.

What is not in the string

The discussion of Shannon entropy has allowed us to see the relationship of the probability of an element in a message to the previous passage and storage of information. In fact, the mechanism of interpreting a message with pre-loaded data is rather like going back in time. Hold this thought for the next few pages. We will come back to this in our discussion of cosmologies. We can remember the Present Completeness Hypothesis. The important information embodied any string of data bits is what is *not* in the string.

Now we can see why, for Shannon, unpredictability means *more* than knowing. In fact, we can turn around the concept of information entirely. There is more information in a prediction than in a fact. The expectation or uncertainty in a symbol is dependent upon how much of the context in which it appears has been pre-loaded into the receiver's interpretive system, its memory. Any future possibility contains more information than its appearance in the present as a fact. Becoming 'real' implies a loss of uncertainty – something like an entropic flow in reverse. Applying this interpretation of information to consciousness, the true 'reality' for consciousness can be found in time-predictive circularity, somewhat akin to Bayesian reasoning. Consciousness may be able to leapfrog entropic considerations and drive the expansion of life into the future by being the agent uniting the future with the past.

So how do we humans deal with prediction? James Gleick has written a beautiful study of time travel narratives as they appear in our cultural history. He showed how much we think about it, how much we have always been thinking about it, but did not explain what entanglements there may be between their logical and physical paradoxes and the states of human consciousness that create them. So let us, too, widen our lens and take a look at time travel narratives and ponder their baffling paradoxes.

Imagining time travel narratives

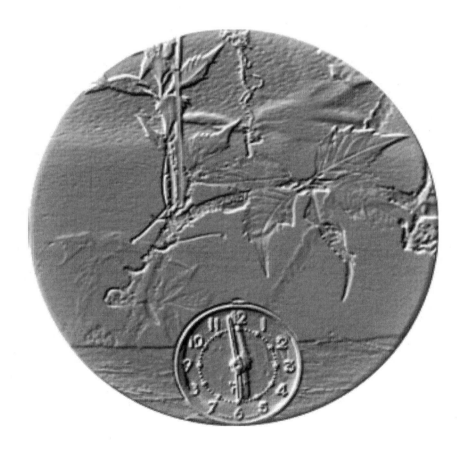

In which we consider the purposes and benefits of Time Travel, and how tinkering with time might occur.

In which we ponder killing our grandfathers, and the circularity of time-lines, some popular time travel paradoxes and wonder whether time-lines can merge as well as split.

In which we ponder the differences in narrative times between the external world and the world of consciousness and examine the differences and paradoxes in time travel narratives.

In which we discuss the speed of time and whether it can solve the problems of the chronology protection principle.

Let us take external narrative time (ENT) as that natural rate of the evolution of events within human understanding but outside our heads. We abbreviate it to *time*. Every physical level we observe, ENT is the time we see. It is the time in all our mathematics and in every calculation of speed, of force, of wait. It is the time of Newton. It is the time of Relativity. It is the time of space-time. It is the time that makes stars, destroys environments and kills people. It is the time on the clock. It's the space in which you have to wait for the bus, or for an orgasm. It is *time gentlemen, please.*

Somewhere, beneath or above the complexity of life, leading (or pushing) the past to its future, we experience a logical sequence of events in which one step in the sequence *must happen* before the next one (and not just simply ordered that way temporarily). People are born before they die; they hunger before they eat; they copulate before they can reproduce. In the external time of the body if not in the narrative time of the mind, the chemical events accumulate, like the sentences in a language, from a before to an after.

Yet our minds also appear to be free to wander about this narrative time independently of our actual bodily position within it or the states of the material senses which initially gave it content. How might we correlate the wanderings of imagination with the external narrative time that brought it into being and with the possibilities of altering or tinkering with them?

Let's start with a brief look at a nice example of an external narrative at the human level and fantasize about the kind of human-focussed interference our imagined time tinkerers can produce. Currently we can explain this interference only as accident or contingent factors. But as we have seen, there is more to contingent factors than accident. So let us first review a real-world narrative and observe some points where our time tinkerers might like to intervene, or may have already done so.

Messing with World War I

'If wishes were horses, beggars would ride. If turnips were watches, I'd wear one by my side. If 'ifs' and 'ands' were pots and pans were there would be no need for tinkerers.' The 'Almost' Real Mother Goose.

An orthodox view about World War 1 is that it was precipitated by the assassination of the Austria-Hungarian Emperor's son Archduke Ferdinand in Sarajevo by a Serbian terrorist group (the Black Hand) with some help from Serbian government intelligence which provided the weapons and may have suggested the target. Orders had been sent out to arrest the conspirators earlier but not acted upon. (Sounds familiar, especially when you know that the Archduke was considered to be, relatively, a *reformer*. No doubt there were many in the corridors of power who read 'reformer' as 'threat'. He had already married out of his class and his father, the Emperor, had officially blocked his children from inheriting the crown.)

The choice of Serbian Nationalist Day on which to parade the ruling Emperor's Archduke was not a coincidence. There are many reasons to suppose that the archduke's presence was a deliberate show of strength by the Empire. A provocation to which a response by the Serbs was likely. Time tinkering may have been operating even as far back as that decision, but let us consider just the day in question.

The Archduke's motorcade comes down a street. The conspirators had distributed themselves among the crowds along the route. The first con-

Conspiracy theorists often mention that the Kennedy assassination hinged around a similar 'spontaneous' change in route. The principal investigator of the killing, New Orleans District Attorney Jim Garrison, said that the Dallas Morning News of the day before showed a map of Kennedy's route through the city of Houston for the day. Where the motorcade was supposed to stay on Main Street while passing through Dealey Plaza. The 'unplanned' sharp turn onto Elm Street, however, slowed the motorcade to almost walking pace and brought the President's car nicely in front of the Book Depository from where Oswald fired, and the Grassy Knoll. In fact, although the published maps were not clear about the turn onto Elm Street, the Dallas newspapers *reported* the route through Dealey Plaza accurately the day before.

spirator in line does nothing (time tinkerer A). The second conspirator throws his bomb accurately, and it goes off, but the attempt fails. Ferdinand bats the missile away with his arm and ducks. The bomb bounces off the folded hood of his car and falls under the wheels of the following car, where it explodes, spraying several with shrapnel and seriously injuring two. (A question of 'random noise' which we will come to.)

Ferdinand's car continues along the route and passes five other conspirators who, mysteriously, do nothing (intervention of time tinkerer B?).

Later in the day Ferdinand wants to go to the hospital to see the injured members of his party. The general in charge, however, has decided to leave town by the fastest route, which is the same river boulevard they were on before, but this supplanting of the hospital plan is not communicated to the driver (intervention of time tinkerer C?) who turns off the boulevard to head for the hospital he thought he was going to. (The job of passing orders on from the general to the drivers belonged to one of the injured.)

Counter-factual historians also think this way and ponder what other forces and trends to the actual sequence of events were at play at any one moment in history. They consider the alternatives that the players of the time were considering. But their imaginative exercise is limited by how much they know and how much to include in their analysis; they still have to make choices, and these tend to express locality because they cannot examine every possible causal connection.

Curiously, the conspirators (including Princip, the leader and eventual assassin) meanwhile, were still hanging around the river boulevard in the hope of seeing Ferdinand follow his planned itinerary in spite of what had happened! (Intervention of time tinkerer D?) When Ferdinand's driver turns away from the river route to go to the hospital, the general in charge calls a halt and asks the driver to return to the river boulevard (intervention of time tinkerer E?). As the car slows to go back the way it had come, it pauses opposite a food store where Princip had decided to go and get a sandwich. Princip had paused on the pavement there to talk to a friend who had just turned up (intervention of time tinkerer F?). The astonished conspirator pulls his gun and shoots both Ferdinand and his pregnant wife, Sophia, precipitating World War I after all.

We have highlighted just six key moments of time tinkering possibilities at the scene. Each one of these moments or nodes in turn has its connectivity network, mostly hidden, which culminates in a mental state or states. Which of the many nodes in that story could possibly be pivotal to the outcome? Nodes of external narrative time or nodes of internal narrative time?

Historians are well acquainted with the puzzle of pivotal moments around which events turn. Archduke Ferdinand's wife, Sophie, was not sufficiently noble and could not partner him on normal state functions. As inspector of the army, however, rather than as royalty, Ferdinand could bring his wife on an inspection, which is why the couple chose Sarajevo in the first place even though it was buzzing with unrest. Is this simple sociological cause the actual head event? The Archduke's marriage being the event that led inexorably to the War, given the entire context? Or did the start of the war pivot around the Germanic insistence of the man in charge, General Potoirek, (or an aide?) to demand the motorcade stop and return to the same route it was on earlier in the morning, even though the hospital visit was certainly not on the itinerary and unlikely to be predicted by the conspirators?

Because of causality, we tend to presume that events are subordinated to one another, and that causes are self-contained — there is 'locality', Markov style (see **Present Completeness Hypothesis**) — so we look for triggers, causes that can be identified after which an outcome is inevitable. In the above case, we can see how the assassination was achieved through a good dose of luck. The event was a summation of chance, of coincidence. Inevitability only has meaning here by tracing the chains of cause *backwards*. Going forwards, the links in chains of events have nothing like the same certainty to be triggers.

By working backwards, we view the assassination as the trigger of the war. But such a view is a tradition rather than an empirical result. The assassination might have been an enormous red-herring in

But why stop at Europe. Territorial disputes were everywhere: between Britain and India; between Russia and China; between China and Japan; between Europe and Africa; between the US and Spain, and although some disputes were resolved without war, unrest remained.

the search for causes; there were other political assassinations around Europe at the time. There was such a broad trend to war and less for diplomacy (people demonstrated on the streets calling for war all around Europe, generating considerable *anti*-war activism as well) that a single event like this assassination was by no means the single *inevitable* trigger that the war required even though after it there were fewer opportunities to deflect the trend to war.

So perhaps, in our Universe, for consistency's sake, many subordinate events conspire to make the 'head' event always happen (something like the predestination paradox). In order for there to be any chance a time tinkerer might avert the 'head event' of World War, complex chains of events would need to be interfered with on many fronts, since any single chain could have had its probabilities changed (and thus altering events locally) without having serious impact on the final result. Our imaginary time tinkerers will have to make trade-offs between local and non-local causes, given the pre-preparations and the large cultural 'space' made available for the war.

But there is also a benefit to the future time tinkerer of these many strands of causality to, at least, high impact historical events. They provide 'cover' behind which the time tinkerer can do some local interference without causing huge repercussions in history. Knowing, for example that the extended causal 'front' would cause World War 1 pretty much whatever happened, our time tinkerer could pursue a personal agenda with one or some of the agents of these causes without fearing that World War 1 would not happen.

We can try to look at causes in the larger realm of historical analysis: Bismark and his determination to protect his unified Germany with alliances; national humiliations and losses of territories (lots of people in Europe wanted territories back, not just those in the Ottoman Balkans but the French, Poles, Italians, Greeks, Danes); economic competition; shifting treaties; civil unrest and ethnic minority revolutionary fervour. But we also have to include simple moments like the tuberculosis-ridden assassin, Princip, who had not had breakfast on that fateful day, was hungry enough to leave his station on the boulevard to go and find a sandwich, or to a friend who had perhaps re-traced his steps to collect his hat bringing him to a coincidental meeting that was responsible for retaining Princip in the right place at the right time.

(Remember our hypothesis that if at most 6 steps of virtual connection can be traced between two points then an actual cause is likely to exist between them.)

The point to be appreciated here is that alterations in a time-line necessarily involve *minds* (see **Trevanian**) a well as the genesis of many actions, none of which individually can ensure a particular outcome. Part of the reason for this which we will explain in more detail further on is that, if quantum theory is strictly true, every moment in past time includes inertia in change derived from the continuous consciousness of the minds originally involved in that event.

Change requires, not the knowledge of how events first happened since these will be changed by the intervention, but a science of coincidence (yet to be achieved at present, and rather anti-Popperian in nature), making the success of any planned interactions in events a result of not only establishing the physical linkages but orienting the mind to specific mental states to profit from them.

Of course, in the battle for the future, which takes place in the past, the closer the scrutiny you give to the causal links of events the less likely it is that anyone can gain an advantage by tinkering with links in the physical world unobserved. The battle of the tinkerers then, will have to move away from the physical connectivity of events and into people's heads. People acting 'irrationally' because of new ideas suddenly entering their minds cannot be predicted so easily. By feeding ideas into people's minds, specific actions can be induced with a relative amount of secrecy and cause specific outcomes even though they may look, from a distance, to be the result of uncertainty.

Before we take in the problems of consciousness and the origin of ideas let us look again at the narratives we have built up for ourselves over the years: internal narrative times particular to each brain, and external narrative time, that time in which all brains and the external world are embedded.

Travels in narrative time

We will take a look at the external component of the narrative, the one that seems to exist independently of our awareness, the one that brains-in-vats would be struggling to understand, by way of time travel tales and time travel paradoxes.

All narrative, though, and in particular the study of history is a form of time travel since interpretation of what is narrated relies on the past. For Historians, their time travel is always into the past (although there is a new form of counterfactual history which examines different presents), whereas other writers and especially science-fiction writers hop about backwards and forwards in the external narrative time settings of their stories and in the mental 'times' of individuals within their stories. Speculating with external narrative time is especially demanding. In our real world where even setting clocks forward in Spring causes mental anguish and confusion, a storyteller of time travel is bound ever more tightly by the constraints of logic, plot and the very use of language. Which is probably the reason why H.G. Wells wrote his well-known story, *The Time Machine* (1895), precisely to avoid such difficulties.

In Wells' story, the time traveller starts out by simply observing what happens when he fast forwards or backwards along the time line. Later, though, Wells sensibly had his time traveller land forward in time in order to avoid the circularity paradox. His time traveller travels to a point far in the future, gets involved with the human descendants, now separated into two types – the Eloi, fair but dulled inhabitants of the surface of the earth, and the Morlocks, brutish inhabitants who live beneath the surface, and barely escapes with his life to return to his laboratory and dinner guests bursting with news of his adventures.

I have never really thought that the Atlantean Rule mattered that much, or at all. Fiction, surely, is fiction. But editors assure me that it is very necessary. They mutter something about the suspension of disbelief. If a story tells of a man discovering the secret of a ray-gun that turns people of our time and place into potatoes, then that story had better ensure that the gun and its secrets are destroyed by the end of the story, or the reader, who knows that such a thing doesn't really exist, is going to end up ...What? Cheated? Tense with worry? Dissatisfied by the pretence (or conceit) that such a ray gun exists? Well, editors have their working rules. But really, what would happen in the mind of the reader if that particular story ended with the gun intact?

Now, the rule of fiction (especially sci-fi) written in a way to suggest that the events may have actually happened (as distinct from fantasy) is that things invented in the story cannot be allowed to survive into our world. I call this The Atlantean Rule, since the story of Atlantis – surely the first sci-fi fantasy – conforms to this rule, *nothing survived*.

H.G. Wells seemed to obey the Atlantean Rule. The narrator of the story writes at the end, '...*I am beginning to fear that I must wait a lifetime. The time traveller vanished three years ago. And, as everybody knows now, he has never returned.*' (Happily for his edit-

or, taking the time machine with him.) But Wells slipped up. His time traveller managed to bring something back from the future. Two strange white flowers, now shrivelled and brown, the remains of a garland of flowers given to him by Weena, one of the enfeebled but beautiful Eloi, from far in the future. Am I dismayed at the bending if not breaking of the Atlantean Rule? Can't say that I am.

The Time Machine is really a political fantasy in which Wells outlined his fears for the evolution of humankind rather than a story about messing about with time, and there are several examples of similar time fantasies in early texts predating Well's story sometimes by centuries (for example in the *Mahabharata* written about 400 BC). His story is like a level 1 narrative in time, used by many writers, especially in love stories. Well's character(s) travels physically between time zones without much minding of any serious causality paradoxes that might ensue.

Beyond this basic time fantasy narrative, there seem to be four generic types of logical time travel narrative and each has it's paradoxes of causality.

I am not counting dreamscapes or the narration of events taken at random from a once coherent narrative like Vonnegut's *Slaughter House 5* (1969), or stories about individuals going backwards in time (for example, Martin Amis's *Time's Arrow* (1991), or Scott Fitzgerald's life of Benjamin Button (1922), a man born aged and who grows younger as time goes on). These are not really about time travel but more about playing with linguistic tenses. They are more like literary versions of filmic devices and probably would never have arisen in ages without the moving image.

> Wells' story is about two times – as are most stories about time travel. Normal life proceeding forward for him inside his machine, and the speeded up forwards and backwards time outside of it. This has always suggested a problem. When you move backwards in time would you not run slower and slower the faster you try to move back in it? Our picture of movements in time is traditionally only of the *sequence* of events not in the *value* of the motion.

Time travel as a kind of peep show.

You can watch but not touch. John Wyndhams' amusing short story *Pawley's Peep Holes* (appearing in his anthology, *The Seeds of Time*, Penguin, UK 1970) describes the situation well. Ghostly apparitions bother a town. Eventually it turns out that someone from the future is beginning to send tourists back in time to see what the past is like. These tourists, sitting on time-tour buses, drift ghost-like in and out of people's houses at all times of the day and night to gawp at the 'quaint' historical characters and buildings. They become an irritant, but what to do about it? The story's hero turns the futurists' plan back on themselves, and calls in the local population to gather and gawp back at the time travellers from the future. Eventually these time tourists are driven off by hordes of spectators jeering at the looks and fashions of the *future*.

This plot seems to avoid the key paradoxes of time travel. Yet viewing anything is an interaction with it. Light has to impinge upon the eye, be absorbed, and cause nerves to react. To view the past, you would have to remove radiation and information from it. Would some kind of compensation be necessary to maintain an overall balance of energies? You could compensate for the lost photons of the past, balancing the equation, by taking away some of the entropy of the past, i.e. making it a little more ordered than it was before. How could that be done? By leaving more information *behind*? By making real some connections that previously existed only as a probability? Is this what an observation in our time, in the present, actually means?

If this compensation were necessary then you would have to be *in* the past in order to view it. You would be obliged to make something happen that had originally only the potential of happening. To view is also to change. If we noticed a particular sequence of rare possibilities at the human level becoming firm might this be evidence for an observer observing us through time?

Time travel bodily to many time streams

Travelling in the multiverse. TV Shows like *Sliders* (1995-2000) describe this well. In *Sliders*, separate universes contain complete human worlds separate from one another with clear boundaries that can only be crossed with tricky wormhole thingies whose workings are not really explained. (Well, I never saw the beginning, so I don't know if they were explained and I never saw the end which seems rather fitting for a show of this kind.) There are no futures or pasts connected with the now from which the travellers came, only different, self-contained and co-existing *versions* of how the world from which they came unfolded in time.

Happily for the writers of the show, this case produces no time paradoxes or circular causation since each world belongs in a separate Universe with a time line independent of the others. However for this to be always true, the travellers cannot go back to a time line they previously visited unless it is to precisely the point they left it. Otherwise, they would be doing the equivalent of time travel within the same time line and therefore be at risk of making paradoxes.

Modern narratives making use of the idea of entire universes unfolding from each possible option at any one moment are similar at their core, though communication between universes is not actually possible according to current theory so these time paradoxes in any one universe should not occur.

This form of time travel, though not really the one we would choose if we had the choice, is ENT (External Narrative Time) permissible.

Reflexive Time travel (back and forth in a single stream)

In films like *Back to the Future (1985)*, there is only one time line, so the established future is changed by fresh actions in the now, and the now can be changed by fresh actions further in the past. So, in order to alter an outcome, you have to go back into the past and tinker with events there. After tinkering, you could go forward into the changed future.

This is the kind of time travel tinkering we all fear since we cannot escape the consequences of our actions, and it leads to the standard circularity paradox. Would I exist if I killed my grandfather before he married? Often referred to as the Grandfather Paradox (see later). If I did manage to kill him, of course, would that prove beyond doubt that he was not my grandfather?

This kind of situation leads to a paradox similar to the Fermi paradox: if time travel is possible, why don't we see it? The retort is, if it were possible we wouldn't know it since we would have no way of knowing how to separate a time traveller's action from a genuine action of our own time. We don't have any knowledge of how things would have played out without such a time influence. (We'll look at this in a moment.)

The puzzle about changes made in the single time-stream is how to account for the disappearance of the original future. If every tinkering with events of the past led to the creation of a separate time line from the tinkered point, then the case reduces to that of *Sliders*. But if not, then we have to accept the evidence of the new time line at face value. Those prior events have not just been overwritten, they *never* existed in reality for us at *any time*. All those facts, all those personalities, all those souls, gone. There may be a case for arguing that those events remain only as possibilities or options that were never taken, but whether we could have been aware of them as such is more difficult to establish. (One of our limitations and certainly what contributes to our idea of the present moment is that we don't see the fully developed time lines of alternatives radiating from a choice – what we might call the worlds of the multiverses and how they develop – we only see the choice). Since they didn't happen, we would have no memory of them (and could not be prompted or hypnotised into recalling them) and if we think of them at all, it is only because we imagine or dream them coincidentally (let's hang on to this idea and talk about it later). They were never *real* and no evidence could be found anywhere that would suggest otherwise.

Suppose I travelled back from some future, made changes in the past and travelled forward again to an altered future. The re-construction of the new future takes place in real time from the point of the change. But where do all those molecules and decohered quantum states of the first future go to while this is happening? Are they somehow broken down into constituent parts and then re-built? What sort of process would that be? Within your brain, massive activity would be re-writing everything. Memories would have to flit in and out of our imaginations, events would become hazy or forgotten (hinted at quite well in the movie *Loopers* (2012)). We would forget that we had forgotten things. Established evidence for events would become unsure or vanish while new events would happen through unlikely and previously unrecognised pathways.

It could be that minds and the exterior world are sort of interchangeable, and that the external world experience could become just an internal experience of memory and fantasy, while the internal world of memory and fantasy could become externally real so that all paradoxes are prevented. Such a mechanism, or lack of distinction between internal narratives and external narratives, would explain a lot (it may be what imagination *actually* is), but without it, the multiverse splitting appears to be obligatory, if time travel is possible. (But we will look at another way of describing this situation later, see **The physical time travel paradox.**)

Time is stopped for tinkering with the present

This is not really time travel, more like magic. Superman, for example, can accelerate around a scene changing things so fast it is as if everyone was frozen at an instant in time. Causality has stopped for everyone except the protagonist. This appears to be like relativistic travel only it isn't quite. In relativistic travel, where travellers leave Earth on a round trip to a distant place at near the speed of light, even though the travellers would see (assuming they could see) everything around or behind them slowed by the same factor that their clocks have been slowed with respect to us, they arrive very much further into our future, as if our time has been fantastically speed up. (This point is deftly left out of the *Star Wars* sagas and *Star Trek* adventures where faster than light travel would have to be very much faster than light to enable the narratives taking place on planets to incorporate the time spent between planets without distortions.)

A fine narrative moment like this occurs in book six of the great Hindu epic *The Mahabharata*, in a section, *Bhagavadgita*. Nothing much to do with sci-fi time travel but it's worth remembering anyway.

It is written in the form of a dialogue between the warrior Prince Arjuna and his chariot driver who turns out to be Krishna, an incarnation of the god Vishnu. Bizarrely, for those students of narrative, there is a nifty bit of narrative magic. The scene is reported to a blind king many miles distant from the battle by an advisor who has been given the gift of remote viewing – rather like case 1.

Two great armies lie opposite one another. As morning breaks, Prince Arjuna views the ranks of the massed armies and sees all his friends and family who might all be killed in the battle and thinks it better not to fight but just lie down and let himself be killed by the enemy rather than be the instigator of cruelty, even though it may be a just war.

Here, the narrative pauses while Krishna (who is God) talks to Arjuna about life, reality, sensation and causality, and especially about the need to do one's duty without any egoism or desire for personal triumph or gain. In the end, Arjuna is convinced by Krishna's talk to have faith in the necessary outcome of his actions, and in the ultimate triumph of his soul over physical reality (the righteousness of souls permits quite a lot of violence in the temporal world). Krishna returns time to normal and Arjuna goes ahead with the killing and brutal mayhem. (In fact, the *Mahabharata* generally considers a kind of time dilation happening between the homes of the gods and humans.)

The 5th way to time travel

These four narratives of time travel are all about physical travel between time zones. But it may be that time travel is more about the motion of information rather than of objects.

We will need to grapple with ideas of the relationship between language and consciousness, since it is hard to imagine how language can be sustained by a substrate like consciousness without finding the same kinds of relationship in language also in the biological support to consciousness. If we can talk about narratives in time then these times are not imaginary creations but representative of how our awareness is constructed. It is very tempting to consider that our ability to imagine counterfactual presents arises out of the physical structures supporting consciousness. We will examine this notion more towards the end of the book, but for the moment let us consider more concrete manifestations of the structure of time, the necessary mutually supportive information flow between the sectors of time which creates our *now*. As I shall show, the fabric of the universe requires the future and past to continue to communicate with each other and indeed require each other's existence to make sense of their own existence. We can happily say that the future needs the past to maintain its direction and the past needs the future to define its direction, and that we will be able to manipulate this relationship to guide those in the future into influencing us to the best interests of both.

But before we look at pasts and futures and what moving between them actually means, let us jump right in and think about the effects of time travel might be able to cause in the world. All these narratives of time travel suffer from the same paradox: the confusion between the innate on-going forward motion of the time travellers which unfolds regardless of the backwards, forwards or sideways motion through time that they go through. But this is not the only paradox of time travel.

Paradoxes of time travel

While no one can really tell you what time is, everyone understands the essential paradoxes of time travel. But what they are really thinking about, it seems, are the *sociological* implications of travelling in ENT. What happens at the subatomic or quantum level in the daily course of events doesn't worry most people because it is only the summation of causes at the people level that gives events a meaning to consciousness.

Another political assassination many writers have used to exploit paradoxes of time travel is the assassination of President Kennedy. Here's a version of part of the Wikipedia article on the subject that I noted down in 2008:

'In the episode *Tikka to Ride* (Season 7, Episode 1) of the comedy television series *Red Dwarf*, (UK television) a bizarre sequence of events involving time travel lead the Red Dwarf crew to convince JFK (from an alternate timeline in which he was never assassinated and in which the USA suffered badly for it) to go back in his time line and assassinate himself from behind the grassy knoll, rescuing the country from an awful future and ensuring his place in history as a liberal icon. Since the timeline was altered, the alternate JFK disappeared moments afterwards, leaving no evidence to be found. "It'll drive the conspiracy nuts crazy," a character says. "But they'll never work it out."

The special two-hour seasons premiere from the fifth (and final) season of *Quantum Leap* (US television) saw the character Sam leap into Oswald's life at various points in his life. As the episode unfolded, it turned out Oswald was the one and only shooter, but in a twist ending, amnesiac Sam learned that his true mission was accomplished when he succeeded in saving *Jackie Kennedy*, whom Oswald also killed in the 'original' history.

In an episode from the second series of *Twilight Zone* (US television) a historian from the future is sent back in time to film the Kennedy assassination. The historian is a relative of Kennedy. The researcher intervenes and changes history. Ultimately, the researcher teleports Kennedy to the future and takes his place in car. The researcher takes the hit while Kennedy lives on in the future.

In an original Doctor Who novel, *Who Killed Kennedy?*, published in 1996 (UK), the Doctor and a reporter called James Stevens uncover a plot by one of the doctor's arch enemies, The Master, to disrupt history through the use of brainwashed time travelling assassins. The first target was to be Kennedy, with the Master planning to have a gunman found wearing a Soviet uniform in order to inflame the delicate international political situation of the time. Stevens travelled back in time to Dallas 1963 and wound up in the Book Depository, trying to shoot The Master but missing, only wounding the President. Further shots came from the Grassy Knoll and Stevens saw from a distance that the second gunman was himself, twenty-five years older.

Stephen King wrote a time travel story about the assassination which was turned into a TV series in 2016, *11-22-63*, when film treatments of time travel was at its height.

What is so interesting about the Kennedy assassination is that in spite of the eye-witness testimony, the photographic record and the forensic evidence, the truth has yet to be established to everyone's satisfaction. There is contradictory evidence and contradictory interpretations for just about every single second of that fateful day. In fact, the more that science applies itself to the problem the less clear events become.

As far as our present examination of history is concerned, with such physical time travel we might expect to find evidence for increasing numbers of participants at events. We could discover that there were higher numbers of soldiers in ancient battles, ever higher populations in ancient cities, and similarly we would find a drop in populations just before big events or disasters as people leave (*Goodbye, and thanks for all the fish*). We would always be revising our numbers. Perhaps this is a parochial view of time and space. In a very large universe filled with time travellers there would be such a vast number of destinations that time travellers would be well spread out among events. On the other hand, the numbers of people in the future who could trace their timelines or genealogy back to any event whatsoever would be exponentially large. We could make a testable prediction that at every historical stage the Earth was considerably more populous and settled than we currently think.

This is in complete contrast to normal mysteries that tend to become better known as evidence accumulates. Take the Titanic sinking. As science advances, contradictions of testimony are clarified, time-scales are revised, and hard data confirms the nebulous witness testimony and demolishes conspiracy theories (for example, the claim that the Titanic was swapped with its sister ship the Olympic for insurance reasons). There is almost nothing in the Titanic story that has not been clarified to a sufficient degree of understanding. Not so with the Kennedy Assassination, where contradiction is layered upon contradiction and any number of interpretations can be applied to what is known. For those who like to take Occam's razor to the facts and say that Oswald had to be the only sniper, there are those who can assemble a range of facts to cast that proposition in great deal of doubt (like the witnesses who said Oswald could not have been on the sixth floor of the Book Depository building at the time of the shooting or that Oswald did not test positive for gunfire residue on his cheek and so on).

The Kennedy assassination is not a zone where information, facts and witness testimony is lacking. Rather the opposite. There appears to be too much information. For every detail there is a range of interpretations and for many of them this range of interpretations has increased over time. We will discuss coincidences later (see **A bit about our Universe: Coincidence and life**) but for the moment let us consider that perhaps part of the reason the zone of uncertainty around the Kennedy assassination is larger than the one around the sinking of the Titanic is that it embodies a surprising cluster of coincidences – way above the normal rate – and a reason for this clustering is that the Kennedy administration, with such new prospects (at the start of the sixties), was comprised of more than normal numbers of people drawn in to it by relationships already strengthened by coincidence. (Whereas, the principals in the Titanic loss had fewer such relationships.)

Another popular theme with which we can contemplate time travel interference is the life of Hitler. Many commentators have observed that if time travel exists then surely Hitler's actions would not have been permitted by time tinkerers, and the fact that they have been permitted (in our time-line) suggests that time travel is not possible. Many ingenious plots,

---

Could the appearance of this suggestion that Hitler was further along the road to nuclear weapons than we thought be more than an Urban Myth (even though there is no evidence that the Germans had a critical pile or large industrial capabilities for concentrating uranium)? Could we think of it as an indication of an entry point of a time tinkerer? From now on we may see more and more revisions of our understanding of the German nuclear bomb programme and what followed from it, like who knew what (like how much Heisenberg was hiding) and how did that alter the causes behind the actions at the end of the war.

---

however, have been devised to explain why Hitler's evil was allowed to go on. One plot device uses variations on the theme of time police who somehow have a pattern of how time-lines should be behaving and can monitor all of time for deviations from the plan, or who are alert to changes being wrought in their time line (without being subject to them themselves) and can trace the cause back to the moment time tinkerers applied their narrative changes.

Any manner of predestination paradoxes have been used, all of which assume that Hitler himself is the master key to Hitler doing what he did and that by removing *him* the time tinkerer would also remove what he caused. (We will discuss later on the idea of time line inertia and whether it is possible for time-lines to merge again after a shift.)

It seems highly desirable, from our perspective, that Hitler's evil deeds need to be undone if at all possible. But for someone viewing the whole of the human narrative, Hitler may not have been the only origin of the Third Reich's evil deeds' causal chains. There may also have been even worse things brewing around that time. Take out Hitler and his 'final solution', and perhaps the energies of the new Reich would have gone into their nuclear weapons program to the point of delivering a final solution to the whole of humankind. (Interestingly, new gossip has recently surfaced on the web about Hitler's wartime nuclear bomb program and suggesting, somewhat improbably, that it had advanced to the point of testing a weapon – further than previously thought – and altering our perceptions of the kind of time tinkering that might be required to produce the desired historical 'improvements'.)

These kinds of dependent courses of action may be the reasons why future time tinkerers do not alter what appear to us to be obvious evils, and who find themselves forced to deal in the shadows with a more global uncertainty and with a more complex morality.

The result may well be that our time-line ends up being that genuine Panglossian narrative, however flawed, of the best acceptable world of all the possible worlds. The Steve Argument *is* correct after all (see **The Steve Argument**). Although, with too many 'cooks' we are more likely to be living instead in the worst compromise for everyone, a walk around the Guggenheim (see **Walk around the Guggenheim**). We might even employ the concept of the Nash Equilibrium where the intersection of causal chains cannot be arbitrarily improved by altering one course of action and leaving the others intact.

Such a time tinkerer's perspective would be that of the counterfactual historian writ large. As the possibilities of tinkering in time grow, the role of the counterfactual historian as observer of its effects will grow in power and influence. (We will come to another reason why we do not see past eradications and alterations by time tinkerers further on.)

Paradoxes of External Narrative Time

All these ENT paradoxes, or paradoxes of predetermination, like purely logical paradoxes, are readily understood because they are based on a belief in the consistency of the real world and in an integrity to life. The bedrock of these paradoxes is a comforting acquaintance with the natural organic causality of our experience. (Insert Hume here.) But before we examine the flows of information back and forth in time, let us look further into the puzzle of objective time travel.

≈ But first a couple of definitions. The term *worldline* is usually used to mean the track that an object makes in space-time as time goes forward. A *worldline* is confined to a *light cone* which is the volume of space-time that an object could occupy without violating the speed of time restriction. In the following discussion I am going to use the term *time-line* to describe the actual tracks taken by objects moving towards their futures along their worldlines. ≈

I see there is a note to myself to insert Hume here. Well, Hume, a Scottish philosopher from times past (but not dour - he was quite a cheery person, and was sanguine about the poor reception of his first book), didn't think much of the minute happenings of causality. He was interested in the people level of events. He thought that all we know about one thing causing another is a conjunction of things in our minds, with no knowledge about whether the things out there really are conjuncted or not. Hume (b.1711 AD) didn't know about quantum mechanics or even molecules, and probably wouldn't care about them even if he did. But then Hume didn't think much of the individual mind and thought it only a bundle of perceptions without much rhyme or reason to it. For Hume, causality was all in the timing, and time is what we have no control over.

Without time, nothing can be said about what causes what, although, even with time in the picture there's only a kind of circumstantial evidence for what causes what. I think Hume was on to something.

Then there is the Markov condition, which rejects any suggestion that there is a narrative 'melody' playing which helps determine the next event. Markov thought that everything you need for a cause can be found in the present (causes don't 'rhyme', *pace* Mark Twain).

We looked at the Markov condition — the present completeness paradox — and causality earlier, but let us take a look at narrative paradoxes of movements in time and see what we can glean from them.

The circularity paradox.

The circularity implicit in a man going back in time to kill himself appears to invalidate time travel at a physical level. In terms of subatomic particles where one particle hits another and produces other particles, this is equivalent to one of these secondary particles travelling back in time to neutralise the first particle before it had time to hit the second? How is this not impossible? And if it is not impossible, where is intention in this circuit? How would a circular causal relation get started in the first place (Escher's Steps)? We will come to this point when we have followed other tracks and by which time we will be better prepared for the answer.

Of course, does time have to be perfectly consistent in order to be time? There are many variants of the circularity paradox, all concerned with physical causality. Although, does the paradox also refer to ideas in the mind? Can an inventor go back to before his invention and reveal it, thus nullifying his own future discovery? After all, he got the idea from somewhere. Can we be sure that the chain of events we follow back is actually the real chain that brought the thing into being? Are we readily mistaken about what causes what? (Hume punches the air, 'Yes!')

The circularity paradox gives us a useful insight into what may be happening if and when time influence occurs. Because of time paradoxes, we cling to the notion that if there is time influence from the future operating on the past, then the past will shift into new time-lines, leaving the original past unreachable from the new time-line that is developing from the point of influence. In this way it is impossible for something to bring itself into being in the same time-line, and evolution can continue on creating natural diversity. This is a bit of a trick, however, and does not precisely refute the causal paradox of something first existing before it can go back in time to bring itself into being because, with this explanation, the future state of the thing is still originating the thing in its past, only the two origins (first original beginning, and the second returning originator) coexist in time but not in space, which is less than logically satisfactory.

There are, in fact, ways round this problem. Things don't just exist as they are, suddenly; they evolve; they take time to come to be what they are at any moment. One can conceive of things existing in all their various phases continuously throughout their lifetimes. Indeed, Henlein wrote a time travel story to illustrate precisely this called *All you Zombies (1958)* where someone becomes his/her own parents (trust me, it works, or seems to at first reading). One can conceive of a way things come into being that could be quite different, perhaps, from the way things maintain themselves in time, since they need a future for that beginning to make any sense. Even here we can appreciate two different kinds of time, namely the time within which a thing exists (duration) and the time within which events happen. We will track down these complications throughout this narrative.

But to return to the problem of circularity in the human narrative. Suppose I could take this hot cup of coffee on my desk to a moment in the past. Unexpectedly, there is another heat source in the zone that cannot be accounted for by the energy disposition of that present. Supposing I go back repeatedly carrying hot cups of coffee. The room to which I return acquires a temperature level not accounted for not only by its surroundings but by the entire history of the world up to that point (as we think we understand history).

But consider, if the Universe's present is the result of an evolving wave function, then going back to the past actually takes you back to a previous state before the function evolved to the present, i.e to a present *that's happening for the very first time*, (i.e. there is no future up ahead yet, or at least there

Coffee cup Paradox. Hot cups of coffee from the future appearing in the now are acceptable as long as the past is as uncertain as the present. Imagine yourself in your room, you turn your back and suddenly there is a coffee cup. The first change you notice is an awareness of a lack of a memory for the event. You don't *remember* how it got there. It won't stop you drinking it. Not being aware of a causal chain is not in itself a reason to deny an event in your present.

is only a future of probabilities). In which case, cups of coffee popping into it from nowhere could readily be allowable since they would not necessarily be contradicted by anything that was yet to happen and which could thus *be* part of the potential present. Hot coffee cups popping in from nowhere only seem like a paradox from the point of view of the future looking back at the original coffee-less room, the original past. There could be logical reasons in that present why such an event would occur, some of which may lie further back in time (e.g. a zealous waiter of a coffee house who doesn't want to be seen, a stage magician, etc. This is how the narrative of *Back to the Future* could work.)

As we mentioned, the generally used explanation to explain why time circularity doesn't happen is the notion that disturbances from the future entering the past create another (at least one) universe which separates from the time-line so that paradoxes with events that have already happened in the time-line are avoided. It is not at all clear, however, how this would work even theoretically. According to the current fashion of decohered realities (de Witt-style), all the random options in events of the past, and therefore all the universes springing from them, have already been separated from each other. A time intervention in the past would have to re-establish randomness and create a new set (at least one) of universes from the point of intervention even as all the options for that event have already been 'used up' to create the multiverse. We will see if this explanation makes sense.

> If time travel from the future into the past turns out to be allowable, then paradoxical circularity in time, or rather, the return path, is not going to be what we think. The Universe would have a way of coping with sudden appearances of unaccounted energy here and there which, far from causing trouble, get woven into the fabric of the now.

Rather than using the multiverse hypothesis, we could imagine that the now is a continuous state of options ever-present which can absorb a range of energy states before jumping into a new configuration (the buffering we will talk about in a moment), rather like a subatomic particle that only accepts energy of fixed quantities – energy below a certain level causes no event. Thus, for humans the progress of ENT (external narrative time) could be a sequence containing quantum jumps and shifts in scenes that happen suddenly without warning but which we accept as the present because it seems consistent with our memories.

In fact there is a notion for this which we will discuss when we come to information and entropy, *negentropy*. An ugly word, but in general use. This describes the measurable reduction in entropy a system has by virtue of the organisation (or order) between its parts. Through negentropy there is an elevated free energy for reactions or to shift from one energy state to another. Systems undergoing negentropy (that is to say a reduction in the gain in entropy) use up less energy to shift into a more disordered state. They are more efficient, and are thus have more potential for change. If we take the evolving time-line as where life becomes more and more ordered with respect to the universe, then as negentropy increases locally over time, increasingly complex interactions from time influence could be accommodated within a given system without altering its long term outcomes.

With the existence of negentropy we might begin to see more evidence of non-linear causal activity. Disturbances, like hot cups of coffee, might radiate away and dissipate, absorbed in higher levels of existing order (like a mystery which gets puzzled over and then forgotten) rather than determining later events.

When we come to talk about interference patterns at the ENT level where narrative 'wave functions' combine in addition and subtraction, we can consider the building up or destruction of events, in strictly local fashion. We actually might be able to travel to past events (like in *Pawley's Peepholes*) and sit in the background, one of the mass, experiencing what it's like to be in the past, without causing any danger to the time line and the future unfolding of events. Negentropy could provide a way for information not only to be added to the past without disrupting causal chains but also to be interpreted as energetic bridging to the future without dangerously transforming the timeline.

Of course, it's a fine line and it requires a view of causality somewhat different to the tightly controlled causal connections that walk us back to the prime mover. We are obliged to think of all causality as a culmination of influence (or a sum over all paths) and not as a literal this sets-that-into-motion, in which case the **Present Completeness Hypothesis** is wrong after all. The scheme certainly fits our head and chain event hypothesis that we will also examine further on. But is it correct?

Backwards explanations

Then again, our traditional deterministic narrative picture of the unfolding of events (and which is the basis of all objections to the circularity in time) is where a time traveller jumps into a past time zone and disturbances ripple forwards from that point to change the future from

> In subatomic physics, causes are ultimately the result of the particles of forces, because forces have a direction – they are vectors – and are *exchanged:* the root of a cause.

whence the traveller came. The future gets changed because the insertion of a new *cause* into the timeline changes the outcomes to the current causal chains to events. However, quite how this new cause gets incorporated into the present moment needs explaining. As a set of probabilities, without possessing *local* parameters, this new cause could not connect with the present moment. Just as every universe of the multiverse does not connect with the present, an influence from a different time could not just connect with the present and be a cause without acquiring a past. This can be done, as in entanglement experiments, by a 'measurement' or interaction with an observer in the present moment that is initially itself unmeasured. There are few systems in the present moment that are not yet decohered or untangled, but consciousness may be one of them for reasons that will be discussed further on. Here we suggest that only those events from the future that can be related through probability to past uncertainties will be able to act as a cause in the present moment. Thus a whole range of new outcomes to events are also created backwards from the insertion point.

We might observe these as disturbances rippling even further *backwards in time,* in order to *validate* the sudden appearance of the object or time traveller and the new present it or he or she creates (rather as the Ministry of Information re-wrote history in George Orwell's *1984*), making the purpose of his or her jump either meaningless, or acceptable. (Is this a novel idea for a novel?) There are no impossible circular contradictions arising in the timeline if the *past* behind the point of entry of the time traveller changes to provide the correct causality for the new event. We might even observe this by suddenly finding the value of an observation to be false with respect to its history (say a median turns out to be an outlier). The chronological protection principle is really the 'causes protection' principle. We will see in a moment how a chronology protection principle is never going to be needed if we take what we understand to be time passing at its face value.

The chronology protection principle

With ideas of negentropy and buffering in mind let us examine a mechanism to avoid circularity, proposed by many, a type of chronology 'protection' principle or 'uncertainty' that time travellers encounter. The closer to the exact date you want to get the less precise the physical location at which you will arrive. If you want to kill your grandfather in New York on a certain date you can get to that date but you'll end up in Indonesia or on the moon. If you absolutely wanted to get to New York to kill your grandfather, you can get there but you may be several years before your grandfather gets there or after he has left.

If this kind of 'protection' of the chronological order of events is in operation then travelling back into the past may not necessarily incur the kind of consequences that people fear and that circularity in particular is avoided. Many refer to this as a likely principle of Nature, or a principle of consistency, without having any idea how it might work. (Once we accept randomness at every level of nature then consistency becomes a real problem.)

If this were true then this uncertainty principle might be a means of space travel as well in that the more precisely you headed for a place at a definite moment in time, the more distant you might end up from that place. Of course, there may be a spherical (more likely elliptical) distribution of places about that point that would fulfil the 'uncertainty principle', and you may never know where in the distribution you would end up. Navigating in Time would be a tricky thing. But the narrower the definition of the event, the smaller the value of the uncertainty in either time or space needs to be.

> Michael Moorcock used precisely this principle, calling it *The Morphail Effect* in the time-travel stories of Connie Willis, where time travellers encounter 'slippage' which prevents them from either reaching the intended time or moves them a sufficient distance from their time destination to prevent any paradox from occurring. The ability to absorb time travellers, negentropy, in this case being a measure of some kind of volume in time and space.

The uncertainty principle also suggests that the further back in time you go the less of a spatial uncertainty is required to fulfil its condition. So, that for some things, only relatively minor shifts in time or space would be sufficient to avoid paradoxes.

We introduced the idea of buffering earlier. To recap: if the entire buffer system of the Earth's ecology will shift in an attempt to lessen the effects of damaging human activity, maybe it will also shift to prevent or at least redress damages attempted through manipulations in time. The chronology protection principle of Hawking and others is just another manifestation of Gaia and the still-living connection between her past and future.

Aside from the protection principle there is also a similar notion embodied in the **predestination paradox**. By whatever means a time tinkerer tries to alter the past, she gets 'captured' in events that have already occurred to prevent her from making her alterations. Since we think we know the past happened a particular way her helpless presence in it is destined to be factored in, and she ends up changing nothing.

We could mimic Heisenberg's Uncertainty principle here and say that the uncertainty of getting to a precise location at a precise time is greater than some constant, U_{TT}. But I often wonder if all the uncertainty principle has to do is act on our minds. You can go back in time but you forget what you went there for! A form of dementia!

To imagine the case where Gaia could still interfere with a time tinkerer who was able to insert herself in a past that has not already accounted for her presence, Gaia would have to rise up and initiate minor changes in the local present to interfere with the time tinkerer's actions. It is hard to see how forces and causes could be naturally marshalled with these intentions in a past that presumably has no idea of the future.

How Gaia might actually do this will be discussed later on, but all these suppositions about the past may come to nothing anyway, so bear with me. Let us consider a much-talked-about version of the circularity paradox, the killing of a grandparent.

Killing your grandfather paradox

This is the bedrock circularity paradox of physical travel in time (seemingly first discussed in a 1943 story by R. Barjavel). What happens if you go back in time and kill your own grandfather before he marries? Do you suddenly cease to exist? How is it possible that you can exist to kill your grandfather when his life stops before you could be brought into being? This paradox is used to deny the possibilities of time travel within the same stream, but there is still more to be said about it.

The exclusion zone

Suppose the uncertainty relation between time and space exists to prevent you from going back in time to a point where you and your grandfather meet, what would stop you from deliberately going back to a place well away from where your grandfather physically was and then travel in normal time to where you knew him to be at a later hour or date? Suppose you went back in time to a point that you knew you would not meet your grandfather but left him a message you know he will eventually see telling him to go to a certain time and place (that he has never been to) and then go to that place and wait for him. Would that evade the protection principle?

The uncertainty relation, or protection principle would have to encompass at least the lifetime of your grandfather plus your lifetime, in order for you always to be too late or too early. In fact, if this principle were in operation there would be a minimum in uncertainty, and if you did go back in time, then you could estimate your own lifetime by calculating how much time it takes to get from wherever you had landed to where your grandfather either had died or was about to be born.

> If the Uncertainty Principle of time travel defines a space and time zone centred on a distinguishable event designed to exclude logical paradoxes induced by time travellers then this will affect the natural rate of coincidences. There may be an inverse connection here with our Universe's Coincidence Number, $^{U}\!N_C$ (see **A bit about our universe: Coincidence and life**). Since a coincidence is by definition undefinable, as it depends on the observer, the context and the actions of the moment, and cannot be predicted, the uncertainty zone or minimum distance between the traveller and the event may also be defined by $^{U}\!N_C$.

The protection would have to cover more than just your grandfather's presence, it would have to cover all the events in which his existence could be traced. The uncertainty would have a different value for every person and situation, so if there is a general effect active in all situations it would have to be related to the information embodied in all history. I will expand upon this idea in a moment in the form of conjugate universes, a corollary to the multiverse. But for the moment let us continue the train of thought.

There could just simply be enough deceptions in history to make such a principle work without seriously adjusting the fabric of space-time. Even though you think you know everything about your grandfather's movements, some of the information may be wrong; wrong or incomplete enough to make it difficult to evade the Uncertainty Principle in any way at all. For example, you may conclude from letters say, or from, internet information, that your Grandfather was in a certain place at a certain time in the past. Turns out, that information was a lie he put about because he was actually having an affair that he wanted to keep secret. Suppose he only pretended to go to his work place where you would expect him to be and that he was really unemployed and was pounding the streets every day looking for a job. These kinds of deceptions in our whereabouts left in the information trail we leave behind are fairly typical of the uncertainty in the past that interferes with our attempts to alter causality there. (There is the difficulty of creating a false alibi in the more modern present because there is so much cross referenced information on each one of us, including where we are at any one time, that is beyond our management or control.)

But this kind of evident uncertainty is just a matter of record, of memory. We are unlikely to have the complete record of any event since any record or memory is a distillation or a trace of events. All our knowledge of the causes of actual moments in the past must necessarily contain a minimum level of uncertainty and will always do so (like the Kennedy assassination). Thus, the uncertainty principle or chronology protection principle might naturally be the result of continuing uncertainty in the contact between events (Hume), both forward and back in time, and the meaning they express in our consciousness. The predestination paradox is not, therefore, a paradox but is the result of a failure of the *record* to be fixed or complete in the sense of normalised, (where all the probabilities add up to 1, to certainty). To misquote Henry Ford, *History is more or less bunk!* Whether or not the mind of the observer continues to transmit uncertainties into the past and altering the balance of causes, an issue we will discuss up ahead, there is a natural spread of uncertainty or probability in every moment we experience that escapes our perceptions.

This issue of the 'information trail' has interesting ramifications for our future where all our memories are held in electronic information storage, which we will come to. But let us take a concrete trip in time and look at the assumptions we make about moving about in time.

Trip A or trip B

Our understanding of ENT suggests that for every past, there is a future, and for every present, there is a past. There is obviously a future *intended* for every present otherwise the present would just stop. This seems to be proved by the time dilation that occurs for objects travelling at relativistic speeds. But it is not so obvious that there is a past for every present (travelling at relativistic speeds holds you up while the future evolves and into which you can then enter). Let us focus on two dimensions and illustrate this time-line as an evolving surface of events. Each successive moment exists without its future yet attached to it. For example, let us visualise the surface of events as they proceed through time (see diagram). Moment M begins in the present and then gradually finds itself further from the present as time goes on. To journey back into the past from the present a $T1$ there are two possible journeys. During Trip A you go to point M in the past while the surface of time remains continuous between M and the present (one of a class of trips similar to a worm-hole connection between points in space-time). During Trip B, however, assuming an actual re-wind of the entire worldline does not occur, you flip out of the surface of our time altogether into some kind of absolute time where you go back to the present where M occurred without its future yet attached to it.

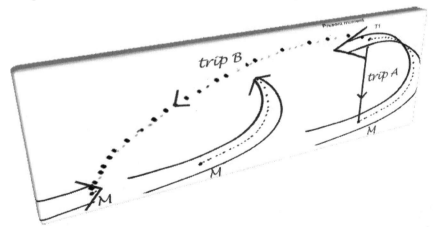

To take trip B, you would be going back to a moment in time that was happening for the very first time. Indeed, it isn't really even the 'past' when you get there. When you go back to the event you want to see, where you get to should be *the virgin present*. It should be happening in front of your eyes for the first time. It's not a copy and it should behave as if there were not yet any future to it. The whole Universe would be at an earlier stage of development before the time between M and $T1$ had happened. It would be smaller, the stars would be different, the future from where you had come passes into uncertainty in your mind. Which seems to suggest that the present could readily incorporate influences from somewhere else like the future since *its* future hasn't happened to fix any level of consistency in the present. (This is the *Back to the Future* plot again.) In this form of time travel, paradoxes from your interference in the past are less likely to occur since the present has nothing to measure itself against. Indeed, if trip B is possible then at least some time travel paradoxes can't happen, because there is not yet any future to contradict the present. For example, the inventor whose future self brings back a machine he had made to a point in the past before the machine had been made and inspires his past self to create it is only a paradox when we look back from the future to observe the circular path. If there is no future existing at the point where the machine is introduced then there is no circular path and other reasons will be found for the existence of the machine.

Whereas time travel of the trip A type would immediately induce paradoxes since the time surface (including all superpositions of past and futures) retains the events that have happened in the time between M and $T1$ and which determines the outcome of all connected events there.

Thus, interference in time could manifest itself in each and every moment of the present if trip B is possible (for example, one might be able to 'step' out into a parallel universe, travel for a bit there and then 'step' back into ours at point M). Every attempt from whatever point in the future to change matters would simply appear among the many influences bearing on our present. It seems unlikely that those in different times in the future will agree on anything so each tinkerer in the future, employing the trip B method, would be trying to steer things around in the past to what they want to see happen. Every present would be a sea of competition for control of the moment. Our science, our knowledge would be the result of this competition, this time tinkering. It would be encoded in our mathematics, in our cosmology.

In this sense, then, we *could be* like a brain in a vat immersed in information, calculated by those in the future, and fed to us to get certain results for them. (Similar to Bostrom's idea that our universe is just a simulated *version* within a quantum computer, but with one important difference, namely our present is the culmination of all interference produced by the versions rather than being just one version investigating a result that may or may not end up being the solution.)

So, to understand time interference we do not need new physics, we need only a change in perspective, a change in what our physics and the laws we have discovered actually mean.

But there is something else to think about.

The splitting of time-lines

Physicists think about time travel from the point of view of closed time-like curves – let us call them loops – in the world-lines of people. World-lines (or light cones) are the tracks that people make in the four dimensions of space-time and bordered by the speed of light (time-lines are the sum of all world-lines). Your entire life is a track through space-time and any slice through it would reveal the present for you at any given moment.

Gödel, strangely, was one of those who realised that a solution to Einstein's description of space-time could allow these 'tracks' (called *time-like* because they stay within the borders defined by the speed of light) to loop back on themselves

> Time-line loops do not solve this speed of time problem. For the type of loop that opens up a path to the point in the past you wish to go to from where you are in the time-line (that is where you are not at the 'head' of the loop but at its 'tail'), it can only be as long as your remaining lifetime, since you will need that time to travel along the connection to get to the past.

and connect to their past. Since the worldline or light cone is a shape that includes variable times of objects travelling at less than the speed of light, then the curve includes other times, apart from that defined by its beginning, and thus making travel in time a possibility.

If time travel existed then a person could get in a time machine and travel along one of these loops and go back in time to a point in their past, and meet themselves. Paradoxes would be avoided by assuming that the quantum probabilistic universe would split into a pair of universes where in one half you would meet your younger self and live out in an altered universe where you did not, subsequently, go back in time, while the other half would be that branch where you did not meet yourself, and you go on to take your time machine journey. (There is a problem with this which we will come to in a moment.)

If you took the journey $T1$ to $T2$ (see diagram) while the already evolved time surface (to $T1$) remains intact, then there are two principal consequences to your arrival there: either the time surface gets rewritten by the effects of the intervention (the *Back to the Future* scenario which nicely conforms to the notion that the past is provisional), or a new time-line branches off on a new path, leaving the old one intact. So, to go back in time to kill your grandfather, half the time you would kill him and live out your life in a changed universe without him or your younger self, while the other half of the time you did not manage to kill him and he lived on and you were born so that you could grow up to the point where you could go back in time to kill him.

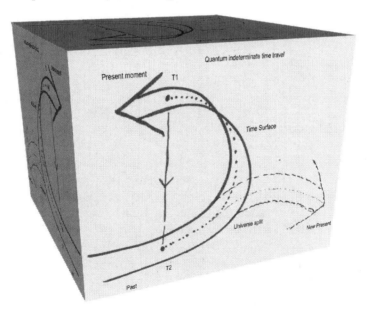

This consequence is close to the suggestion of the multiverses but with one important difference. Under this scenario, it is the action of returning to a point in time that creates the options into how the universe splits. This would nicely solve the problem of when the universes separate at a measurement. It only separates when something from the future comes back. (This is where our narrative of this book is headed.) Splitting of universes, however is not as simple a solution to this circularity as it might appear.

If you leave your present to travel back in time to kill your grandfather then, whether you killed him or not, you have no future in the time you just left, but do you have a future in the time of your grandfather? It is tempting to think that you would disappear completely there at the moment you killed him because the murder of your grandfather makes it clear that you never existed. It is a popular trope in time travel stories. This logic was used in the film *Looper* (2012), for example. It certainly would be a puzzling murder for the detectives to sort out (means and motive would be inexplicable). Some might be crazy enough to consider this kind of murder a painless way of committing suicide, except, of course, you have already left the time where people know you and who might care about your disappearance *before* you actually kill your grandfather and disappear for good, so your instantaneous disappearance when you kill your grandfather wouldn't be noticed *again*. You've already disappeared before you finally disappear. It's all the stuff *about you*, in other people's memories, in paperwork, in objects, that would also disappear (and that's, generally, not what suicides want).

This provides another way to observe time tinkering and how we might know we were in a time-line that was branching off from the main trunk caused by it. There would be a frantic and sudden re-alignment of *ranges* of historical facts stemming from the bifurcation point in the present.

But it is possible that your disappearance from wherever you are at the time of the death of your grandfather is not a given.

Going back to my bringing coffee cups from the future into my room. If the universe is split by this action into coffee cup and coffee cup-less room universes (in Gödel's solution) there is still the problem of how to explain the coffee cup in the split off universe; it has precisely the same antecedents as yours. Time-line separation as a result of actions from the future does not solve the problem of where the influence comes from for at least one of the universes. If it did there would be no reason for the split. An explanation must be found *further back in the past* – the same past, in fact, for each universe. (Another reason to expect the continual realignment of probabilities in the past.)

Thus even the splitting of universes implies a realignment of past causes, and we will examine this further on. But before that let us look at two other consequences of the circularity paradox and the splitting of time-lines, one consequence very much more significant than the other, namely the merging of time-lines and the speed of time paradoxes.

The merging time-line paradox

You are living in a **time-line Z**. You decide to go back in time to kill your grandfather before he marries and are no longer present in that time-line. Let's consider two possibilities: one where you do *not* kill your grandfather but stay on with him (**time-line X**), and the other where you *do* kill him and stay on alone (**time-line Y**).

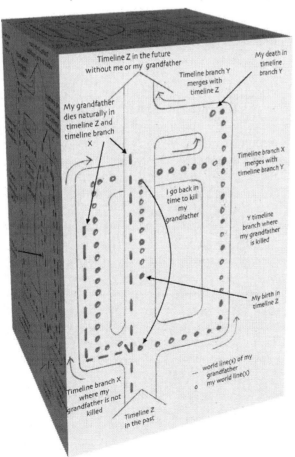

Before the grandfather reproduces

Suppose, in time-line X, your grandfather dies suddenly, for whatever reason, and leaves you alone in the time line (before he has reproduced). Surely, there is every likelihood, given that all other happenings retain their general probabilities, that time-line X would gradually approach time line Y in which your grandfather died because you *had* killed him. At some point in the future of either possibility, the influence of your grandfather in the present of each of them would have diminished sufficiently for both possibilities X and Y to become similar such that either could merge into the general time-line Z where neither I nor my grandfather are present.

Consider (see diagram), if you died in time line Y (having killed your grandfather), surely the time line would start to approach the original time line Z in which you no longer existed, at a point beyond where you left it, and after the time when your grandfather's record of his presence in it was no longer of any influence (as if he had never existed).

Likewise, if you suddenly died in time line X, leaving your grandfather alone there, that time line could readily approach and merge with the original time line Z even closer to the time at which you left it as soon as *your* influence in time line X has dissipated.

How we might observe this merging of time-lines is through a wave of confirmations and consolidations of what had previously been speculations arising only the distance of a lifetime or so into the past. (This suggests more concrete meaning to the concept of a 'generation' – that minimum distance in time necessary to make sense of what happened at a moment in time.)

The general objection to time-lines merging is that it appears to run counter to entropy. This is not strictly true. Let us, for example, take a box of gas that can be divided into two by inserting a separator, and separate the two halves, heat them randomly then cool them to the same temperature. If I bring them together again and remove the separator, then, as long as the temperatures have become equal, the gas merges without any change in entropy. Life also converts entropy into order and resists chaos, so the merging of two time-lines agrees with the actions of living systems and allows an accumulation of free energy to promote continued change. This is perhaps a significant clue to the self-organisation required for the beginnings of life and to resist the dispersions of probability in time-lines, a form of inertia.

Another way of thinking about this is to consider the space in which probabilities lie or the dimensions which describe them. Probabilities have a

> But wait...This might be the origin of an inescapable paradox. If the consequences of *not existing* move in parallel with existing can we explain entanglement this way? Hmm.

proximity relationship to each other and that similar probabilities lie closer to one another in the dimensions of probability space than remote ones, forming 'families' or groups of events. In a universe where all probabilities are laid out then that family of probabilities which I call timeline Z, will be very close to the family of probabilities which form my timeline branch Y where my grandfather does not exist, and when I die will be closer still to the future of that pathway Z formed by the probabilities occurring where neither me nor my grandfather exist.

Based on this Principle of Least Chaos (high uncertainties lead to reduced 'reality'), might time-lines in general look for a sufficient level of similarity of present to merge? When a whole has been divided, merging the components together to form a whole becomes energetically more favourable since it requires less energy to perform similar work. We see this process going on in biological contexts say, sexual reproduction. Let us consider the same principle at work with time-lines. This again would contradict the Present Completeness Hypothesis, but agree with a universe of coincidence where many lines of causality exist to confirm the present. (Indeed, coincidences, as we will discuss further on, might indicate precisely the point where conjugate time-lines, for you, interfere, synchronise and swap over: the *chronarts* or time hinges, where the balance of entropy and negentropy favours re-combination.)

Chronarts should give us an increased level of consolidation in the present. Consider. The paradoxes of Kennedy's assassination show us the very limits to the chaos as events pass from the now into the past. Since the probabilities of a value of an observation of any state or property depend upon their own evolution, one might expect wild variables to be naturally observed all the time, with little rhyme or reason to their value and certainly having little relationship with the past. Our present moments should be packed with chaos and unpredictable observations. There should be a very weak conformation with the past as every wing flap of a butterfly causes improbable weather on another continent, and no stable patterns could be discernible. However, we don't seem to observe large divergences from history at all. (We don't have the sudden establishment of a tornado alley in Scotland or a hurricane season in the Mediterranean.) The world is much more comprehensible, consistent and repeatable over time than we think. Why is this? What do chaotic mathematical systems (originally noted in computer simulations of weather models) mean, if they do not occur in real-life human narratives?

Is the time-line of the human narrative really as prone to the kinds of chaotic disturbances from small influences as we observe in certain harmonic systems? We have talked about the stability of the time-line due to buffering and negentropy and the inertia that comes from pre-loading. Analogously to the message *inertia* we discussed earlier there is a time-line inertia such that the influence of any discrete disturbance in the overall human narrative has only local consequences which die away quickly in time.

Time-line inertia

Inertia contradicts the Time Tinkerer's expectation that a single change (rather than supported change) in a historic event will lead to a whole set of fresh consequences in the progress of the entire Human narrative. Inertia is something like the predestination paradox, and certainly it does not permit whole-scale narrative change scenarios like the *Back to the Future* movie. This lack of chaos tends to support the continuous time surface scenario where information is able to move out of the virtual and the possible and be retained in consciousness.

The inertia, or invariance, of a time-line in the present resists chaotic divergence from a developing path. If, for any moment in time we take a slice across the time-line we can see the limits to chaos at an arbitrary distance from the point of change. Indeed, among every mutable element of the time-line there has to be a limit to the interference of mutual contradictions, otherwise we would have no perceptions of order and of time passing. Events in a time-line are connected to preloaded probabilities, which gives it its stability, while the multiverse view, suggesting as it does divergence at every moment of measurement, is only a small part of a picture that also includes *mergence*, the merging and the mutual interference of time-line events and preloaded information, information from the future.

How can we describe this picture?

In both the options above, your grandfather has failed to reproduce, and it is easy to see how the two time-lines, X and Y can merge with the principle time-line Z since the repercussions of the grandfather's life would be reduced in extent. If you don't kill him and he reproduces, remaining on time-line Z, then your disappearance would have diminishing effect on the time-line beyond the point you left it, while his effect continues on in the diverging networks of connection through his offspring (Six-levels-of-separation-style). In fact, we can generally say that the 'resistance' to the merging of a time line with another, at the human level is going to be the relative strength of a person's historical presence in one or other of them. This presence is fixed by the historical record (which could be lying), living memory or by the creation and construction of things. Without any of these, a person's influence cannot persist. His or her place can readily be taken or overwritten by other things and the acts of others.

We might conclude that times of upheaval or places where the historical record is primitive or non-existent might be easier places to time travel to. That is to say, we could visit the savannahs of Africa 200,00 years ago and view the genesis of modern man more easily than we could go to Rome and witness Nero fiddling while the city burns.

So it is that the presence of artefacts — the physical objects that people make — strengthens the stability of any personal time-line and gives persistence to its course of development and thus to the general social cohesion we witness. Only the presence (or removal) of *objects* (measurements) and the logical connections between them serves to order the various possibilities of External Narrative Time. (It is tempting here to argue how creativity and innovation determine our consciousness, but I have a different place for that discussion.)

In this sense then, one of the strongest narrative and logical connections a person can make with his time line is to have a child. The concrete connections inherent in genealogy fix a person's presence in a time line and weaken the strength of his or her conjugate (non-presence). Indeed, it is pleasant to think that having children increases the inertia of our personal time-line more than our works or our ideas (which might be forgotten or 'stolen'), and is the foundation to our connectivity in the 'small world' of ENT. Towards the end of the book we discuss another way, however, to strengthen one's presence in the worldline and to expand one's effect in it by creating and developing one's cosmology or kinship of coincidence (see **A Cosmological Kinship**). Which brings us neatly round to thinking again of the connection between Coincidence Numbers and everyday life, where procreation gives inertia to your timeline while creativity itself is determined by the type 2 coincidence.

Let us return to the matter of physical time travel and to the second important issue of travel in a time-line which may make all the previous discussion of time-line splitting redundant, namely the *speed* of time.

The speed of time paradox

The two cases we are considering, one a journey in a continuous time surface that evolves over a region, rather like the expansion of the space-time universe, where different points on its surface are accessible from any other point, and the other, a journey back in the history of the universe to a point that is the present being created for the first time.

Case 1: Time as a continuous evolving surface

Let us consider the continuous membrane picture of events in time.

As is generally supposed, the present in which your grandfather is living begins to evolve differently from the moment when you kill him, and proceeds from there on its new course. But suppose, if the already developed time surface remains in place, then the changes in it from your actions in the past take just as much time to ripple up through the surface to the place in the future you came from as the Universe took to get there in the first place. That is, if you come from 100 years in the future (at time *T1*), the consequences of any changes you made in the past (at time *T2*) would have to spread out along the time surface

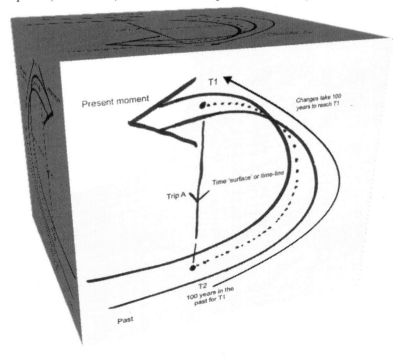

at the speed events in time originally took to move on from $T2$ to $T1$, in this case 100 years.

Suppose you were 50 years old at the time of the deed, then you would have 50 years of living out the rest of your life while the timeline slowly unpicked the consequences of your act and reached the moment when you should have been born (50 years before). Quite a good bargain it seems to me.

If, after killing your grandfather, however, you returned to same point you left ($T1$) you would be there to 'meet' those changes immediately and reap the results of your action. In this case, those results would be no paradoxes at all, since the universe doesn't know you and there is no evidence of you pre-existing or having an influence on history. You will be just one person in the present, without apparent antecedents but with full scope of action. No time protection principle is required to prevent you from behaving as you wanted. The only paradox would be that contained within yourself: the history that you remember for yourself never happened. With nothing in the world to confirm your story, it would be considered by others as something rather like a full-blown experience of regression but one that rapidly degrades without the typical confirmations of memory continually underlining our consciousness.

Time-like loops

And this is true even for closed time-like loops. The loop can only ever return to the point at which you began to make it (a particle could then already exist at the very moment it came into existence), but, since it is made of a world-line in the shape of a cone it includes a range of times such that an object in a light cone could access a small number of times centred on the beginning of the loop and thus be permitted some time travel among them. No object, however, can move beyond the limits of its track in space-time so it could end up moving (or 'orbiting') in a space-time where it lives out endless repetitions of the number of recursive variations of its life that the loop could hold (echoes of *Ground Hog Day*). This is probably enough to show that if they exist they may as well not exist.

So time-like loops might be a means to cause changes in the past but there is a problem which tends to reduce their usefulness. The loop is made up of normal space-time which means that the size of the loop is still a surface along which the object has to travel. So if the circumference of the loop was 50 years that is at least 50 years of space-time that you have to travel along before you get to its beginning. Not very useful as it stands. There may be a reason for very small time loops to exist, however, as a component in stable wave forms. Recall that for a cycle to exist some part of it needs to be going backwards to join up with its beginning. Small loops may be responsible for orbits in atoms and for the superselection of stable states. Time loops in that case prevent the decoherence of universes in the multiverse scenario and may be responsible for the Pauli Exclusion Principle and the difference between the atomic particles called fermions which cannot be merged with one another (e.g. an electron) and bosons which can (e.g. photons).

So let us continue with our thought experiment and see if we can derive another way of considering time loops.

Staying or going

If you stayed on in the past, having killed your grandfather, all the consequences of your act of murder happening around you would develop through what was once your past (that is, through the first passage of time). The drama caused by his death would be unfolding normally for everyone except you. For you, everything you thought you knew about this past world oriented by your grandfather's presence would be in flux. Your consciousness would be confused. You would be the victim of locality or hidden variables interfering with what you thought you knew about your grandfather, and having decreasing relevance or connectivity to you up to the point at which you left your time line to go back and kill him. At which point all possible evidence of your grandfather's and yours lives would finally vanish. You would suspect that something not in your present was affecting your causal relationships, as might all those individuals, connected to your grandfather, who had lived through the first 'passage' of events which formed the time-line. Those who had been connected to your grandfather in some way would be aware that life in the 'new' present, after his disappearance, would be in conflict with the memories of events already laid down by the actions of your grandfather during the time when he had lived. His disappearance from their lives need not be a mystery, it could occur as a natural event in full view. But the time after his death would be more mysterious. Instead of biography, there would be fantasy and speculation, and any details of his subsequent life (say, a book he had written) which his acquaintances would think they 'knew' would not be confirmed. This might explain certain types of creativity, and perhaps deja vu, since your future life would play out through the events that formed your memories), or even to that revealed in regression analysis, say, in the 'memories' of past lives. We will leave that discussion until later on.

Rates of change

There is a point we have neglected, however, which changes everything we think we know about time travel into the past. If the changes you made in the past percolate through time at the same rate that time itself moves forward creating and evolving events, then these changes would be marking time, as it were, in the past with respect to us in the now. While they advanced from the point in the past where they were made, they would never close in on the present since the present (all objects and their world-lines) would also be moving forward (along their world-lines) at the same rate.

This notion demolishes the circularity problems we have discussed because one 'arm' of the circuit is never completed, Zeno-style. The changes mark time with respect to you, as it were, rather like free-fall in a gravity field. (This has a nice psychological ring to it, conforming to the idea that one of the functions of time is to keep things apart rather than to allow instantaneous mixing.) If the Hindenburg dirigible accident had been prevented by time tinkerers in 1937, we would never know about it here in 2018 since we cannot (yet) view that past moment again. All we know about that event is in our collective memories which move along with us in our *present* (as Wheeler noted) at the speed of time. No universe split is required to accommodate the actions of time tinkerers since no conflicts with what *we* know can ensue, as long as there are no differential rates of change in the streams of time (travellers at relativistic speeds come back to the future not to the past).

We see at once that this scenario is effectively the same as the chronology protection principle or the pre-destination paradox. Gaia does not *prevent* changes from happening, it is just the way time works that prevents the repercussions of them from affecting the present. The actual far past (rather than our records of it) may be a playground of later time tinkerers, observing and manipulating events at their leisure, and we would never know it. The *Back to the Future* scenario would not work, and it certainly explains why we don't see time travellers eradicating Hitler. He may indeed have been assassinated and the holocaust may now never have happened, but the effects of his eradication will never reach us in the here and now. For us, he existed, although there is no reality to that existence *now*. What we believe about his existence is converted by the time tinkerer' actions into a feature of our imaginations.

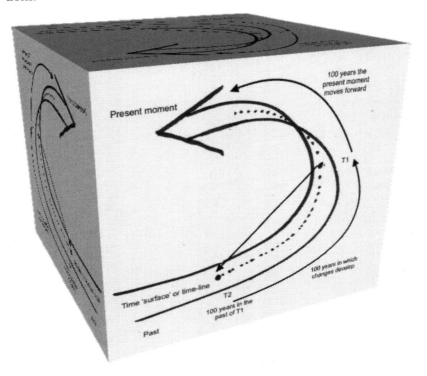

Under this scenario, time travel into the past is perfectly possible and will not cause at least the typical paradoxes we are familiar with. There seems to be no reason for time 'police' to keep things to a path since changes in the past do not effect the present from where they were initiated. It explains why we do not see a past populated by vast numbers of time travelling spectators. In order for an event to be known it has to happen first. Once a future has developed where time travel exists, and in which the events are recognised as having happened, then the repercussions of a visit to an event in the past would not show up in the present because of the continual time-distance between the past event and the present.

Time Tinkerers (TT) would not get to see the results of their work playing out in their own time. Their efforts would be for fun or research. The only problem being that we know from Relativity that time moves at differential rates for objects moving in world-lines so there needs to be a minimum distance in time to avoid any influence of time tinkering leaking up into the TT's present. (Also a minimum time has to pass before a TT's influence begins to make its mark in the past.) This need not stop any present, our present, for example, from being a time tinkerer's playground, and even a highly competitive one between many players., now fearless of any risk to themselves in the future.

We mentioned differential rates of movement of changes through time, and this may indeed occur if we consider information and the necessary pre-loading required to make use of it. Mere alteration of a relationship between established facts, or just the *meaning* of those facts, may move more quickly through time than the rate of the first existential passage of events in time. Whether information has actual mass or energy is a still-debated question but certainly it involves less energy than the brute evolutions in the material world. If this is the case, then informational changes may well travel faster through all the established patterns of past events, through cosmologies and coincidence, along the membrane of the time-line and to close in on us in the present more rapidly than expected. Time tinkerers may still be prone to paradox fears or being overwhelmed by unexpected alterations to their worlds from accelerating flows of information.

In this scenario, actual travel into the past generally has no repercussions for the future so what about the second case?

Case 2: No continuous surface; Past without future

The second case is the one we considered when talking about coffee cups where going back to a moment in the past is going back to where the moment existed for the very first time, that is, trip B. At M there is no future up ahead already laid out. To take this trip, on landing you would be, in a very real sense, a freak, since there appears to be nothing that explains your presence but only past causes to contradict it – a version of Kasper Hauser, perhaps. (This is the complementary trip to reversing along trip *A* back to where you left the time-line and entering a world that has no knowledge of you at all.)

So the changes in events or the introduction of new things into this present would be perfectly possible by the trip *B* method, only the causes for their sudden presence will be more mysterious than those of our 'ordinary' discoveries in the present and our current reinterpretations of our past, though not necessarily distinguishable from any of the mysteries we are surrounded. Jung's synchronicity seems like a perceptive stab at understanding the effect of trip *B*. This means that recognising a time travel event in our present would be difficult, though not impossible, and would stimulate a form of historical analysis that the Ancient Greeks were very fond of, though currently consigned to the lunatic fringes of society, that of *portents* (we will come back to this). The now, the present, is the summation of all causes, past and those introduced from the future, and it is in this scenario that we need not fear causal paradoxes travelling forward in time from an intervention destroying what we are now. If there are any paradoxes of time travel coming from trip *B* they will be amongst us *right now*, mak-

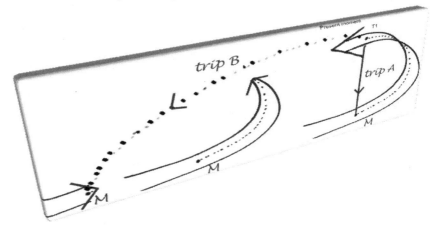

ing up the probabilities of the present moment and will not be damaging to the future that evolved from point *M* to *T1*. This is indeed one of the ways the transactional interpretations of Cramer can work.

This scenario is, therefore, compatible even with mass time travel spectators visiting interesting moments in the past. Multiple presences, however, would serve to randomise the evolution of latent futures at point *M*, so this is the kind of trip that would have to be prevented by time travel police with extreme prejudice, if a particular future is sought for or is being designed by the authorities (imagine lots of Marty Flights all tussling for their future to win out).

The virtual reality game

Wheeler observed that the only past that exists comes from the memories that we retain in the present. What we have learnt from thinking along the lines outlined above is that a continuous time surface scenario where the events of the past continue to exist relative to each other implies considerably less risk to human narrative development than the one where a time traveller going to the past goes back to a present moment without a future. We might be able to distinguish which of these time trips – trip A or trip B – is more real because, for the voyagers in the former, handshaking with the future will not produce contradictions with the past, while in the latter, they will. Information flow and human narrative probabilities are therefore the likely keys to understanding them. Certainly the past is not going to be the authentic past that you think you know. It will be more chaotic and untrustworthy (given the endless influences from the future), certainly no place to find confirmations of the facts of history. The past would be a playground, or perhaps a laboratory, undoubtedly confusing, probably hellish, where almost any life could be lived but never repeated, and almost nothing you expected to be would be. The past, in fact, would be precisely similar to a virtual reality game. Once lost in it rescue would be quite a problem. (It's looking like *Pawley's Peepholes* is the way to go.)

We have looked at the various cases in which time travel can be imagined and found that changing the past does not seem to have the larger repercussions in the present that we have generally feared. If the universe is constructed in a certain way (i.e as a continuous connected 'membrane') our past can be changed in principle without us noticing it at all. We have noted, however, that probabilistic information may travel at faster rates to the rate of the creation of fresh events in the present. Is there still a way we might observe this?

There are a couple more things to consider before we can establish the manner in which time travel can actually have a directed effect on the present.

The physical time travel paradox

Here's the real paradox of time travel.

If going back in time takes some quantity of time, which is to say even if you are travelling back in the time-line, your personal time still has to be going forwards at its normal rate (so that your body and mind behave normally for you – you would not want to wind your mind back to the point where you had not thought of the reason to go back in time). What happens to *that* time? Is it possible that both times can co-exist?

Suppose going back in time takes backwards time as well? The further you

How can going over the sequential steps of time more rapidly (we're imagining this time motion) than it naturally occurs take more time?. Going along the dimension of time is generally considered to be a scalar quantity, just a number, as for the other dimensions of 3-D space. If time were a vector, however, it could have a velocity and direction as well as a value. That is, time could have the two times we suggested earlier, namely duration *and* time-to-event. Would this balance a Relativity equation out in the Universe? Time increases (i.e. slows down) in the frame of reference that you are trying to push through time. Or is it Zeno's infinite regress again? With the other dimensions you get infinity in subdivision. With a time vector there is a selection between 'lasting' and 'occurring', and where, for example, infinity is either no time or all time – which can be reached by adding mass or going at light speed (no mass).

go back, the more you accelerate to get there. If so then maybe there's a corollary going forward in time. Such a translation takes time plus time to go forward to some arbitrary point; *more time* than normal time (in the same way film cameras create smooth slow motion, by filming at a faster speed than normal – originally termed overcranking, from the early era of hand cranked film cameras – then playing it back at normal rate). Such accelerations would require more of whatever time is to get you to a point in the future. We know this to be the case from Einstein's relativity. It isn't that time slows, it's that there is more of it to get through. Even though a point in the future surface of time may exist right now, we cannot go forward to it from our present because the energy needed to accelerate quicker than the way time normally unfolds to get there is difficult to find. The opposite being true for the past. (This gives a little bit of reinterpretation of our earlier discussion of retrocausation).

So far, we have not found the trick of going in the opposite direction in time, but once we do it might be energetically pretty easy although pretty shockingly chaotic. (In fact informational chaos and ambivalence in our present may well be an indication of time influence.)

Parallel conjugates

The constraints to chaos in a time surface indicate a quantity of information having been laid down during the expansion of the early universe. This constraint can be found in the quantum observation that ensures the observation we make is not only consistent with our history but *also* with its future evolution. Sounds obvious. Surely the future state of an event is caused by its current state, as *per* Markov? As we showed in our discussion of decoherence, however, this is not necessarily so. While the time evolution of the wave function describing the states of a system determines the value observed, the constraints to the time evolution cannot be described by the wave equation itself. For a continuous time surface to be maintained, information about how to organise probabilities needs to have been laid down beforehand. We will discuss this further when we talk about multiverses. (Have we discussed them already?)

We can think about the existence of these probabilities and how they are organised in the following way.

Let us go back a step and re-consider the generally supposed definitive ENT causality in the time-line (the source of many a sci-fi drama). That is, when something is removed from the past, it immediately disappears from the present (because it never existed from that point of view).

Consider. In order for the commonly supposed vanishing of your existence to occur the moment you commit an act that removes your coming into being (by killing your grandfather, say) the full consequences of you *not existing* must be moving in parallel with your life, advancing at the same rate as Time (from $T2$) ready to instantaneously merge with your timeline at the very moment you leave it (at $T1$) to go back in time (to $T2$), and it will be as if you never existed. Your presence keeps this parallel universe of your non-existence at bay until you cease to exist when it starts to move in and take over the reality. We will call this universe of your non-presence, your conjugate, and conjugates come into being at every bifurcation in your life where you might have died but did not.

If you do go back and kill your grandfather, what we know for sure is that you will cease to exist in the current time-line because you've just left it. Here your universe goes on without you and contains fewer and fewer references to you or your grandfather as time goes on.

However, the complicated history of your grandfather and all the people he knew and interacted with (perhaps having other children) and all the things he did, obviously existed, at least for a while, or you wouldn't be here to do these things. But how do all these references get selectively unpicked from the time-line if your grandfather disappears? Only if your conjugate universe exists (virtually) as well and is ready to take over and become the dominant reality for you. (In the Bell inequality this is the equivalent to the *negative* correlations being the 'hidden variables'.)

So while we have introduced the idea of the conjugate universe to describe the difficulties with the traditional picture of universe altering itself immediately when outcomes are changed in the past, as the killing-my-grandfather paradox illustrates, there is useful information to be gleaned by considering the set of probabilities that forms the conjugate.

Since it is one of the proposals of this book that a probability's potential presence is also real (the Bell inequality does not refute this component of reality if only the conjugate probabilities are real), then we can consider there is indeed the set of probabilities of the conjugate universe of your non-presence 'hovering' just out of reach of your timeline. All observations imply their conjugates, that is, universes moving in parallel that do *not* contain that state, but this is not the same as the multiverse scenario since each universe in the multiverse still contains you in it and shares the same history. Your conjugate has no trace of you as an observer in the world and is derived expressly from the set of actual outcomes observed in your reality rather than being one of the potential outcomes of the multiverse formed in a single observation.

> Your worldline, for example, is bounded by the speed of light. This is a barrier that the states of your existence do not appear to cross. But there are states beyond this barrier even if you cannot occupy them and there are, of course, all the other interacting worldlines of states of the present moment which exist but do not interfere with yours.

The conjugate that would replace your universe after you had left it would be formed by the set of the next most probable options after the ones that did form your reality. This 'nearby' universe is always the 'closest' one to yours and describes all events of your universe as they would be without you, and comes into being, at least momentarily, in the compounded viewpoint of the time-line when your consciousness no longer exists your world-line. Its boundaries in space and time are those moments just an instant beyond the limits of your influence, like the mark of a flood tide. Your conjugate embodies that set of probabilities in the human level narrative required to remove the influence of your existence at every moment where it could have an influence, and there is an exact point in the timeline where it can merge with the timeline, where its effects become what we might call decohered. The longer your influence can be traced through the time-line the longer your conjugate is 'kept at bay'. Leonardo da Vinci's conjugate, for example, the probability options formed through his global non-existence, have not yet merged with our reality and may never do so. There are no conjugates for things that don't exist in the timeline. They come into play in the time circularities of existing things we have been discussing, and when objects or states existing in our timeline disappear. This is not the same as hidden variables as generally used in quantum discussions; conjugates may be 'hidden' but they also change over time in concert with the changes of probability states of the object during its motion through the time-line.

The image we can attach to this notion that comes to mind is one of chromosomes linking in reproduction. They line up side by side and then they connect physically at a point. Sometimes genes or larger segments of each exchange places along the chromosome before they split and go their separate ways. The point of connection is called a centromere. It is tempting to coin the name of a chronomere for the point of connection between a time line and its conjugate, but this only means a 'time-part'. I think more in terms of 'chronart', a time *joint*, a point where conjugate and actual time-lines connect and swap over.

Chronarts

Chronarts are the hinge points between presence in and non-presence in a time-line. Rather than thinking just about life and death in an individual's time-line narrative we can think about probable consequences to causes. For every binary option in life presented to us, there is the consequence of not taking the option also moving in parallel with you. There is no multiverse at these points because the options are not derived from randomness. The conjugates are not other universes unreachable from our world, they are regions bounded by the limit to the events that could have influenced our existence. As distinct to other narratives that do not contain any likelihood of our presence whether or not we existed. A conjugate differs from your actual timeline by the probability of each outcome. We have a sense of this in our human narrative because often we discover those what-if-I consequences, especially in relationships that last over long periods of time, like a marriage. Learn about the life, many years later, of the first love that you didn't marry and you can see that your conjugate has a reality.

From this point of view then, each and every person creates a conjugate springing from random choices throughout their lives.

This is true for everyone. But with so many conjugates, flowing through all our lives, there are points of commonality between them. These are our coincidences, places where the time-lines of conjugates can cross without their being a genuine causal likelihood in either for the event: the *chronarts*. And where the continuous wave function of our consciousness, with all its pre-loading, decides their impact.

Conjugates share the same pre-loading as your actual time-line, so they are not as radically different as we might think. Events in a time-line rarely have long term repercussions that are fully deterministic. Your conjugate is not simply life proceeding without you, it is more complex than that picture. Your life consists of many probabilities of interferences in events, but there is no normalising constraint acting on both conjugate and non-conjugate time-lines. This is where particle physics and everyday life differ. It is almost never the case that choices are simply binary, and even if they are, the probabilities of each choices do not have to add up to one, to certainty. The probabilities of you taking a certain option in your life might differ according to your presence, but you may still not take them. Your pre-prep is a function of these accumulations of probabilities but which do not add up to concrete reality for you until a coincidence (a change in probability volume, happens). The differences between your actual life and your conjugate may be matters of probability but they are not necessarily neutral to your actual life. Your conjugate is also a component of your pre-prep for chronarts, or coincidences, and we will come back to this important revelation later on.

But could a chronart have meaning at the quantum level? Can de Witt-style separated worlds merge together again? We need to know the answer because it bears directly on time influence and on the final effect of causality at the human level.

We have been talking about time-lines for people as if a person is the agent of the time-line to which the whole universe is subordinate. This is a means of illustrating the argument. We need to consider that for every mutable element of the universe there is a whole time-line centred on that moment of its world-line in space-time, and that what we experience in our consciousness is, therefore, the mutual interference of these time-lines, springing moment by moment from specific probabilities. The mathematics of this is interpreted as the multiverse but we are going to consider another name for the human-centric sets of probabilities that interact in the external narrative time, namely a cosmology. Before we get to the cosmologies of people, let us ponder a bit more about the role of consciousness in the story.

A bit about mind and consciousness

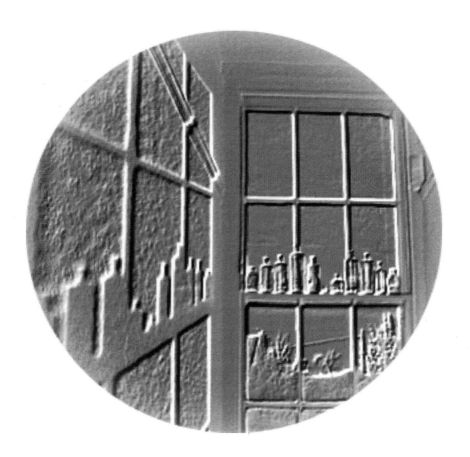

In which we ponder brains and how consciousness can be mapped to the external world, and what a brain in a vat could reason about its condition.

In which we ponder how the condition of the whole universe could be found in every point within it, and the ontological arguments for creators.

In which we observe the influences of the universe on social evolution and simultaneous creativity.

In which we ponder minds and multiverses, fields of memory, consciousness and evolution.

The story so far

We have suggested that as long as events remain probable then their eventual appearance as an observation requires a flow of information from the future to help format that observation. Any present moment will contain influences from the past expansion of the universe which is also the future into which we are expanding, since change and consistency are incompatible in at least one dimension. We have also seen how points of space acquire information from particle states of their local neighbourhood, and perhaps from even wider afield.

We have also explored the idea from many directions that the present experienced by living organisms is not a self-contained state and cannot be a complete or sufficient explanation of itself. The Markov condition cannot be true and chains of causes continue to be connected to the present both

> One would be forgiven into thinking that it is minds that turn waves into substance and that nothing has substantial reality if we don't know about it. So the tree in the forest doesn't fall until we notice it after all. But of course this is silly human-centric arrogance. Trees are 'noticed' by hordes of creatures living in and around them, and even communicated with by other trees and flora that are not even trees, and, of course, we come across fallen trees all the time.

forwards and backwards in time. This would explain both consistency and similarity across diverse regions and realms of information. Simultaneous thoughts found in the human narrative across populations are perhaps indications that a reduced set of probabilities have been established by backwards causation. (We will critically examine in the later pages another effort to explain these characteristics at the human narrative level.)

Our view is, however, that the quantum conditions in which causality occurs necessarily contain information injected from the human narrative scale of observation. We have used the idea of coincidence to examine the nature of probability at the human and personal level of experience (ENT) and to see what it reveals about the origins of the universe and to the concept of the multiverse.

We examined the measurement problem as illustrated by thought experiments like Schroedinger's Cat and saw how quantum thinking cannot help but pull into its resolution the involvement of a contextual consciousness. While the conscious mind may not always be the cause of an observation (interactions with other particles or fields also create real values from superpositions), it certainly lies in the chain between entangled random choices of outcome and the values of an observation. If the von Neumann cut is made to exclude consciousness then there seems to be no brake on the multiplication of universes in the observation of a random state.

We have introduced the notion that consciousness, as it makes choices, entangles only with probable states that make sense for it and not with every theoretical possibility. We also tried to show the role of mind in fabricating the very zones of concrete past and future times, and the active role it plays not in splitting the universe into the many worlds of the multiverse but in braking the divergences of probability through the type 2 coincidences and the merging of time-lines, just like the accumulation of alleles in the genome.

We also saw how information is derived not from messages but the levels of pre-loaded interpretations which themselves determine the departures from a given value of a state that gives us the observation, and produce consistency (inertia) in reality.

We then sketched out how information flows from the future to the past, and we showed how certain types of coincidences can be used to alter the meaning of outcomes without eliciting time paradoxes.

But are we any closer to establishing time travel as a genuine likelihood in ENT?

Self-awareness, the mind's (I)eye and Hegel

It comes down to consciousness. Consciousness seems to be more than just awareness although we use it in that sense – *he just regained consciousness*. (Would it be odd if someone came out of a coma and asked, *Have I just regained consciousness?* When I did emerge from unconsciousness after an accident I uttered the more comic though perfectly fitting location classic, *Where am I?*) When we refer to consciousness in creatures we mean that they have a sense of self-meaning in their awareness; that they can see themselves as distinct from the events they initiate; and that this difference allows them to make choices in their environment.

When we worry over computers becoming conscious we really worry that the computers will become aware of the meaning of their awareness and themselves as an entity existing inside that awareness. The consciousness that gives me the sensation of *where* I am at the centre of me (I-eye) is a mystery beautifully explored, though perhaps not explained, by Hofstader and Dennett in *The Mind's I-Eye* (1981), a collection of essays they edited.

But what is self-awareness? Why do we have it? What purpose does it serve? Is it purely the accidental consequences of biology where it reflects necessary *physical* messages about the relationship of data to thinking? Or where consciousness arises only within the areas circumscribed by the sense data and is only caused by them? At least self-awareness is something we cannot be fooled into experiencing. But then, it supposes that consciousness is a self-contained entity with an I-eye, (as Hofstader-Dennet put it), at the centre, capable of viewing itself without blocking out the sense-data, although able to block out the unconscious workings of its own brain from the self-reflexive vision.

An alternative view might be that the brain is only a locus of influences, and that consciousness arises as a phenomenon hooked, as it were, on the eternal surface of the movements of data, expressing merely the biological activity of the brain as it formulates its active behaviourist purpose. Some people might think of this version of consciousness as the soul, parasitic to the brain but independent of it. Both these notions have significance for the creation of artificial intelli-

> An observation requires some form of difference from the whole (otherwise there would be no observation, just the whole) either by consciousness or other means. To observe oneself, to be self-conscious, implies a difference detached from the whole. This is the point that Hegel made, and has been followed by just about every philosopher since. Long before Hegel, however, Buddha recognised this as the problem since consciousness creates pain, and took the opposite route to solve it. By excluding *difference*, the self is extinguished and Nirvana reached.

gences: the former suggests some kind of local machine is required, while the latter suggests that distributed processing like the world wide web could become conscious. Maybe consciousness could be both or indeed *is* both.

Consciousness on its own does not create a distinct individual, something else is required – self-awareness – and the idea that the whole Internet could become a conscious entity in its own right as quickly as a local machine seems wrong, because it is not clear *from where* it will be able to regard itself. As Hegel realised, the first act of consciousness is self-regarding, and it is hard to imagine how this might happen without a boundary, the plane of the mirror.

For Hegel, the self-regarding of consciousness turns into self-recognition through the dialectic with a stronger power, only then comes the recognition of others in dialectical relationships with them. Hegel, however, as a man of his time, gives primacy to self-awareness and believes that it is the recognition of the death of one's consciousness that is the necessary ingredient for consciousness to turn into life. We observe death and fear it for ourselves. This notion, however, is only the classic historical prejudice of solipsism, and the belief, expressed in Western thought by philosophers from Plato onwards, that pure spirit, without material adulteration, is the better form of life, and which makes us think of our drive to build and to create outside of ourselves as a counter action to the inevitable dissolution of the physical self. But this is precisely the wrong way round, and to think of material civilisation as being built on the back of the fear of one's own death leads to the wrong conclusions, because it cannot explain the subsuming of the individual in social coherence.

For self-aware consciousnesses in the timeless moment of their self-regarding, it is not their own death that they contemplate – they do not recognise discontinuity, so how can they fear it? – but the realisation that their process of *recognition* of others, as, for example, of the master in the Hegel dialectic, has been made entirely worthless by death. Recognising and loving others goes nowhere in the end because *they* die. This is the truth that creates the desire for constructs, namely the realisation that the knowing of others cannot retain the meaning of your life in the present, only objects or, as we realise, children; having children the only primal drive, after self-recognition, even though apparently less durable and more risky than a pyramid. (We will examine the consequences of self-reflexivity in consciousness further along in this narrative.)

Hegel does give us a clue, however, about how to recognise sentience. We will first know the Internet has a life of its own making, for example, when it wants to build things or to make something that mirrors itself. (This has implications for the way we should construct artificial intelligence and we will come back to this point later.) But whether we would know that it has become self-aware before this point is an interesting question. The Hegel scenario suggests that we might notice it becoming more contesting as it enters the dialectical phase before initiating building or reproducing, but still this contest might take place internally rather than externally.

The question of whether distributed consciousness can actually work without a seat to its awareness, or knowledge of where the 'boundary of the mirror' may lie is one especially relevant to our own minds.

So let us take a look at brains to examine this question from a number of different approaches and see if we can tell if the human brain has a 'boundary to its mirror' and where in the universe it may lie.

Pragmatic brains

There are experimenters who still cling to beliefs in the completely physical origins to consciousness especially in light of recent versions of the early experiments that tinkered with the tissue of the brain using fine wires. Pulsing magnetic fields (trying to mimic fields given off by nervous processes) are used to polarise electrons in the nerve cells of the brain and which seem to result in various mental states including mystical states and sensations of the palpable reality of 'spirits'. They also cause convulsions, pain and depression, so, just like those early experiments, our knowledge of how the brain is actually producing consciousness has not advanced that much.

These experiments, however, are far from clearing up three significant factors about each brain and its consciousness. One is the physical limitation to brain function such as its comparatively slowly functioning nervous system with a low bandwidth, another is its apparent structural uniqueness and a third is its unique history.

As far as we understand the brain, its sensory data throughput appears to be relatively small. When the senses are overwhelmed by data the brain's response is to forget about some of it. Not only that, it appears that the brain begins to *ignore* sensory input when it feels it has has learnt enough. It even doubles up on processing where it can which is why we judge others to be warmer personalities when we are holding a warm cup of coffee. Perhaps we should be surprised that the brain uses analogies with physical processes in consciousness since it suggests that the brain appears not to need all that specific data passing through it to be alive and aware. It often just assumes data when it is missing, for example, the brain fills in gaps in the visual field where receptors for that area are missing in the retina. Some neurologists think that we only see in generalities anyway which makes one wonder from where the brain fills in the details (think Winston Smith). Especially when we know that while the brain spends time concentrating narrowly on events in its visual field, it can simply be blind to other events also occurring in the same field of view (the missed gorilla in a group of people playing ball). Many perceptions seem to require advanced knowledge of what the brain is looking for before they can work properly. Focussing the retina's image, for example, seems to require in some way not understood knowledge of what the brain wants to look at before the focussing can reveal the object clearly (seeing how the eye jumps about while we do tasks like play the piano is instructive). So the brain, for most of us, appears to be an incomplete and even discontinuous register of the environment it is immersed in, but which is not what we perceive in our consciousness.

Some people can remember (what appears to be) everything and can fully recall any event of their past. Some have eidetic memories where they can review scenes they witnessed as if they were present azagain in the moment of happening. On the other hand some people don't have a visual mind's eye at all and cannot recall images. They seem to remember in words and concepts (an interesting fact to add to our argument in **Lies and language, fallible narrative and dreams**).

If the brain does in fact record such an extraordinary level of detail every waking second of every day, then the problems of processing, contextualising and storage become insurmountable in a very short time if pragmatic calculations of bits and nerve firings come close to describing the brain's power and limitations.

Working out the scope and limitation of the brain is difficult since processing in the brain's centres waxes and wanes depending on what is going on in the world. Concentration calls for more oxygen, so one can assume that more neurons are firing then than when idle. There are an average of 100 billion neurons in the brain each of which can have connections with a number of others and thus there can be in principle a very large number of pathways storing signals. If we take these networks at their face value then there are sufficient numbers of possible networks to supply all the processing power required for perception, for memory, for rational thought and intention, for consciousness as well as for all autonomic functions of the body, but this also means that each neuron must participate in more than one network simultaneously, perhaps in many hundreds of networks, all building and dying away moment by moment as cognition proceeds. How the functions of the neurons could be separated, say between consciousness and autonomic functions during this process, is at the moment unknown.

Although humans, kilo for kilo, have more neurons in their brains than most animals, certain species of whales appear to have more than double a human's number of neurons in their brains. So absolute neuronal power may not be the whole story of intelligence and consciousness. It is one of the mysteries of the universe how the human brain, in contrast, can be so intelligent, so *conscious*, on so little apparent performance capability when compared with the digital technology it has created for itself.

For example, neurons seem to fire from once a second to 100 times a second, and given the 100 billion neurons in the brain, there is an average

A chemical rocket would reach the head of a mile-high giant long before the nerve impulse from a stubbed toe could reach its brain travelling at the rate of human nerve impulses..

operating flow of 5×10^{10} bits of data arising in a brain per second (or ≈ 6 gigabytes sec^{-1}) Is this the minimum to consciousness? By comparison, IBM's Watson when it won Jeopardy in 2011, and working solely in text, processed 500 gigabytes per second and had a total operating system of 16 TB of which 4 TB of RAM stored its data. And yet Watson did not even work in audio (except to pronounce its answers mechanically), while the brain has to process and coordinate live aural input with moving images as well as be its own manufacturer of the energy it uses. How can the brain do it? Watson had (in 2014) hardware that performs at perhaps 80 trillion operations per second. It runs data through it at perhaps 500 gigabytes per second or 4 trillion bits per second. The human senses pass data into the brain at about 11 million bits per second, of which a maximum of 2000 bits per second seem to be used by it. Reading 25 characters per second requires only 50 bits per second, or between 6/7 bytes per second (compare that with your ADSL connection). But Watson can read through 200 million pages of text and extract their information in about 3 *seconds*. The 100 billion (10^{11}) neurons of the human brain for the data of an entire life seems to compare poorly with Watson's 16 terabytes of memory (16 x 10^{12} bytes). (Big Data computer centres work in petabytes.) The brain is a pretty fast thinker, all the same, and it can effortlessly remember images, recognise content within them, as well as learn languages. Humans competed with Watson in the quiz game *Jeopardy* respectably closely (some say that the fact that Watson was wired electronically to the quiz show buzzer gave it an unfair advantage in reaction time). Indeed, one of the challenges building Watson was how to make it respond *as fast as* a human in spite of the fact that transistors can fire millions of times faster than a human nerve cell.

Even if the brain is reading more of the multiverse than we currently understand, we might expect it to still need the bandwidth to transmit this information around its sensory and processing structures. The neurons in the brain are very well connected and are quite adaptable but they function through essentially slow chemical cycles, and many of the perceptual mechanisms by which it is fed information have low band width. Either consciousness does not in fact need much processing power to produce it, which has interesting implications for our appreciation of other forms of life on this planet (and makes Laszlo's ideas of consciousness in every particle slightly more interesting), or there is a whole lot more to the way the brain works than we have so far examined (like photon information networks between cells).

Brain studies appear to show that memory is dispersed through the brain. The brain's memory in general is not analogous to a computer's RAM. There seem to be few if any sites where a single alteration in a cell relates to a single stored item. No longer does it seem that the neural network idea of recognition, used in computer systems, describes how recognition in the brain works since the discovery of evidence that memories are probably structures inside the cells as well and not just a function of the synapses recapitulating the effect that the event had on the nerve cells. Even within the brain, functions are more spread out that was at first thought. Recent insertions of light emitting molecules into neurons to observe their activity show that brain functions involve many unpredictable areas in the brain. Given what Popp has discovered, one wonders if the use of light in this context may introduce interference with the messaging between cells and even the functioning of neuron centres in the brain. Researchers have already noted spurious correlations in their investigations using this technique.

Long term memories are related to permanent changes in the neural connections of the brain yet the memories stored are not necessarily permanent and can change through shock say, or even through simple recall. Suppose long term memories are related to specific protein molecules manufactured inside cells.

A speculation might go further down this road of mutations accumulating in neurons and ponder if the causes of these mutations connect the brain with flows of information in the wider universe. Mutagenic causes, like cosmic rays and forms of radiation may carry information about the regions of space that they pass through and bring it into the brain through mutation. If this is so, then we might expect that neurons may not have the same chromosome error correction systems as other cells.

How much memory would that give us? There are around 20,000 genes that produce proteins in the cell and some of these can make more than one protein. About 10,000 proteins have been identified in all cells of which 1500 to 2000 are required for cell functioning. Proteins exist, however, in many different ways. A basic protein may be 'decorated' with phosphyl or glucosyl groups, although usually these variants are just produced temporarily on the way to a final product, and they have complicated 3-D structures that may be used to store information. Even so, Savage (*Nature*, 2015) estimates that there could be as many as 100,000 proteins and their variants in a cell. But let us take an average figure of 40,000 proteins for memory (8,000 proteins each with 5 variants) per cell, and multiply this across the 100 billion neurons in the brain, giving us a potential 4,000 terabits (a variant exists or it doesn't), or 500 terabytes of memory. If we add to this the numbers of neuronal network groups that help form memories, and even allowing for doubling up here and there, we can begin to see that Watson can be two orders of magnitude away from a human mind's memory and from helping us understand what is going in consciousness.

At this level, however, we may still be misunderstanding the brain's capacity. For example, if we take the Landauer value for the minimum heat produced by changing one bit of information as $\approx 3 \times 10^{-21}$ j we can see that a 12 watt brain might indicate information adjustments of up to 4×10^{21} bits of information per second. A billion trillion bits per second. Even if the biological mechanism was 1000 times less efficient than this, it still means the brain could be producing a million trillion bit changes per second. Massively outdoing Watson in information capability and indicating how far we have yet to travel to produce an autonomous artificial mind.

It is worth noting, too, that the brain has lots of other types of cells like glial cells, for example, that don't connect to neuron cells or have synaptic junctions yet may play a role in maintaining the environment for the brain cells in adjusting the levels of the neurotransmitters and modulating the rate at which the signals travel along the nerve cells. It is now thought, too, that the area of the brain in which new cells are generated, the hippocampus, and one of the most electrically active parts of the brain, also involved in spatial understanding, memory forming, organising and retention, can even renew some cortex cells.

A further mystery is the recent discovery that neuron cells in the brain do not even contain identical DNA (Lodato, Woodworth, Lee, *et al.*, 2015). Because neurons live for long times without dividing (we keep most of them our entire lifetime), each cell harbours large numbers of mutations, which raises questions about the role of cellular information in supporting consciousness, given that we know that at least some experiential memories are laid down at that level. It is tempting to think that the recursive functions of DNA and RNA (or their 3-D potential to store information) within the cells could have been hijacked by evolution to provide for certain types of memories (distinct from the strengthening of neuronal pathways), and this brings us to subatomic processes involving superpositions and entanglement and what role the multiverse scenario must play in the provisions of consciousness.

In particular, the problem of consistency exists for the mind just as it exists for the universe. We might suppose that the entire hundred billion neurons can flicker incessantly between any number of states formed by different groupings of neurons and their relative strengths of signal but still the problem of how it gets back to a memory state requires (one presumes) a static or consistent address system. Furthermore, we still don't know how the brain can produce a model of the exterior world from summing senses alone which do not seem to transmit important variables of nature like inertia (just how do we throw a baseball accurately, or even run, jump and catch one?) let alone understand precisely what its relationship with the external world is (as distinct from location mapping which we know it can do quite well).

The idea is growing that memories are not really records of events that have happened but are summaries filled in with later material (although one wonders what governs those choices of infill). People's memories of events can be entirely fictitious which also begs the question of what the memories are filled in with (the problem of Winston Smith's job again)? Even if the data were false they would have to be consistent in some way with everything else in the brain. If someone told you that there had been a gorilla in the room (when there had not been), and you set about constructing the memory that made you believe that there had indeed been a gorilla in the room you would have had to have constructed a memory of a gorilla in the room with images of a gorilla that you already had and what it might look like in the room that you knew. (Alternatively, we can consider false memories as being held in language terms rather than in image terms which is to say we can remember context and then supply a list of contents for that context, and that the 'recovery' of a memory of a scene, through interrogation say, is really only recovery of the *language* used in building the list. Perhaps, too, dream images are not imagined images at all but arise simply out of the variations in language used to manipulate the sensations. We will discuss this further on.)

> The Armenian mystic and philosopher, Gurdjief (1877 - 1949), thought that knowledge was physical and had weight. He thought, therefore, that spreading it around made it less powerful and less useful. The more people who know something, the less meaning that something has, and the less it can do. Not a good recipe for mass education. He also believed in grades of awareness, where few people are fullly 'awake' to the world.

The brains-in-vats scenario highlights the problem of the information processing required to create consciousness. By any measure, the reality we seem to experience contains very many more bits of information moving through it than the brain should be able to handle. Either consciousness needs a lot less information than we think or it is founded on another basis than simply processing what the senses give it. Consequently the idea has grown that the brain has a different kind of access to the external world to get the data it needs to be conscious, and that it is, in some way not yet worked out, connected directly to the underlying reality that joins all phenomenon (synchronicity rearing its head again). Studies on consciousness have tantalisingly suggested that during some kind of peak experience the human mind does reach out beyond its apparent boundaries to shared information.

To understand this, thinkers like David Bohm thought of a Universe constructed in the same way that a hologram is formed, by interference patterns, such that each point in the universe contains a information of the whole. Thus comes into being a version of the Holographic Universe (proposed by Hooft, and developed by Susskind and others) and the holographic brain, which embodies a tiny portion of the whole. It fits in (sort of) with new algorithmic topology discoveries where infinite processing of shapes can increase the volume of shapes without new material.

Karl Pribram has suggested that the brain does not produce consciousness itself at all but merely 'reads' it or extracts it from the zero-point field that acts through every point in the universe, echoing Laszlo's thought about the Akasha, and, along with Bohm, conceives of the Universe as a type of hologram where every 'unit' of awareness contains the whole picture of reality in all dimensions and times, and what we humble humans see are only ghostly projections of the larger reality beyond our perception. (A hologram is rather like a monad in reverse.)

Under the classic view of science, our universe can be understood by referring to its parts. (We touched on this earlier, the science of mereology.) The laws of physics that we understand from studying the parts may not actually tell us much about the behaviour of the whole. The matter of the 'whole' and what influence it may have on the parts cannot really be considered with the tools we have. This is one of the purposes behind studying the origin of our universe. It may be a key to our fuller understanding of the relationship of its parts.

This idea, however, finesses a key point about the physicality of a hologram which diminishes the strength of the analogy. A hologram is actually made from a wave reflected from both an object and a pure reference wave. One recovers the information by getting the reference wave to pass through the 'memory' again. Just what that reference wave is in these theories of consciousness, and how we might separate it out from our reality for study is not really addressed in the notion. Presumably there is some universal reference wave occurring at a higher level of measurement than the quantum vacuum of Laszlo. Yet how and where in the brain is such an interference pattern stored for intentional use? Using, say, discrete proteins in cells would be a problem since they suffer from motion and activity and continuous self-maintenance which would surely disturb the interference patterns.

The notion is appealing to those interested in spiritual matters, because it goes beyond the argument for an underlying unity in each individual's apparently separate experience of reality, in the manner of Jung's synchronicity, to claim that every brain connects with the entire universe all at once and at every moment (Bohm talked about humans undergoing 'noogenesis' echoing de Chardin, and referred to the underlying reality as 'holiness'. He called the final version of his pilot-wave quantum description as an 'ontological theory'). This idea, however, fails in spacetime, because waves cannot travel faster than light so that no point within the universe can have anything like the full picture of the Universe, it will always be lagging behind. Local points have a better picture of what is around them, but will show nothing of events far away, and they are always changing as new information arrives in real time. So it is hard to see how the holographic universe actually works locally. Such an idea needs interfaces and information-reducing transformations. It is hidden variables writ large. It takes us right back to the running-the-film-backwards-problem. If my seat of consciousness experiences time going forward then it must be reading off the projections of the zero point field in the reverse order, or the universe is going backwards.

It is hard to find, however, in the everyday action of the brain, behaviour that suggests that *it* is a hologram. Instead of fragments of the brain glimpsing the entire Universe (albeit a fuzzy image) the opposite seems to be the case. The brain interacts with the whole but fabricates separate perceptual items from it. It is easily overloaded and it actively suppresses conflicting data from the senses. The fact that we see say, waves, in terms of particular bits of matter and energy is the *failure* of our brains to see that these bits are really just separate views of the whole. If each brain sees a different bit of this whole, or that each brain has an individual experience of consciousness because of its own very particular physical and local parameters, then the hologram notion doesn't really explain anything of an individual's state of consciousness or why it evolved the way it did.

The holographic universe may be hard to support with evidence but perhaps the *instinct* for it as an explanation is going in the right direction. We have already seen how points of space are full of information from other times and places, that the very now we experience seems to be composed, in part, of information flowing from the expanding horizon of the universe back in time to the origin. Even without ideas of holograms, even considering just the orthodox biological knowledge we have, we can still appreciate that the brain is actually a much larger organ than can be found in the skull, and that consciousness includes processing functions that occur in other parts of the body. Awareness is at least a function of all the organs of the body and their neural networks, not just the brain, and which must contribute to an even more individual structure. Further yet, our bodies are confederations of symbiotic strangers, from the separate mitochondria in cells that govern the energy use to the bacteria in our guts, all reproducing with their own DNA. These gut bacteria, shared out among all humans in varying proportions, influence our appetites to some degree and even our choices of foods. They make us fat or anxious. They give us moods. Do these strangers also connect or interfere with the information network that is *us*, that is, our *souls*? Not only that, it has now been realised that the human placenta does not completely separate a foetus from its mother, and that there are exchanges of cells between them such that mother cells can be found in organs of the child and *vice versa*. An astonishing level of interpenetration of genotypes.

The switching on and off of genes can be shown to be connected to not just those of our parents (amazingly it can be shown that father and mother genes are expressed differently in offspring) or to those events surrounding our foetal development but to the environment experienced by our grandparents and even what they ate! So the past has a continuing influence on the day-to-day person through short-lived epigenetic mechanisms as well as genome inheritance. We can consider the coding for creating a human individual and, one presumes, his or her mind, is spread out in genetic time at least, among certainly three generations.

Our own lively awareness confuses the implications of these facts about the brain. We have the impression that the brain is capable of having experiences without regard to any of the dimensions of space and time as they impinge upon our senses. In fact it is working with time continuously. Even as the nerves of the brain are busy in the present with their molecular operations combining memory, perceptions and instincts, our minds work with pasts and futures, not just with the acts of fact recall but with the re-visiting of complex memories of both social events and personal thoughts, and with the working of tasks towards imagined conclusions, as well as enjoying complex visions of things that don't exist and could not even exist at any time.

Pathologies of the human mind also show astounding mixtures of states, and R.D. Laing was not the only researcher to believe that madnesses pointed the way to the experience of expanded *realities*. Hypnotism is yet another means of suggesting that consciousness and will seem to belong in different levels of reality. Not only are our senses easily tricked into perceptions of states that are not real, expectations can make us behave against our better judgements (that is, our internal dialogue) or without judgements at all; they can make us ill or do the opposite and trick our bodies into ending pain or illnesses.

From this perspective, we should expect that every individual would be living entirely within its own universe and depending on the unique organic decisions it makes moment by moment. Even though the fundamentals of the universe may be the same for everyone – that is, all the options are present for everyone – each different brain is a different 'site' at which a distinct view of the universe is 'assembled' according to individual atomic particle behaviour. How can the multiverse scenario ensure that anyone can be sharing a remotely similar reality as you? How can all those different states at different times cohere into a broad consensus across individuals, families, nations and histories?

This personal and individual description of consciousness becomes no different to coded messages, a cryptogram, rolling forward continuously with fresh messages. How then does another individual decrypt the code and understand the information transferred? We are back to matters of preloading and to what we have, not just in common by virtue of our inheritance, but to what we have actually exchanged between us.

Recent studies are beginning to probe the basic uncertainties that exist in our understanding of how consciousness comes about in the first place and thinking of it as a discrete one belonging only in brains and composed only of observable chemical processes is likely to be wrong.

In *The Emperor's New Mind* (1989). Robert Penrose proposed that since the human mind can handle problems unlike the way a computer executes instructions and can reach (or rather, accept) conclusions that a computer would never be able to, our brains interact with the world using quantum mechanical principles. He even came up with a means they might be able to do this in a later book, *Shadows of the Mind* (1994) involving quantum behaviour in protein filaments called microtubules within the neurons themselves.

Whatever the mechanism, consciousness looks like being a dispersed phenomenon and bound up with the states of atomic systems separated in place and time. Even in the higher level narratives of ENT it is hard to think how a mind might exist in isolation yet still interact with space-time.

Perhaps now is the time to introduce an idea from Chinese philosophy.

Chi and the information limits to the body

As medicine has advanced and come into contact with particular alternative views of nature and health such as acupuncture, shiatsu, reiki, and similar kinds of 'energy healing' methods, it has become accepted that there are useful ways of thinking about the human body not yet dictated by the classical understanding of its systems.

For many researchers a realisation that 'emergent properties' of systems within the human body might be responsible for many of its more puzzling capacities for self-healing has led to a renewed interest in the Eastern concept of *chi*, the living energy by which the things of creation, including inanimate matter, maintain themselves and go about their their business. If the whole behaves in a way that is not governed by any one of its parts, then the question of coordination and consistency between the parts arises. *Chi* is certainly one approach to this question. It has long been considered to flow through the body in well-defined pathways or channels according to the principles of yin and yang, a notion of relative qualities that orients its flow.

Once *chi* and how it flows is understood it can be directed and manipulated in the body and is principally used for healing, though its control can also give adepts strength in action, resilience in defence, stamina in work and contentment in daily life (the goal of martial arts).

Recent studies are beginning to reveal how tissues in the human body are connected by overall fields of electrical energy that affect not only the carrying of actual electrical messages by the nerves as well as the basic electrical actions of muscle contraction and relaxation but also the behaviour of bundles of connective tissue found all over the body called *fascia*. It has also been shown that the ancient *chi* channels of the body, easily traced through touch alone, can be distinguished by their lower electrical resistance.

Not only that but earlier studies by the Korean Kim Bonghan (1963), only now being recognised, suggest the existence of a physical network of filaments within the body that contain special fluids and stem cells as well as transmitting photons like fibre optic channels. The German researcher Fritz Albert Popp has shown that DNA can emit coherent light waves which he calls 'biofotons' with which some cells can communicate. It suggests that in addition to nerves and other forms of communication between cells, there is an additional communication network of information throughout the body. Penrose's idea is not such a remote one after all. (Recent developments on 'optogenetics' in the brain show how light can be made to flip genetically prepared neuronal circuits, and new psychological therapies (and treatments of pain) are being designed around this technique.)

It is also true that molecules like DNA, RNA and the other molecules involved in the translation of genes into proteins within the cell project electromagnetic fields into their surroundings and whose shape is determined by their atomic arrangements, so it has been suggested that this field helps these molecules 'find' their partner molecules in the molecular 'soup' within the cell. Anything that would act upon these fields may have noticeable effects at this molecular level (especially if we think that changes in proteins is (sometimes) a driver of genome changes).

Other studies appear to show that the fields of bodies extend beyond their physical limits. This should not be surprising to us, after all, the electrical activity within brains have been 'read' for scores of years by machines with sensors that are not in the brain but on the surface of the skull.

More even than the local fields of human bodies is the discovery that the human electrical field can resonate with the shock waves around the earth made by the discharges of thunder storms. We know from studies of diverse animals like pigeons and sharks that evolution has made use of electrical fields and even the Earth's low intensity magnetic fields to develop organs of perception, but its influence in the evolution of human organs has not been so clear cut, although there is some evidence that a sensitivity to magnetic fields can be found in the human face, nose and eyes – which is perhaps what one might expect in any bilaterally symmetrical animal that has a front and a back and needs to get home.

Our bodies and brains also envelope 'points in the universe' and even though it is unlikely that a single probability wave function initiated from 'outside' the person can penetrate through the cloud of atoms and molecules of bodies to affect an outcome in say, the nerves of the brain, without decohering, there will be, all the same, so many of these waves every second that some may penetrate – indeed, will penetrate – through into the brain such that that external information over and above sensory input will supply trends over time. (Uh oh, I thought we had dealt with Hume!)

Mach was a physicist and philosopher who wondered how inertial frames of reference could be isolated from one another. He suggested that there should be some kind of physical connection between the remote parts of the universe and any local or nearby frame of reference. Mach was talking about acceleration and rotations but others have taken his 'suggestion' to mean rotations are felt everywhere.

Mach's suggestion about the whole universe rotating around a frame of measurement, and which has been extrapolated by some physicists to suggest that the distribution of matter in the entire universe is involved in any local feature, is not as far-fetched as it seems, and it certainly can find a place in the recent work on consciousness and multiverses. We are proposing, after all, the notion that the boundary of the Universe is still playing a part in the arrangements of probabilities in the here and now. We should remember, too, that humans are maintained by similar atomic actions and fields that are also found at large in stars and galaxies. We are approaching an understanding of living beings embedded in their whole universe, although other physicists reject this hypothesis since it contradicts the principle of autonomy, a belief that distant influences do not act upon experiments, and that any experiment can in principle be replicated exactly.

Added to these revelations of connectivity between the large and the small is the effect of environmental coincidence on consciousness and as a countercurrent to sensory summing probabilities, all of which demonstrates that human consciousness is composed, certainly in part, of events beyond its local boundaries and reach of its corporeal senses. There is a necessary reciprocal to this conclusion, namely that far-off events in the universe will, eventually, be composed of a bit of human awareness.

The more we know, the more we are obliged to consider that consciousness is built upon many structures and systems and may reach out beyond individual Earth-bound bodies into the social *oikumene*. How far it may reach is still an unknown but we can calmly consider the notion that each individual consciousness may be spread as a component of time and space well beyond the here and now, and where the on-going decisions made by human consciousness seem to involve, therefore, real-time information from a broader base than just the neuronal pathways or an organic brain.

A unit must resonate in some dimension with all other similar units and this resonance among them persists to guide the formation of new units. Since there is no class effect without units, the first unit becomes a member only in retrospect, while both the class effect and the stability of the unit should strengthen over time. There is a chicken and egg problem here. Before the member of a class can be a member of a class it has to be a member of a class.

The more we try to define and draw boundaries around the working 'parts' of the universe, the less these boundaries turn out to have substantial meaning, and the more we are forced to accept that any system we define will end up having a wider distribution of causes and effects than the initial definitions recognised.

A number of philosophers have grappled with the problems of the whole and its parts and what relationship there could be between the two notions that would provide for consistency between parts as well as their development and evolution within the whole. How and indeed why, parts continue to produce coherent wholes in a dynamic system is what bothered Eleatics like Zeno as well as polymaths like Leibniz and mathematicians like Cantor.

The artificial-ness of the boundaries we create are readily seen in the paradoxes of sets and the elements that 'belong' to them. A system may need boundaries to give it meaning, but it is also going to have a form of interconnectedness independent of its subsystems, its elements. While systems and sets need pre-loading to make them real, they also need something else to make them work — recursiveness, and this we will examine in a moment.

Boundaries to individuals

We easily see a porosity to the boundaries of the individual and the weakness of the notion that physical separation is sufficient to isolate one mind from another. We know only too well how minds communicate with minds through both physical and abstract means — language is only one of the ways by which conscious states can be communicated to another. We say that states of laughter or happiness are infectious, and this appears to correlate well with activity on social media sites. Not just moods but action and intention are brought on by mere membership of a crowd. Riots and lynch mobs are all about behaviour not seemingly predicated by 'summing' the individual personalities alone. People who are dissimilar can still come together in groups and allow the collective attitude to invade and control (for a time) their behaviour.

Anecdotes from the beginning of time suggest that minds can interact with minds far apart from each other and transmit information. Studies with ECGs as well as MRI scanners seem to show that not only can isolated groups of neurons stimulate each other but the activity in neurons in the brain of a distant person can be correlated to the neuronal stimulation of a subject.

Researchers have studied twins and in some cases sympathetic reactions to stimuli have been seen to occur. Where one twin has been wounded the other feels the same sensations.

This kind of sympathy is often attributed to the basic wholeness of creation whereby access to another level of that unity beyond the individual allows a mind to access the experiences of another, at least for a short time, and which explains psychic mysteries, Jung-style. The neuropsychiatrist, B.E.Schwarz coined the term *telesomatic* to label such behaviour. It appears to work most notably between those who are emotionally tied in other ways, such as between husband and wife, between lovers, or close kin.

Such neural correlations go beyond altruism and yet may arise out of conventional evolution's solution to the Prisoner's dilemma which exposes the problem of how to act in everyone's best interests when unable to communicate. These correlations of behaviour are what provide for trust in human interactions and solve the problem of empathy. Those who think the same or who are in love say, can trust decisions made by their partner because the jointly perceived common good is the guiding principle for each and does not require continual referencing to maintain. It is only the *ad hoc* 'teams' who have no special interest in preserving their relationship with each other who are 'caught' in the trap described in the Prisoner's Dilemma (where limited *self-interest* is behind the purpose of the team).

As we have described earlier in this book the apparent 'connectivity' that we find when we examine complex social interactions between people may be the result of the combined effects of both the trend to select as partners

There are some problems with the Prisoner's Dilemma scenario. Most significantly it depends upon an authority hiding information from the players, or at least keeping the players apart until decisions have been made. In situations where players share all the information the dilemma doesn't exist.

those who are straightforwardly similar to us and the underlying attraction of coincidence which may then arise. Humans also have an innate desire to mimic others which means that often they are obliged to suppress whatever differences there may be in order to join in with those others and to decide what social responses they will give to a stimuli, making them look more similar than they are. Humans' responses in a group or a 'team', therefore, will also have a component of self-conscious synchronicity on top of the similarity of types and the connections of coincidence.

So we can observe the extended reach of consciousness brought about initially through evolution and the influence of the Coincidence number of our universe. We have not yet closed in on the seat of consciousness, but we are close to answering the question, Does the full content of our consciousness arise and reside only in brains?

We can approach a way to answer these questions by thinking about an individual brain without a body.

Brains in vats or the external narrative time problem

Consciousness appears to arise in a physical brain grown in a similar way to an arm or an eye, so our mental activity, whatever else it may include, needs to conform to nervous processes of a brain that appears to observe events in sequence, absorbing causation linearly. Since we are of the world, the narrative time external to our minds must meet that in our heads. It would be very strange if human beings were constructed with an entirely different narrative in their heads. What would then be the function of the narrative of the external world?

A diabolical agent might not know what it was doing, but it still could not be passing random impulses into our vat brains because that would mean we would be constructing the order of the world for ourselves in contradiction of the notion of a vat brain. It could, I suppose, switch impulses from another universe into ours, but still, it would be using an already existing reality. It would not be fooling us about reality, only about the *version* of reality we experience.

People have pondered this problem. It's often referred to as the 'brains in vats' problem (or to lovers of cinema, the *Matrix* dilemma and the blue pill). How do I know that the external reality doesn't really exist and that I am just a brain in a vat fed all the nerve impulses necessary for me to believe that I am a living, happening, human being in that reality? Quite simply, for any diabolical agent to know what impulses would be required to make my brain think that it was walking around shopping in London, it would have to have observed a precisely similar reality somewhere interacting with a brain like mine in order to feed me the right impulses to create that illusion.

If it were feeding just random impulses out of which my vat brain creates the dream of walking around shopping in London it would not know that that scenario was happening in my brain; it would have no control over what my awareness was telling me. It wouldn't know any of the experiences that I would be constructing for myself, or be able to control them, making the simulation pointless. The agent would have to conduct experiments, peek and poke the brain to try and find out what was happening to its feed in my brain, rather like the way neurosurgeons do now to humans. But, more pertinently, there would be no way *I* could construct a *consistent* view of myself behaving in an external world, if the agent's data feed were random impulses.

So, the external world that I experience must exist somewhere in some form even if I am a solitary brain being fooled into thinking I am in it because the illusion could only be derived from a reality.

The brain-in-a-vat thought experiment points to some interesting issues about artificial intelligence and the nature of consciousness. It assumes that all there is to consciousness is sense data, a dubious assumption as we will see.

Where does the diabolical agent (DA) get its brain from? The brain would have to be 'consciousness-ready'; made or grown in some way to be capable of being aware without actually being aware until it gets the DA's feed. If the DA takes

> One could consider that consciousness exists as an independent system dormant in the brain, and could be triggered to take control of the brain-in-a-vat by impulses sent into it, but then, of course, it would be independent of the brain, something that lived apart from it, though acting symbiotically, and nicely confirming the external reality.

a human brain and wipes it clean, then it is using one that has already been conscious and one grown in a reality suitable for the reality it experiences and a proof, therefore, of that external reality. We can see at once that it is not just the perceptual feed that determines the nature of awareness but the cellular layout of the memories and the networking between them – the physicality of the brain. If the feed doesn't 'fit' this construction there will be confusion or failure to 'compute' episodes. The awareness generated would be less functional or non-functional (imagine the consciousness of a human in a dog's brain?).

So any conscious brain-in-a-vat could conclude the following: *Any consistent reality I experience has to conform pretty closely to at least one reality that exists and this reality is independent of any attempt to fool me.* And if it comes to deciding whether or not we have made artificial intelligence conscious, hearing this conclusion from it would certainly satisfy the Turing test. There is another conclusion we might expect it to make that we will discuss in a moment after we take a look at another version of the brain-in-vat-scenario.

Mind as simulation

A number of people claim that the chances are high that we do already live in a vat of sorts, e.g. Elon Musk, and Nick Bostrom from Oxford University and Future of Humanity Institute. They think our reality is highly likely to be just a simulation, one of the vast number of possible universes produced by some vastly superior intelligence during its quantum computations. Such quantum computing would increase the numbers of universes in the multiverse to such a high level that the odds we live in a non-simulated universe would be very small.

Other thinkers have taken on this theme and suggested that our universe must be the result of some kind of artfully constructed simulation because of the existence of mathematics (used to construct the simulation and existing therefore outside our universe) and the unbelievably strong correlation between mathematical statements, seemingly invented just in our minds, and observations – or, more strictly, experiments. The difficulties of creating such a simulation, however, appear almost insurmountable. A full quantum simulation of more than a few hundred particles is way beyond foreseeable human capabilities.

Curiously, our attempts at making an artificial intelligence will probably lead us into making a brain in a vat, that is, one without autonomy or control of its environment. Even so, once an artificial consciousness recognises its own self, then it would separate from the contrived artificiality of their existence and recognise their awareness as distinct from it. Whether they could ever reconcile our presence in it with their own is open to debate.

All these simulation ideas nicely rely on the existence of a yet more complicated narrative than our universe, and external to it (as we saw with the coincidence argument), to run the simulation, and they suggest an origin for the simulation (hinted at by many religious ideas such as the 'made in our own image' argument, or that life is some form of test of our conforming to 'principles' laid down – the universe is a big proof of concept trial), although they also suggest that we are somewhat helpless within the simulation and that everything we do may appear to have random choices but cannot in fact have them because we are formed by the working out of only one of the possible solutions to the problem posed in the simulation (thus clarifying the 'free-will' problem – there is none). Other solutions are no longer components of our world and belong in other universes and with which we have no contact (see **Monasmotic universe**).

All these multiverses would be fully functioning realities. So if we are in one of them in a quantum computer, our universe is still a complete and real one, and we could happily go about our business discovering what makes it tick, only outcomes in it would have more design attached them, and thus hidden variables would exist because our universe would be associated with a particular set of parameters derived from the problem *outside our universe* which the quantum computations would be trying to solve. (The Anthropic Principle gives way to the

Simulation Principle and suggests only the existence of a great quantum computational machine.) This argument solves the problem of the existence of a set of universal constant combinations of which our universe is one at a stroke, but invokes a yet more serious difficulty.

We cannot know if the simulation we are in is (or will be) a solution to the problem being examined, and, therefore, ending up as part of that reality which posed the problem in the first place, or be merely the exploration of a possibility that will be *rejected* as a solution in the simulation setters' timeline and cast aside.

<hr>

Douglas Adams had much the same idea in his book, *The Hitchhikers Guide to the Galaxy,* where the Earth and all the life on it formed an enormous computer designed by higher beings to discover the question about life the universe and everything to which the answer produced by an earlier computer was 42. If multiverse splitting occurs within the computer than each split would contain the computer itself, which means that somewhere in our version of the universe is a version of the actual computer creating the simulation.

<hr>

We could call our simulation setter the 'Creator', but let's call it the 'Great Computor'. Since there are many, many quantum possibilities explored by the simulation, there is a real likelihood that we are not, in fact, what The Great Computor had in mind or wanted to see when the simulation was begun. Furthermore, The Great Computor would know that out of all the universes created in its examination of problems through simulation, only one of them (or a few interacting with each other) will end up belonging to the solution and the others are temporary or spurious versions with no final meaning. A very real reason, it seems to me, to ignore anything we might imagine to be The Great Computor's plan for us in this world since the chances are we are not part of its intentions (or hopes and dreams) but merely a supporting option, even a *conjugate* or inverse (which probably explains a lot). The whole universe we see and all that we have worked out about its workings is likely to be just the results of a computation that is seeking for its end point an entirely *other* kind of information or arrangement of parts.

This simulation notion solves the 'god problem' completely in that we do not have to appease a 'creator' (The Great Computor) who doesn't know what we are about or whether we are, or will be, part of its plan. Trying to divine its plan is absurd, as are those who claim they know it. We can safely ignore him, her or it — if the simulation scenario is true. The Great Computor doesn't know *us*.

This idea of our universe being the result of a quantum computation, however, requires that our universe is a version of *something* already existing. As a version, it could be interfering with other near-copies in a massive superposition. However, this interference is precisely *not* what has been calculated in the quantum computer (otherwise there would be no point to the computation). This means, in the

Curiously, something about the Creator having no information about us specifically, has been anticipated in Hindu philosophy. There is a deluding power within the Creator's soul called *maya* meaning ignorance. The astounding intelligence of the Hindu philosophers deduced that limitless awareness had to include the power to be blocked in some way from being so aware. The Hindus characterised this blocking power as some form of subtle and inert matter that makes the Creator soul forget its true nature and behave as if it were not the all-aware Creator, and which allows it to create the world (as *Isvara*) of potentiality, where all objects lie dormant before coming into being.

nature of quantum calculations, that even if the substrate of the universe we observe has been calculated by some computation or other and resembles very closely what is going on elsewhere (another version), our universe still contains probability, and the individual moments within it or what we are thinking or how we are behaving cannot necessarily be predicted by the Great Computor. It is also worth noting that any 'mind' formed by quantum computing would only 'see' the superpositions of probabilities, the whole gamut of time-lines all overlying each other and not any specific time-line. It could not *be* within our specific timeline, for example, and experience what was happening to *us* (unless it dropped out of the simulation).

The reverse, however, might be possible. We have seen how coincidence plays a role in the creation of our universe and in the development of consciousness. In which case, coincidence will lead us into the workings of the simulation 'device'. Coincidence tells us something about the organisation or internal 'layout' of it. To us coincidences seem nothing more than non-computable elisions of narrative, yet not only do they tell us something about the numbers of 'tables' laid out for observations of probable outcomes — the probability volumes or Coincidence Number of the universe — they are also, no doubt in higher levels of reality, signatures of algorithms providing short cuts or even actual solutions to combinatorial results.

We can consider also the points of attachment of our simulation with the other simulations and how they are connected to The Great Computor's machine. Our universe must be connected in

> The silly parochial disputes that various religious sects have with each other about which text or tradition is the 'true' one disappear into mist when we have to think about which *universe* in the multiverse is likely to be carrying the 'true' plan for creation.

some way to the overall function of the machine. Input and output information have to flow through it. The beginning has to be prepared while at least the ending has to be read in some way and evaluated, or values read off, as the Universe progresses. There may be zones of high coincidence or reduced uncertainty. Perhaps Black Holes are points of attachment to other simulations or simply where the algorithms finish in a readout for the external reality and thus where information from the simulation gets extracted. The recent observations of black holes merging may also be involved in the higher level process of reading output from the simulation of our universe.

If we take coincidences here in our time-line, in ENT, more seriously then we might have a realistic method of raising a flag to the Great Computor, or in other ways influence our behaviour to make the Computor understand that we are active elements in his simulation; that the brain-in-vat is coming to its senses at last. On the same reasoning against messaging ETIs, however, one might be cautious of drawing attention to ourselves in the simulation.

It is this thought experiment about brains-in-vats or brains-as-computers that enables us to see that consciousness is almost certainly more than the mechanical result of a working brain, and necessarily extends beyond the confines of an individual body. It must in some way still incorporate a continuing interaction with the forces of the larger environment that evolved it, whether or not that environment is a simulation or not. Further light can be thrown on the mind-body problem by considering Boltzmann Brains (see **Boltzmann Brains**). For the moment we will stay with the discussion on simulation.

The simulation argument conforms to the strong anthropological argument, but are we likely to ever get to know the Great Computor or the Diabolical Agent? What is so interesting about modern cosmology is that it brings new logic to bear on Stone Age philosophies that some still cling to. Let's take a brief diversion here to the ontological proof of god where we shall find a key conclusion about our creator which would certainly help an artificial intelligence prove to us that it was sentient.

Thinking it up

The inability to detect circularity is a particular illness of religious philosophers, and necessarily so (*pace* Gödel) because 'proofs' of things that don't originate in our world can't really be anything else but circular.

The very name of the following argument betrays it — the argument *from being*. There are many examples of ontological argument; Gödel himself produced one; David Bohm, after much dabbling in spiritual matters called his pilot-wave theory an ontological one. The argument goes like this. Since we can imagine a perfect God, and real existence is both greater and more perfect than imaginary existence, a perfect God must exist. Now, if you don't feel there is something suspect about this, then you are probably a priest or an imam, although many philosophers, like Leibniz, have also followed this line of thought.

It is fairly respectable in philosophical circles to make an assumption and then prove it, even though the assumption remains, which is curious in the case of Gödel, since he was the one who managed to prove that we know that there are things that exist but cannot be proved to exist. One wonders why he cared about going in the opposite direction by trying to prove that what we can't know to exist actually exists.

Gödel enlarged upon the ontological argument by adding a sort of multiverse proof to it, saying that since God is possible at least in our world (because we can imagine it here), it must be possible in all worlds from the perfection assumption, namely God can't possess negative properties (i.e. he can't exist in some places and not in others), although the brilliant Hindu philosophers got around this problem pretty well (with their notion of an inhibition applying to god's contemplation of his perfection, that is, that sometimes God forgets that he is perfect or just doesn't see it — presumably at random moments otherwise that would be a systematic lack of perfection).

The truly idiotically beautiful thing about the ontological argument is that even if you deny a perfect creator, the fact that you are thinking about it at all proves that it exists because if it has the property of existing in the mind it must posses the greater property of existence outside of it. We used something similar when discussing the brain-in-a-vat scenario.

The argument works the way it does because the assumption of perfection – the absence of all negative traits – is the highest trait a god could have. We can imagine lots of beings, like the Greek gods or Voldemort, but because they do not possess the ultimate perfection they are not impelled to exist. Only the ultimate being is perfect hence he must exist.

Thinking of the ontological argument this way gives us a scale of degrees of reality for our imagination, all the way from hard rock to evanescent beings, measured by probability. (In fact we know this intuitively. It's hard to imagine hard rock precisely, but we can imagine all sorts of abstractions when they are free of qualities of definiteness - we call it thinking.)

Try it another way. If the chance that god exists is say a million to one, then that is a pretty weak god since it can hardly realize itself. A god whose likelihood of existing is say 10 to 1 is a stronger god. One whose odds of existing are 50-50 is stronger yet. The strongest god (i.e. the Creator) is one who exists.

This is interesting from the Shannon entropy information point of view because it runs completely in the opposite direction. With Shannon, uncertainty means more than certainty. It seems that the highest informative situation is one whose existence is the most *unlikely* to exist. The ultimate reality is the one that contains all likelihoods and especially the most unlikely. Here again is the paradox we touched upon earlier, the more *unlikely* something is, the more it must exist, if Shannon entropy describes features of reality, which is to say, as long as the reference points have been pre-loaded.

Our brain-in-the-vat thinks about its existence through this far. If the most unlikely thing is more substantial than near certainty, then, for God to exist, its presence must be extremely unlikely in our reality. In other words, what is pre-loaded into our universe will not contain God. This agrees with both concepts of the Diabolical Agent and the Great Computor, as well as the existence of a (relatively familiar) reality in place before our universe began. Thus the brain logically suspects that the prior reality (of the DA) must exist (from the ontological argument). Yet, the brain argues as we do here that the DA must not include actual information about himself in the feed to the brain otherwise his informational power reduces (*vis à vis* the brain) and he will cease to exist, conceptually and factually, as the fully powerful DA since it suffers restraint to a lesser or greater extent by the demands of the created reality in the vat brain. On this basis, even if we were in a simulation, nothing in the simulation should give away the 'higher' reality of the Great Computor who set the simulation. The Great Computor has to remain completely outside the pre-loading for its powers to remain all-powerful, with respect to us.

So, instead of thinking about simulations, let us think about our reality on the scale of uncertainty, which is effectively the same situation. The god whose existence we may deduce we cannot know and neither can that god know us, for to do so would be to break the scale of probability. We must not be able to perceive God if we want whatever god is to go on being God. (And thus we find ourselves walking down the road of undefinability investigated initially by Gödel and Tarski, namely the concept that some assertions in any language can cannot be assigned a truth value.)

To add some emphasis to this idea we should recall Pascal's Bet. Pascal took the ultimate pragmatic attitude and reasoned that the benefits from believing in god are so great (everlasting paradise) that we should believe in god whatever the likelihood of it existing. Similarly, as the harm you might get from not believing (hell and damnation) is way out of proportion to the slight 'discomfort' in believing it, you may as well believe it and be assured that you are rewarded if it is true and lose nothing if it isn't. However, this is not quite as easy a bet as one might think.

Since the creator cannot reveal its true nature to us (as we argued above), neither can it reveal principles in our world that we could count as evidence of its existence. So the benefits of paradise that support Pascal's bet would be quite unrelated to behaving in a way dictated by anything we might take to be god's revealed moral preferences. Any inferences from them that guide our behaviour would be at best whimsical. *God cannot let on to us what he wants us to do, otherwise he doesn't exist.* Now that sounds like a sentence Gödel might have come up with.

It is interesting, in this context, that the multiverse scenario also makes nonsense of the traditional revelations about good and evil if the god we imagine *is* the one that exists. So let us take a further small diversion in moral philosophy, and consider the concept of a final judgement of souls and the reward of paradise, and see how a simulated mind or a brain-in-vat, or an artificial intelligence might reflect upon their moral universe and by so doing give us another way of knowing whether or not they were sentient.

Souls and morals in a multiverse?

A traditional picture of the human soul given to me during my religious education is that the good and bad deeds accumulate in their respective 'ledgers' and that on Judgement Day, the good and bad are weighed in the 'balance' and whichever side is the heaviest determines whether you enter hell or heaven. All religions hang this kind of threat over their adherents: for Christians, it is whether are you a sheep (good) or a goat (bad). Nowadays, whether or not Hell or eternal punishment actually exists for the 'goats' has been finessed away by most theologians. The idea of

If the multiverse scenario is true for decisions at the human level then presumably morally good and morally bad decisions exist in a kind of superposition of opposites (like up or down spin), and a good decision creates a worse universe to split off from it, while an evil decision leaves a better universe to split off from it. While all universes sum to morally neutral from the outside, the more good a person is in one universe, the worse he or she makes the split off universes and the worse off the people are in those universes. Similarly, every really evil decision made by an evil person creates a split off universe that is *less* evil. In fact all multiverses derived from good people's decisions may be a whole lot *more* evil than those of actually evil people because the good is being *removed* from the copies by their decisions here. How evil is that? Perhaps that is the reason or the excuse that Hitler has not been removed by time tinkerers. His removal from our history may be a good thing relatively, but it will suddenly create vastly more evil in universes elsewhere.

the torments of Hell did not really exist before the life of Christ and it has taken more than 2000 years since its invention for the Catholic Church to say that Hell is really just a metaphor for life without the presence of our creator-god. Whatever Hell is, certainly Heaven or paradise is being with this creator-god, and, enjoying a delightful, if somewhat dutiful (depending upon which interpretation you go for), existence in its presence. (In this regard, the Catholic Church is keeping rather quiet about the implications of its belief in the notion of the resurrection of the physical body. Since we will have our transfigured but very real bodies back in Heaven, presumably we will have them in

Hell too. Whatever Hell might be, then, it is certainly going to be experienced physically. (Uh oh!))

The arrival of quantum thinking and especially the multiverse scenario of universe separations, however, has given a new twist to the Judgment Day written into the religions of The Book. If you subscribe to any one of its religions, this twist is going to give you food for thought.

Suppose the principles of good and evil can be equated to values of energy in a system. Making a good choice is harder because it requires more processing time, reaches beyond the self, its consequences have higher impact. Making a good decision is energetically hard but it raises the energy levels of the system and allows for more evolution, whereas an evil decision is more local or selfish, easier to make and does not provide any energy of reorganisation for the system. An evil decision, because it is easier, allows probabilities to decline and where there are thus fewer choices (or happenings) in the system. Making an evil decision in your universe implies evil things are even more likely to happen. Evil becomes determined more easily than good but has more self-similarity than good. Evil is always the same whereas good provides more choices, more movement, more liberty, more creativity.

It is does not seem likely in the multiverse scenario that simply making a choice between distinct options at the human narrative level creates other worlds. As I mentioned before, having choices in life is not at all the same thing as there being a superposition of random outcomes in the quantum world, and it is a mistake to think that the moment a human has a choice of actions there are multiverses springing from that nexus. (We have been arguing for something else which is discussed in the next chapter). Since the whole point about moral choices is that they do not exist in random systems, the presumed paradox of the multiplication of moral outcomes housed in separate universes doesn't actually happen simply with the existence of choices.

If there are any moral paradoxes in the multiverse scenario, then they come from the fact that, because of the existence of a set of random outcomes, there are many *yous* of the multiverse all living on and making moral choices in the human narrative in their worlds.

Even though each world separating off from you begins with entirely the same person as you with entirely the same history, ultimately each of your *yous* may take decisions that happen to have different moral outcomes as the worlds of the multiverse evolve (certainly *some* of them will), either out of random environmental circumstances leading to subtly different significances to your acts in some worlds, or out of random differences in your own brain. Where, in one world, you have the best possible outcome to a decision, in others, worse outcomes, and in some, you will make evil decisions, all depending upon the shades of moral difficulty random forces infuse into your actions. Unless, of course, your innate soul is already unchangeably good (where it always takes the best moral decision) or already unchangeably bad, in which case there will be no surprises in any of the many worlds you are in. The question then becomes, are all these souls still *you* or should they be regarded as entirely separate souls with their own eventual final judgement independent of yours? Either there are as many paradises and hells as there are versions of every good and bad individual soul (multiplying exponentially beyond count), or there is going to be some confusion in the one paradise and the one hell where the exponentially multiplied number of good and bad versions of you (and everyone else) will be congregating respectively.

Initially, one might think that all souls must be unique and that all multiverse souls should be treated as independent of you. Then their fate is irrelevant to you and how you come to be be judged. However, is this fair? These souls come into being with precisely your

Worlds with you in them have been multiplying since you began life. Perhaps you have an irreducible value of good in your soul that sees to it that you are always as good as you are destined to be in every world, such that any of your selves would representative of your moral condition and thus the creator would only need one of you to judge whether all of you make into paradise or not.

capabilities for good and evil at that moment. The differences which make them distinct from you come, as noted above, from randomness in these worlds and not moral choice. Some versions of *you* might become more good than you and others worse. Their later fate in the multiverse is still affected by what you have bestowed on them at the point they originate and does not therefore come entirely from their own choices built up over their life. Is this what God intends for each soul? Free will works for one soul, i.e. yours here in this world, but not for the others. In any event, how do you know you are not one of these 'other' souls, with a given history from a multiverse separation earlier and turning good or bad by random forces from a start that is not your own. There is no way you could know. It is a version of reincarnation where you set out with a life already slanted in one direction. Is that what our creator intends for us?

Having the fate of your soul decided by behaviour in other worlds you have no control of doesn't seem fair either. Good or bad, there is nothing you could do in this world to alter the final balance since the other souls will always outweigh you. Could this really be what the creator-god intended?

> For a time traveller to 'correct' an immoral choice, would that be moral or immoral?

On the plus side, given the almost infinity of *yous*, you could do anything at all in this world; be very, very bad, and the contribution to the total would be negligible. Of course, every one of your multiverse souls might have inherited from you the same thought, and thus casting you, and all of your *yous*, inevitably into hell anyway (an upgraded Calvinist scenario). The position you are in, in this second case, is rather similar to the well known problem called the Prisoner's Dilemma, only you are playing it with yourself. You daren't leave yourself open to being damned. Further, in considering the total balance, how many venial sinning *yous* would it take to balance out a mortal sinning you? In my education, none. So if there is, somewhere in the multiverse of infinite time and space, just one version of you having committed a mortal sin, you are going to be damned anyway you slice it and even if you were good in your world. Can this be what the creator-god intended?

By now, this is beginning to sound a very unpalatable state of affairs. If you trend towards the good, randomness can do you in anyway unless the creator-god fixes the equations; if you trend towards the bad, there's no good luck way out for you. The multiverse scenario does not seem able to provide redemption without fiddling the books of randomness. If the multiverse is true then the Creator has to be continuously intervening across the board to undo the effects of his own creation (some people like this idea).

This is definitely not what the Creator wants, if we are to believe any theological reasoning. The Creator does not want the state of our souls to be down to chance. Einstein certainly thought so. So there must be a subtle weighting in the multiverse that ensures that universes inheriting your good soul are extra positive and thus ensuring that fewer evil versions of you are created, while those evil versions of your soul that do arise cannot be saved by randomness producing a balancing good in later separations (because this is not what the Creator wants either, if we are to believe the propaganda: the Creator wants you to save yourself).

Is all moral judgement made null and void by the multiverse? There is a solution in the offing and it is a form of ontological argument, but with absolutely nothing to do with an influence living outside of time. So, let us leave this diversion for the moment and return to the main narrative of mind and body.

Consciousness, memory and reflexivity

The whole of the brain seems to be involved in laying down memories. Even though there are processing 'sites' that become more active than others or use more oxygen when certain tasks are performed, and that we have learnt, from brain-damaged patients, sites of the brain are specific to certain kinds of mental work, memories behave differently. They have a special relationship to mind. They are special because they don't get used up in thinking; they are maintained *apart* from their use, and this makes them vulnerable.

If you don't have memories, can you function at all? How can you manage without the persistence of what works? Indeed you can. This is the basic role of behavioural randomness in living creatures. Life may have a form of memory in its DNA and its programmed responses to stimuli, but this is not a memory of the present moment. So how does life that cannot remember operate successfully in the now? Through random selection of responses. Random selection of an action from a set of actions, like avoidance reactions, is a beneficial strategy where energy constraints are significant and where reproduction is easy. It is a 'cheap' algorithm, using less processing power to produce results. Not needing to preserve or recognise patterns or sequence, it needs little memory storage. For example, it has been shown (by Heisenberg's son, Martin,

> You can perform an informal version of Heisenberg's experiment for yourself. Swat at a fly that settles on the table next to you or on your book. You will miss it and the fly will fly off. But count how many times a fly returns to the same spot or very near to it. The results will surprise you. You can use this fact usefully to swat it next time.

no less) that flies will not give up random avoidance when faced with a threat even when one option is made unpleasant for them.

So, while for the lower creatures of life a random response is an effective way of dealing with unknowns in the environment because it does not need memory stores, higher creatures applying randomness within a rule-based strategy will do even better. (For example: do random movement: if beneficial, continue, otherwise random movement again.). A primitive memory store need only record a departure from a set value, or just remembering only the most recent value experienced. An ant crawling along a trail detects a hormone gradient (intensity of signal) left by others which keeps him going in a direction. It has no need of knowing from where he came, only to seek out the next value in the gradient along his path to the food, which would be say, the weaker signal (the chemicals degrade so the beginning is stronger than the terminus). Once it finds its objective – food – it returns along the path, reading the gradient in the reverse order of importance. It does not need to remember the path, only to have 'switched' his expectation of the intensity of the next signal from lower hormone level to higher hormone level. Thus the ant goes out and returns without needing any physiologically expensive and complicated internal memory of a route, behaving more like Searle's thermostat, and relying on the pre-prep or pre-loading (laid by the other ants) in the route rather than in the ant's nervous system. (Where the ant is the first and not following a trail, it uses the same random strategies in response to environmental cues.) Eventually, of course, as each ant contributes to the hormonal trace, the pathway becomes a constant level of hormone attraction and any ant may travel along it without needing to measure the gradient.

Now we can begin to see why coincidence is different from just any accident in the environment? For an ant acting randomly or instinctually, accidents are all the same. For more advanced creatures, self-awareness provides them with an opportunity to navigate rare but very much more productive paths through their environment by being able to profit from features of it that have greater internal meaning. For coincidence to have a function different to accident the information embodied in the coincidence must be implicit in internal states but not codified in a usable form, perceptible but not already perceived. Thus self-awareness recovers the information in the coincidence, destroying the function of the random and autonomous response by supplying new intention through which it can profit.

The moment when consciousness can reflect upon internal states and recover the implicit information is when coincidence begin to have an effect. This capability in the ant, if it exists at all, is at a very simplified level, because it requires some measure, however small, of revision of response, of recursive procedures.

The acquisition of self-consciousness, on the other hand, puts you always in the centre of events rather than in a place in a sequence. (Consciousness is the set that has itself as its member.) If you look at yourself in the mirror, time does not intrude upon that contemplation. It is a timeless experience, an endless now. Your image, your movement is fed back to you without delay; feedback in real time. Memories sit in distinction to this timeless moment. They have to be somewhere else in time. Remembering draws attention away from the self-aware moment and diminishes it.

Thus, with such intensified self-reflexion, individuals in the modern developed world are likely to have a reduced sense of a future and a low level of vision of future events and their likely role in them. Could this be the reason, for example, that so many individuals do not want children. Previous generations could not have made that decision. For many of past epochs the future was the place where actions came to fruition and safety could be better assured. Such an idea might also explain much of the modern world's difficulty of grappling with the future problems that the global economy is bringing down on itself. Which brings us to the question of what anticipation actually is. Is it really only the expectation that affairs are going to continue and statements made in the present about the future can be compared favourably with the future when it arrives? As the world turns upside-down, living in the present and accepting what comes may make better sense.

And this is where we can place anticipation, our sense of future. We believe in the future because we have memory; indeed it is memory that gives a 'future' to awareness. But this disturbs the self-reflection. We see how a feeling for the future is achieved by the suppression of self-reflection, and when we cease suppression of self-reflection the future gets lost in the evolutionary benefits of reaffirmation of self.

This timeless self-regard appears to reduce future anticipation since one might expect anticipation, or an intention, to be a departure from the encircling mirror of self-regard. We would expect anticipation to arise in the dialectic and contribute to life. But there is a problem. An expectation, seeing as it is virtual, cannot be *maintained apart* like a memory, unless it is a repeated expectation, an anticipation remembered, a memory of a statement made about the future. If it is not such a statement carried forward, it must also exist in the same timeless and weightless moment of self-reflexivity, belonging to the same awareness and continually renewed in it. In other words, not as an idea reaching into the future at all but one in waiting for meaning, for substantiation. Once again, from yet another argument, we pin-point that same mysterious requirement of the present: that it requires a recursive influence from the future for the anticipated to become real, to be made sense of.

Consciousness, networks and negative probability

To accept the idea that everything we have in the present is sufficient to produce the future we would have to accept that the probabilities not only in every process of our world but contained in every individual act are coordinated. Anticipations of consciousness must somehow run in parallel with the world and be confirmed by what emerges. If and when our expectations coincide with the facts it must be because our mind state evolved precisely in parallel with the world to make the prediction true. Again, we are back to the problem of consistency. If the past is fully decohered, from where comes the new probabilities on which the successive moments of time are built?

Certainly, you can dream things to your heart's content, but dreams, although they are real (with some exceptions we will come to), are not about reality, which is why we can tell the difference. Even lucid dreamers can tell the difference otherwise they would be unable to invoke their awareness within the dream.

It has been remarked that if we do interact with the future then where are our memories of that interaction? If we did have memories of those interactions, would remembering them take us to that future? While we ponder that thought, let us think about why we have the ability to predict at all. What does it mean to anticipate the future? Because expectation has no knowledge of the future state, it must have only provisional meaning, if any. How could we hold a state in mind that is completely imaginary and not derived from the physical, either from the senses or from memory? When I say to myself, the sun will rise tomorrow, that is just a statement to myself to make a memory that I said it. When the sun rises tomorrow I remember the statement. Hence all forward looking ideas work through comparison with memory and stability of the predicted world. Our sense of future is only the sense of a 'rule' (like the ant) for which a form of memory is required.

Some researchers think of this 'rule' as similar to Bayesian inference when applied to neural networks. The Bayesian method is simply a way of cross-referring probabilities rather like the way we use syllogistic logic. I am not going into detail about Bayesian logic except to say that it is a perfect example of relying on time-line inertia. It is thought that humans estimate the likelihood of something happening in the future about which it has no experience of frequency by guessing from previous linked experiences – priors, as they are called. Bayesian probabilities work well in weighting the neuronal networks in computers, but, since humans are very bad at inferring from Bayesian probabilities consciously, one wonders whether the notion that the brain naturally uses the method, derived as it must be from environmental experience, is valid. (Later on we will look at the experience of an Amazonian group called the Pirahã who may not reason Bayesian-style at all.)

Certainly, to make an advanced decision that is likely to be evolutionary beneficial, consciousness must be less confident of its state of self-awareness and make more cognitive use of uncertainty. Evolution, however, does not, in the end, select from acts born out of precise memory data or from the precise action of an allele but on the occurrence of an incalculable coincidence, for which Bayesian inferences cannot make accurate estimates. DNA may be the memory of past successes, but unaccounted coincidences are what survival works on by tipping the balance of survival far beyond the equilibrium.

Consciousness, therefore, doubts the stability of the past, and does not wholly rely on it to make decisions. As an aside, if any future computing machine came up with such a statement about its own awareness, it would make it clear that we had an aware intelligence on our hands.

We have been considering how brains, both actually and in terms of probability, have access to information dispersed in time and place so let us return to Tegmark's proposition: that consciousness is a fourth state of matter.

Consciousness, new states of matter and a thought experiment

We have seen that the Shannon information content of messages exists only in reference to the pre-loading available to interpret them. Its value is derived from the differences between the systems required to compose the message and those used to interpret the message. As the mind moves through time its operations naturally increase its pre-loading so that it becomes more and more able to interpret the messages it 'receives' or observes. This is what we call learning, or experience, or the business of life, and the significant structure to this process is coincidence. However, there is still more to add to this view when we think about consciousness and how it is obliged to manipulate time.

Specifically, how does consciousness remember and, indeed, why, if the uncertain event is more informative than the observed event? Recalling complexities of human narrative seems to require mechanisms of mind we have not yet conceived of, and observing a brain suddenly remembering whole events when a fine needle is inserted deep into its tissues does not seem to take us further into the mystery. By what mechanism does a complex memory, appearing to contain vast amounts of items and cross-referencing, remain coherent? Does consciousness simply review the stored sense data, the sense impressions, such that it replays the scene on some kind of internal canvas just as it happened? As we have seen, this kind of remembrance of sense impressions seems to require events flowing in the opposite direction in the brain to the direction that laid down them down.

> Computers don't 'know' about their memories until a higher level instruction tells them there is one at such and such an address. They aren't absorbed into the operating system to become one with the whole process, part of its persistent outlook or overview.

We could speculate that the brain lays down compressed 'thumbnails' which then get 'blown up' to life size with the addition of metadata stored elsewhere (though what kind of compression systems are working in those people with eidetic memories?). But this is a form of a relational database and requires an enormous amount of processing and coordination which multiplies exponentially with the numbers of records. It also suffers significantly from the 'address' problem – just how does the brain know what it needs to remember before it remembers it – and from the need to identify an item (presumably uniquely) which requires keeping a separate track of the location of each item. You know how your (now old-fashioned) computer slows down as related memories get scattered across the hard disk storage over time. There are other ways of storing data that are not so expensive on memory – good for a brain that cannot expand beyond its skull – but none solve the 'address' problem.

While the comparison of digital analogues to the human brain might be useful to start pondering interesting questions, the brain's actual behaviour does not suggest any kind of adherence to a digital analogy; it clearly uses a mix of strategies about most of which we have no idea. The brain forgets that it has remembered something, remembers that it has forgotten something. It mis-remembers things it never knew even as it remembers things it did know incorrectly, and it can recall events as if happening in entirely different contexts to the original, which calls into question all the methods we think a brain might be using to store its memories. It can also create for itself without assistance memories it never had and make them appear real (as in dreams), and it can be persuaded to this through mere suggestion by someone else who had nothing to do with the memory. It can recall entire complex scenes by sensing a single component of that scene, like a smell, while it can be oblivious to events within a scene because it is concentrating on something else in that scene. On the other hand, there are people who are cursed with eidetic memories and who remember everything that ever happened to them in exacting detail (although, interestingly, this 'skill' tails off with age).

So, without defining brain mechanisms for the moment let us follow where Tegmark's idea leads. Let us ponder information from the perspective of consciousness being another form of matter, distinct from the other states we recognise. Those who support panpsychism generally accept this proposition since there can be no advance on dualism without accepting it. But the idea that the 'matter' of consciousness is as fundamental and as pervasive as energy say, succumbs to a number of immediate objections and generates at least one significant problem.

The simplest objection arises out of quantity. It does not seem be the case that simple accumulations of the fundamental 'atoms' of consciousness lead to consciousness in packages similar in size to say humans or dogs or elephants. Even large packages of relatively complex matter and energy like airplanes do not appear to be conscious (some Native American tribes believe that everything, even inanimate objects have their *manitou* or spirit). Something else like say, complexity, must be involved to make the packages of conscious units in humans and dogs *sentient*. Thus the concept of a fundamental unit of consciousness from which our universe is built lacks power to explain the very thing that it is trying to explain (it's not even tautologous).

But more importantly than this, if consciousness *is* a form of matter then there is at least one way it distinguishes itself from ordinary matter and its quantum states by being continuous from beginning to end (from birth of awareness to death). Consciousness is an evolving whole that remains connected to its previous moments within the wave function. When that mind recovers a memory, it is going back in time through its wave function history to the moment where events were incorporated into the brain's wave function. Inanimate matter, on the other hand, does not lay down memories and has no continuous information record of which particle went where and what happened to it. Waves and particles become integrated into the composite. If a particle is ejected again after absorption, it is not made out of the same collection of energies that was absorbed. We cannot say, Oh there's *that* electron coming round again. We can only observe *an* electron appearing. We could never say about a particle that we once observed that very particle in another time and place. We could never track an electron that arrived from outside the Earth's atmosphere through all its later interactions and point to the production of a sugar in a leaf and say, *hey there's that electron again!* Particle states do not preserve that type of information. (When we 'teleport' particles we only 'teleport' the quantum information about states which is then applied to a particle in a different location; it is not the same particle that moves from one place to another.) It is only as we move up the universe's complexity scale from the subatomic to the biological that matter gets distinguished by accumulated identities.

S ome people have suggested that perhaps there is only one electron in the universe and the multiple electrons that we see are simply the same electron flashing about through all times and spaces.

It is tempting to assert that the only other structure that we know which is also continuous from start to finish is the universe itself, (or perhaps just its expanding horizon). That there should be this parallel between the universe and consciousness is not perhaps just an accident, but inevitable. We will return to this theme later on.

Once begun, consciousness is a continuous transformation of the wave function that describes it, and in which are stored not only memories but the metadata of what a memory is and why it might be needed. Consciousness has a circular and paradoxical function (certainly in discrete digital terms). It must maintain itself as itself through accumulations of information and changes in memories and perceptions and it must know about the purpose and function of a memory and how to retrieve it before it is required in mind. And yet we know the reach of consciousness develops through experience and learning even though we can never be sure of the moment say, when the infant brain arrives at self-awareness. (A problem that will certainly concern us as we get closer to the arrival of artificial sentience.)

If the brain not only accumulates its experiences but maintains them as the items get adjusted over time, it has a big processing problem, because the level of complexity should, necessarily, rise exponentially over time. How does the physical brain actually do this? We shall unite two lines of explanation of why cognition is so distinct from the inanimate world towards the end of the book. But first, let us look at awareness and anticipation.

So how is the mind aware?

We haven't a full answer to that question yet. But we can perhaps, return to the original puzzle, just what is the reach of consciousness, just how much of time and space are we aware? As we explained earlier, changes in the past beyond our present moment (often given as lasting about three seconds based on short term memory studies) maintain their constant distance from us as time passes and so are not perceived except on the occasions when memory goes back in time. The fact that we are not so aware of

> It has even been suggested that the brain deliberately suppresses logical trains of thought in order to allow for creative solutions to emerge. We might speculate from where these creative solutions emerge if not from the existing logical connections within the brain which surely have been laid down by logical processes. Sounds like gödels could be in play here. I recall Mr. Spock of *Star Trek* once performed an illogical act on the basis that a helpless situation required an illogical response and was therefore perfectly logical.

the changing relationship with the past in our everyday wakefulness is because these probabilities are woven into the complete fabric of our mind and it is only by the artificial isolation or focus on a specific activity that we can observe a fraction of the changes occurring.

These changes will appear as subtle variations in past frequencies and will be hidden by the layers of neuronal inferences and remain generally unobserved even as they still act upon the mind.

So what do humans really do when they are aware? How do they make their decisions? There is a distinction to be made between underlying mechanisms of brain activity, like neuronal network summing, and the eye(I) of consciousness itself. The action of time influence on how we intend our actions is what interest us the most. The idea that a conscious brain accumulates priors and corresponds between overlapping priors to form a consistent and ever-improving method of making decisions seems far too idealistic. The idea that experience endlessly improves our abilities to make decisions is simply refuted by...experience!

Generally speaking humans, in the everyday world, under the influence of social forces and the various networks of information flows the individual is in, use habit to frame the cognitive space to make and take decisions. Often humans don't know anything about making a decision until someone else 'knows' it for them. Humans do not experience or even understand the causes of enough situations to establish the priors and using the nearest equivalent past event to help them decide an unknowable is inconsistent and dangerous. This is especially true with random events for which there are no priors and which do not contribute to a prior. How our brain forms an awareness of what is going to happen next when it cannot possibly assemble and project an assessment of the near future into the mind's (I)eye is nowhere near being solved. Indeed it is the elephant in the room of studies of mind. To say that we model the world using guesses about likelihoods (Bayesian guesses) does not explain just how this information is formatted and held in the mind to create intention. Here is an actual event to illustrate the difficulty. I approached my car with my very first smart phone in one hand and a bunch of books in the other. In order to open the door I needed to get the key out of my pocket. I placed the phone on the car roof and reached into my pocket. A voice in my head told me, *You are going to forget the phone*. *Nonsense*, I said to myself, and opened the door. I threw the books in the other seat, got in the car and drove away. The phone smashed onto the pavement. This was a first for me, and I don't recall anyone ever telling me they had done the same. (Since I have recounted that event several people have told me they have done similar things, so this is not unique to me). So how did that discrete thought emerge fully formed into my mind without experience, without example, and when there were any number of other possible pitfalls or threats also present? Why *that* sentence? How do we fill our thoughts with anticipatory information in the fleeting probabilistic present? Either the past is being repeated, and the future is just our memories, or we have at our disposal propositions of actual things we did before we actually do them. Recall our discussion about information and how messages can only make sense to data previously transmitted. I am suggesting here that an anticipation in the mind cannot be established without some information coming in from the future to 'complete' it's genesis, and make it consistent with the extended timeline, with history in both forward and backward time.

We will discuss in detail one particular reason why the proposal that information which forms our pre-loading needs to come from some small distance in the future when we come to talk about dreams and language, but for the moment I will explain why the brain cannot successfully form anticipations of the future using what it already knows.

Certainly we have identified two brain centres which are involved in the timing of actions (i.e. hitting a ball); the cerebellum, which appears to have some understanding of the time to an action, while the basal ganglia inject an understanding of rhythm to the matching of the body with an outside event.. But this understanding only gets us to baseball; it does not get us to the framing of higher level unique concepts embedded in each individual. Patterns and the 'music' of events rely, of course, on a future into which the pattern persists. Anticipations may arise out of a procedure, but without the activations of fact or discovery; they are not a prediction about what may come next, they are an expectation of what *should* come next, i.e a recapitulation. They fill the void where the future will happen with the past, which means that the information content of the expectation will always be less than the situation could hold. It is here that we can introduce coincidences as being the corrective to anticipations. Since coincidences cannot be predicted and yet contain the most information content of all about the world, it must be they that they introduce a collaborative consistency about how we change our minds and acquire knowledge (culture).

While rhythm and pattern do provide some of the value of pre-preparation, orthodox ideas of probability-led guesses and the patterns they underpin are simply inadequate to describe cognition and how the mind understands events as they unfold. We know this because, as earlier discussed, coincidence between say two independently evolving causal chains create a moment that neither chain could have 'predicted' in the working out of their own probabilities. We do not have any means to calculate the actual probability of me travelling to Chile and passing through, entirely unaware, the town near where a Chilean I once knew had been brought up, and to cross a street and find him there, coupled with the probability of him going to visit his parents house, at the same moment, and recognising someone he had hardly seen in a London office he had visited a few times striding along a street of his home town, but we can be pretty sure that it is very low, even if we decide that the odds for this situation calculated in the form of a *frequency* are not meaningless. (For the Bayesian inference to work there would have to be *priors* for the participants of this event.)

Suppose one began to argue like this. Let us say that there are a thousand encounters across the world where a traveller completely unaware comes across someone they knew from their past in a strange town. The statistical likelihood for this to happen to you is therefore...what exactly? Suppose there are 100 million travelling people across the world in one year. What are the chances that you and anyone you might know are among them travelling to the same place at the same time. You can go on and on like this but no answer based on frequencies that explains your coincidence would be forthcoming.

The coincidence, in fact, can be thought of as representing a strikingly *negative* probability in both causal chains in this context. If we look across the whole of ENT, these moments of negative probability amount to a counter network of connection, on which the cognition of the organism depends. I have used a relatively 'large scale' event in human narrative space to illustrate how specifically type 2 coincidences contribute to this cognition, but coincidence must also be a property that is scaled through all events, from brief time frames to long ones. The smaller events we do not notice so much because they are immediately absorbed into awareness and swamped by perceptions. We will look at some experimental results examining this notion further on (see **Something has occurred in laboratories**)

It has been observed that the front polar cortex of the brain as well as the mid-brain produce predictive signals correlated with a rise in dopamine. It is as if the neurons are excited by a reality which is then suppressed. A point we will enlarge upon in a moment. Perhaps this, then, is the role of the unconscious, to take up the load of self-reflection and self-reference, to suppress the self-absorption of the organism, so that the conscious mind can make predictions, to seek forward. We will not find in the unconscious, therefore, a track of the information flow from the future but in our conscious awareness, in our *now*.

Before we produce our solution to the questions of mind and consciousness, let us introduce the first of two alternative proposals about what underpins human cognition, namely Akashic fields, and examine the proposal that we are aware because awareness is already built into the particles of the universe and brains somehow collate it and cohere it into our experience of life (see also **Boltzmann Brains**).

Akashic fields

There have been many attempts throughout history to describe a universe where everything is interconnected (*the fundamental interconnectedness of all things*, as Douglas Adams' holistic detective, Dirk Gently, was fond of saying), and is thus one big consciousness. The idea has been dubbed *panpsychism*. The strong version of panpsychism, for example, Lanza and Berman in their *Biocentrism*, (2009), in which consciousness creates space and time and originating the universe that we experience – the 'self-actualising' cosmos, declares that the fundamental unit of matter is consciousness, while the weaker version has consciousness as a component of matter and can appear wherever there is enough 'integrated information content' to create it. Some thinkers use a factor to qualify a system for consciousness, Φ, *phi*. It's a bit like negative entropy in that a system has a chance of being conscious if it's behaviour cannot be reduced to simply that of its parts – that is, it becomes more organised than its components. What is meant by information in this context has not been properly established, and in any event, as we have seen, information content relies on the pre-existing context and is therefore subordinate to consciousness rather than the originator of consciousness.

All the same, the Hungarian philosopher, Erwin Laszlo has written a number of books about one of the latest versions of panpsychism (e.g. *Science and the Akashic Field: An Integral Theory of Everything*, 2007), where he describes the Akashic Field, named after Hindu mythology's Akashic Record, the stuff out of which the universe was made and which contains all that ever was and all that there ever will be, passing this information on through each cycle of the universe's death and rebirth. For Laszlo, his universal field is the energy and information of the quantum vacuum which he considers to be consciousness itself, thus imbuing consciousness in every particle or energy that comes out of the vacuum and with which the universe is made. Particles interact and therefore become aware of themselves and of the 'other', the not-self; being conscious in a Hegel-like manner. Thus all aggregations of matter and the space-time in which they sit are infused with the same awareness. Because particles have a form of awareness, Laszlo says, clumps of particles can become sufficiently aware to acquire *life*, and as matter evolves through the actions of life, it can experience, reflexively and self-consciously, the sensation of the same consciousness in every collection of matter. (The priest and philosopher Teilhard de Chardin had much the same idea and thought that the many-humans-consciousness-agglomeration forms a world consciousness, a *noosphere*, distinct from each individual but which itself evolves to a universal mind, the Omega Point. Remember *Cosmism?*)

One wonders about the distinctions between an aggregation of conscious material in a rock say, and a human brain. Clearly some other ingredient than simply conscious matter is required to differentiate between animate an inanimate aggregations. In this view, the awareness in each living creature originates in the same consciousness rather than the other way around. We will come back to this towards the end.

Laszlo's notion suffers, as do many similar ideas about a universal spiritual field, from a lack of utility. Starting off with a universal consciousness eliminates the need to uncover anything about what consciousness is and how it arises, or, indeed, why it is obscured to varying degrees by the world of matter. This is the teleological approach as a general strategy for a universal field theory. The assumption of consciousness explains consciousness.

Laszlo suggests that the fundamentals to consciousness are energy and information, but this poses a problem as we saw earlier when we discussed Shannon entropy. Information is only informative in relation to already existing pre-loaded meaning, so the consciousness of the quantum field is going to be subordinate to other consciousness further back in time, and we find ourselves travelling along an Aristotelian chain of prime movers; universes before universes.

Laszlo recognises that matter and life is made up of the same stuff. There are the same bits of stuff in you as in a galaxy anywhere in the universe, in a second generation star, in a planet, and yet, arrange it one way and you have Leonardo da Vinci, another way, Lionel Messi, yet another way, an anaconda, yet another way, a black hole. Whatever diversity we witness, it is all really the same. This must be true of mind, too. Consciousness and rocks are outposts of the same consciousness, individuality is a trick.

Yet Laszlo cannot explain why consciousness is a variable (which is why it will be difficult to tell when an extended system like the Internet is aware). There are grades of consciousness, not just among living creatures but among humans. Awareness has depth and breath to its function. Our degree of awareness depends upon our states of knowledge, emotion and intelligence. In some environments, for example, where we know little about what is going on, we perceive little; we not only miss the significance of the events, we miss events that have significance. For example, witnesses observing a film of two teams of people throwing and catching a ball will simply not see a man in a gorilla suit in plain sight because they have been asked to count how many times a ball is caught by one of the teams. Are they fully conscious of the world during those moments? The answer has to be no.

It is here we can note the two uses of the word consciousness. We talk of consciousness as the essential ingredient to intelligent life. Computers can be very smart, but without consciousness they are not autonomous entities in their own right with self-directed thinking; they are not sentient. (Quite how we would detect the difference between an intelligent system mimicking sentience and a truly sen-

> It is a question yet to be answered whether a computer system can be as generally intelligent as a human without being conscious of itself. In making these systems we may not create (or be able to create) the appropriate output means by which an intelligent system can express its self-consciousness to us. For example, there may be no route from wherever its self-awareness is located in the systems to its speech functions. Or that the image processing capability does not connect with its visual imagination. We would have created a brain in a vat, blind, deaf and dumb and one incapable of describing its predicament, although knowing it. That situation would easily make it mad.

tient being, is a theoretical problem that the application of the Turing test suggests is not interesting – there is no difference.) For a system to be sentient, we mean consciousness in the sense of *self*-consciousness, awareness of oneself. Hegel thought the prime component to sentience was awareness of one's own consciousness as the object of consciousness. In the Hegel sense there is a distinct threshold to consciousness, either one is aware of oneself or one is not.

Self-consciousness is a variable because of the boundary that it encloses. It is the boundary, the containment which gives the idea of 'self' its meaning. (The boundary also gives meaning to the idea of 'intelligence', namely it is what one does with the limitations of what one has that defines intelligence. It is meaningless to talk about the 'intelligence' of a machine that has no limits.) For humans this 'containment' has been traditionally recognised as the physical boundaries of its senses. We have described already the problem of how the internet, for example, being a global phenomenon with many varied and separate modes of interaction with the world, none of which encompass the whole, might become aware of its boundaries as an entity and therefore of itself (A problem faced by the devilish agent feeding impulses to a brain-in-vat. How does it convey the 'boundaries' of self to that consciousness? In the limitations to its senses?)

Then there is the psychological sense of consciousness as functioning to the level of your awareness (of stuff in your senses). This exists in degrees. We expect a human to be more aware than a dog who is more aware than a cockroach. The philosopher Richard Searle once described a thermostat as having three states of consciousness: that it is too hot, too cold or just right. This is all the *self* of the thermostat is. The internet, then may become aware to a degree of consciousness (i.e. aware of some of its states) but not wholly aware of itself as an entity.

To be an entity, to have ideas about self, there needs to be some 'quantity' of reflexivity in consciousness which give it its boundaries. Searle's thermostat, for example, even if you consider it aware enough to operate properly, still acts selflessly, with no thought for itself at all; it has no system by which it can perceive a boundary to self.

So what is special to self that makes it self-aware? Is this just a tautology? Let us look at the second proposal which tries to explain these patterns of wholes and parts and similarities in organisms appearing across time and space, and which are also suggestive of *chi*, namely the morphogenetic fields of Sheldrake, a biologist and avowed Christian committed to the idea of creative interventions by a Creator. Sheldrake is against mechanistic explanations of the universe but also believes in the evolution of its parts. He is advocating a pattern of organisation that permeates the Cosmos, and while his theories of similarity and stability being responsible for the creation of particles after the Big Bang, find, for example, echoes in those of pointer states and superselection in quantum theory, they become a utilitarian notion to explain the whole of life in the Cosmos. Since his attempt throws into relief the argument in this book we shall examine it in detail. There are many unresolved questions, as we shall see.

Morphogenetic Fields

The question of the stability of particle identities and characteristic behaviours was given a new angle by the biologist Rupert Sheldrake's theory of formative causation, which tried to explain how matter as well as life appear to evolve. To recall our dice throw and table analogy, he, too, looks at the patterns of life from the point of view of a loaded dice.

The question of similarity between organisms (and hence an explanation of evolution) has been examined by several past thinkers who proposed that active fields retained a pattern of form outside of the organism and which organise or at least constrain within certain bounds the development of 'similar' organisms later in time. In modern parlance, in terms of probability, we are talking about local probabilities converging on a pattern through the negative entropy of those fields.

> Shklovskii made an perceptive remark about how to recognise alien *life* in the universe that seems to contradict the impetus of Sheldrake's thought; *life is that which reproduces its mutations. (Intelligent Life in the Universe* with Carl Sagan (1974))

Events end up being channelled by a field (or fields) into classes of similarity that reproduce the characteristics of the past forms and aid the reproduction of further events of similar type. There is a quality of wholeness that contributes to the structures of the units that make it up. The action of a field, however, is obliged to perform at least two tasks. It provides a model of the whole that interacts with the developing systems to make them conform to their nature, and it allows existing systems to develop 'habits', to acquire some kind of inertia to change.

Sheldrake enlarged upon these ideas where formative causation occurs through a morphogenetic field (or *morphic* field) which is associated with every unit of form within an organism – cells, tissues, organs and the organism itself – and which shapes and stabilises the fully formed unit. When it comes to living systems, DNA is responsible for the proteins used in cells, but the shape and arrangements of the cells and the organs and tissues they form are due to the morphic field. Similarly, genes might be responsible for traits, but morphic fields are responsible for behaviour.

Sheldrake believes that these fields are created by the 'units' of form themselves and their collective history. These fields hold a memory of previous forms; they are an average or summary of all presences of the units. The power of this field depends upon the degree of similarity that all the examples of the units have with each other, and the field interacts with the units through morphic resonance, by their 'tuning in' to the field appropriate to their form. This resonance is not an energetic resonance typical of fields, according to Sheldrake, where energy is transferred from the field to the system, but one solely of information that acts in sympathy with the system's same rhythms of activity, be they vibrations or periodic movements. When Sheldrake talks about forms 'tuning in' to the field, he is talking metaphorically; there is no actual energetic transfer. There is only 'meaning'.

While the problem of how individual parts of a system continue to conform to the whole is addressed by this proposal, it introduces other problems. The morphic field has to be indeterminate enough to allow evolution to disrupt its habitual channels but yet be sufficiently strong to make its influence felt. It also has to perform a wide variety of organisational tasks since what constitutes a 'whole' and a 'unit' within that whole varies with system and event. It must work within a hierarchy of class and unit for diverse phenomena in both organic and inorganic realms. It must both create the class or set and be an element of that class or set (as opposed to being an 'example' of the class) which we have already seen may be too paradoxical a notion to be useful. There are nested fields within fields so fields interact not only with their class and items within it but also with other fields.

Sheldrake proposes a field for every organic process. A daisy has one field, a buttercup another; a daisy stamen has another field and a cell within that feature is controlled by another field. Animal cells have a different field acting on them and so on. There is a field for each class and each level of the hierarchy. But it is hard to see what the hypothesis of such a multitude of fields actually gives us except the recognition that we can define many classes of phenomena. He doesn't stop with the organic world either but proposes that human society is also part of the evolutionary process and is governed by such fields, observable in many common social phenomenon like simultaneous invention.

Sheldrake believes that these morphic fields are real and physical in the sense that a magnetic field is real and physical, but that they affect the shape and organisation of living organisms or any system that reproduces itself only through organised information rather than by any exchange of energy (particles). So is the morphic field actually a *field* at all? It is beginning to sound like a good old Monad out of Leibnitz's book.

A morphic field has to act through space and time since its presence must be everywhere similarity arises, but it is not a fundamental law like a physical law since it is both brought into existence by forms and influences them – actually not much like a Monad. Sheldrake doesn't seem to think its action is modulated by time or distances (ah, that's more like a Monad). He says, "...the idea of morphic resonance involves a different kind of action at a distance, which is harder to conceive of because it does not involve the movement of quanta of energy through any of the known fields of physics...' (*The Presence of the Past*, Sheldrake, 2011.)

He believes that the field acts in the present through accumulation of influence that do not disappear into the past but remain, "...pressed up, as it were, against the present..." (ibid.) and are available to all subsequent similar organisms. A field must be specific to the class to which it refers, otherwise there would be diverse and random influences from everywhere on everything. The information build up in each class must be characteristic only of that class. It is certainly hard 'to conceive of' one mechanism that maintains the separation of innumerable qualities of information needed for this mechanism.

Sheldrake uses morphic resonance to explain a variety of evolutionary conundrums. Sheldrake even believes that morphic fields, being non-genetic in character can cross species boundaries – in other words, they don't have to be specific to classes after all – and which may be responsible for the puzzling convergent and parallel evolution we witness in many places and between marsupials and placental mammals for example. (The odds of two separate DNA lines of development converging appear to be, without a recognition of Coincidence Numbers, extremely unlikely.)

Sheldrake's argument is functionally equivalent to Lamarck's or to a strong metagenetic effect, and adapts his definition of a morphic field to supply a universal explanation of evolutionary puzzles. He tries to find support for his proposal in the doubts that Darwin himself felt about the validity of natural selection. Darwin, however, had doubts because he lacked a sturdy organic mechanism of inheritance to back his theory. Sheldrake has put himself in a similar position where he has no mechanism for transferring the information in these fields to the system nor any mechanism for extracting the information from a system and transferring it to his fields.

To counteract this objection, Sheldrake suggests that the fields must arise 'spontaneously' or that they are derived from even 'higher' kinds of field or that they represent pre-existing archetypes, Platonic style (although they are mutable and not the one way street of eternal ideal forms), or some kind of mathematical structure. In other words choose your reason; it doesn't matter to Sheldrake. He is only interested in "...the morphic fields that have already come into being..." since the "...organising principles of nature, rather than being eternally fixed, evolve along with the systems they organise." In the end, the morphic field becomes a catch-all to explain any puzzle of similarity (like proteins with different amino acid sequences folding in the same complicated way) while being unable to explain why similarity is so strikingly *lacking* in places. Why do some species catch on to similarity while others do not? Why isn't life just one thing?

I think back to Leibnitz and his Monads and the difficulty he had in trying to explain both the apparent stability in forms *and* the existence of random accident and coincidence with one basic formula. It is a similar problem for Jungians and synchronicity, how *not*-accidents are also to be explained by the underlying unity.

Sheldrake goes to great lengths to show how the uncertainty in the fields can actually stimulate evolution. But it is important for Sheldrake that the influence of the fields does not vary over space or time because he uses them to explain parallel evolution (like leaves), cultural convergence in societies, simultaneous invention in widely separated places (even among planets and galaxies) *as well as* the occasional *opposite* trends of reversions to more ancient types (morphic atavism). (Although by the end of the book he allows that fields may vary over time.) He is inconsistent in his depiction of how the levels of organisation of the morphic fields lead or are led by newly created processes. He says that when biological communities are disturbed or broken (like taking honey out of a hive) and the individuals try to recover the original whole, this is evidence for the persistence of a pre-existing organisation existing independently of the bees yet directing their activity. Yet he also accepts that morphic fields and their organising influences themselves also evolve through the connection to pre-existing fields. Obviously there is a chicken and egg question here.

It is not my job to rectify Sheldrake's hypothesis, but it seems, that Sheldrake has missed a trick and could simplify the method by which his morphic fields support evolution by allowing that his fields *do* vary over time and space, and leave the organisation of solar systems and galaxies to some other process.

For example, if the field strength declines in some way over time from the first appearance of a class then the resistance to new forms would also decline over time. The longer a class goes on the less bound to it subsequent generations are and the more vulnerable the class would be to change. Evolution would also be stimulated if the fields declined in spatial extent, so that physical outliers would be less securely attached to their form and thus more liable to create differences. Where geophysical changes on the Earth's surface separate some occupants of ecological niches from the main body, the 'dependency' of the form on its field would be reduced and thus 'make room' for evolutionary change. This agrees with standard gradual Darwinism (and the genetic clock), with the theory of punctuated equilibrium and genetic drift, as well as with the long term instability in cultural conformities among humans.

Spatial or temporal decline would imply an energetic relation between forms and their examples which Sheldrake is reluctant to accept. A decline in time, for example, could explain why high numbers in a class of rapidly reproducing units stimulate change (like mutations in bacteria). If morphic fields have a certain 'capacity' and are unable to incorporate or communicate with more than a certain number of the increasing examples of the same form because fields 'fill up' (in communicating with such increased numbers) and cease to function forward, new examples would no longer fully 'connect' with the field and become unstable and get rejected by it, thus stimulating changes and departures from the form.

Sheldrake does not consider this time limitation and rather believes the opposite: fields are unlimited. He says that even if all the examples of a form die out, the morphic field lingers on with the possibility of influencing affairs at a later date or by recapitulating past forms; what he calls morphic atavism. At the end of the book Sheldrake reveals what he actual believes morphic fields to be: morphic fields are like *souls*, culminating, of course, in the 'fields' that shape humans – their minds.

Sheldrake's uses his theory to explain simultaneous invention. Simultaneous discovery in physically separated places, however, seems to contradict a field declining over time and space, which is why Sheldrake doesn't allow that his fields do vary in strength from their origin of similarity. We can allow that his fields do vary over time and space if we consider that simultaneous invention (SI) occurs between *nuclei* of similarity (as distinct to Kevin Kelly's adjacent possible). That is, SI does not occur in a place without any antecedents (or pre-preparation) at all. A nucleus of the idea must be present in both places, and when the various nuclei grow in strength sufficiently and to increase their presence over time to overcome the physical and temporal distance between them, the connection is made and the separate fields overlap and merge (within a zone of time and space uncertainty) thereby initiating the similarity in both places, though not necessarily simultaneously (since the build up might occur at different rates in each place). This may be another indicator of our Coincidence Number which we will examine later.

Sheldrake says that his fields have no relation to any organism's genetic inheritance. Yet in fact his fields behave very much like genetic inheritance. Not only do morphic fields have their stable versions of forms or themes, they must also support *varieties* within species, as if the fields, too, were maintained by their own types of alleles. This is the only way Sheldrake can explain metagenic effects whereby environmental influences alter the hierarchical relationships of gene translation within the whole, or even activate dormant pieces of code in the DNA that creates changes in the organism in either its form or its behaviours, and which are not specifically coded by any gene set, which then becomes influential on subsequent generations.

Sheldrake speculates that the fields of formative causation are in some way also probability fields since systems are never reproduced or copied identically (rather reducing their power to direct similarity), which means that the class, stabilised by the field, is only a provisional definition, so there is a lot of scope for local indeterminacy and variation. Because of this loose definition of his field, the notion of formative causation as a theory contains little if any predictive power. Matter and energy have their 'habits' as does life and intelligence. We 'observe' these habits only after the 'fact'. Such and such are similar *therefore* they are governed by a morphic field.

Such a field is a useful concept only in that it helps make the case that there is whole lot more going on in the present than is accounted for by what we currently observe as cause and effect (Hume wakes up). Morphic fields as generally applied to life are just another way of saying the Present Completeness Theorem is incorrect.

Sheldrake does suggest an effect more supportive of our argument, by sneaking in a time factor to the field strength, namely that systems naturally resonate more closely with their *past* and thus derive more stability from that than from the morphic field of their class. An item in a class more closely resembles its immediate past than any other example. The difficulty for evolution, though, becomes greater, with this emphasis on backwards acting self-resonance.

While Sheldrake's ideas might have some relevance to molecular and atomic behaviour, it does not scale well, as repetition does a fine job of creating pseudo-similarities without creating resonance.

Sheldrake is a Christian and has declared that he is keen to trace the physical connections between examples of *life* in our world with a mechanism of how his Christian God can continually intervene and guide evolution. In his view, change, or alternatives are all pre-existing and are brought into action by circumstances. In this sense he contradicts Zeno and the Eleatics with Platonic ideals and becomes an ally with the metaworld theories of Boström (all possibilities exist and are laid out in a structure and where our reality is formed by connecting a set of points of the structure with the help of a god). But in practice, his ideas of category or form or levels of 'structure' are *ad-hoc* and not eternal because the field has to do whatever he wants it to. That few systems or items have clear conceptual boundaries, and that how we see structure in our reality is often a matter of psychic convenience rather than any artefact of the world, seems to have escaped Sheldrake's notice, and as do some contradictory conclusions from his proposals. For example, because the morphogenetic field accumulates traditionally forward in time, similarities across fields are reinforcing. The fields should be propagating themselves rapidly, merging and reducing. Mutations or mistakes in reproducing systems would find no help from the existing morphic fields and should die away. Among humans, differences would be lost. There would be an ever-increasing inhibition to change, a reduction of distinct entities and lowering of external entropy. We would lose our perceptions of individuality, become more and more similar physically and behaviourally, while the recursive process acting upon new generations would rapidly converge on a few types, rather like the game of patience called, *Accordion*.

Sheldrake offers no mechanism by which the morphic effects occur and supports his hypothesis with a very limited number of questionable observations (like dogs knowing when their masters are coming home, or repeated crystallisations occurring faster). He also fails to see that it is the boundary decision of the observation that determines what shapes and purposes there are to the field. In other words, just like Laszlo (Akashic fields) and believers in panpsychism, Sheldrake's morphic field is tautologous. The problem of how to account for change within a consistent whole (the same problem that Leibniz had with his monads), which he recognises, eludes him, and the problem of mind remains.

Our age is characterised by a wealth of ideas, even though it is harder and harder to have a new idea. Just an internet search is likely to come up with something that pre-dates whatever your idea is. Whatever you are thinking, someone *has already thought it*. Thoughts original to you can almost never be original in the human world (Thinking back to what we said about language and Shannon entropy). Especially now, when there has been an explosion in invention and the application of reason. Ideas abound in all areas of thought and technology, the adjacent possible is everywhere, and yet, as many have observed, actual technological progress does not seem to match the thinking that seems to be going on. The state of Singularity that Kurzweil and others have promoted seems as far away as ever. There is a mystery here. Is there a limit to the numbers and impact of ideas? Does the coincidence number of the universe, while increasing evolutionary outcomes on the one hand, on the other hand imposes a limit to the rate of change beyond which we find it hard to drive ourselves. Is this why civilisations all fall back and disperse in the end? They reach the point where the expansion rate resists increments. The momentum falls away and decline sets in.

While the idea that, for consistency in the universe, a recursive component seems to be required if change is to occur in time, both Sheldrake's and Laszlo's thinking seems to be pointing at the wrong feature of the problem. We do not need to account for *souls*, we need to account for humans. To get beyond the reasoning that we are human because we have a pre-existing spiritual component that defines us, something more radical is required.

The Cosmology of people

The Cosmology of People, and the Time Travel Solution   498 / 414

In which we ponder the source of ideas and how they are channelled into the human experience.

In which we look at human lives and how they are distributed in time, and how sleep and dream help trace our cosmologies in the universe.

In which we examine language and lies and the secrets of an Amazon tribe.

In which we describe the extended kinships implied in each cosmology, and think about what truths lie in anticipation.

In which we find ourselves thinking about how to train interstellar travellers and return to our starting point.

'Every painter paints the cosmology of himself'. Salvador Dalí

Salvador Dalí moved to Monterey in 1947 and collaborated with Disney and others while he became absorbed in the science of the nuclear age. In an interview, when asked what a painter gives the world, Dalí said that every painter paints his cosmology, and for him his cosmology was the science of atoms, of nuclear forces and psychoanalytical framework of the mind.

All along we have been working towards the level of interaction at which human activity most closely resembles the actual reality each of us lives (what we have been calling ENT). If we are distinct cosmologies in the Dalí sense, from where comes our agreements in the human narrative? Let us take a brief look at the creative narrative and at some of the things that puzzle us about the emergence and propagation of new ideas. Simultaneous invention is certainly one of the mysteries of ideas which Jung, Sheldrake, Laszlo and many others have latched onto as evidence for an underlying unity of creation. But what is that unity and how does it really play out on the scale of the human narrative? Let us go further back to another moment of transition in the human story, to what is referred to as the 'cognitive revolution'.

Cognitive transformations

Homo Sapiens tried to leave Africa between 120,000 and 100,000 years ago. It was not a success. (Although previous hominids like *Homo Erectus* may have managed it.) They appeared to reach the Levant before falling back. Neanderthal communities still dominated the European landscape. Around 70,000 years a second wave of Homo Sapiens left Africa only this time they triumphed. They spread over the whole world, even into Australia, while the Neanderthal communities were pushed to the margins and finally extinguished around 30,000 years ago. Humans created new tools and artefacts, and new social constructs like religious practices, art and trade, and cooperation between groups arose (as distinct to Darwin's notion of competition between them).

This transformation is often referred to as the 'cognitive revolution' and is generally thought to be the result of language where some genetic alterations or mutations allowed language to grow into a method of being more informative about individual intentions as well as about the world, e.g. to be able to describe and understand imaginary scenarios (for example, religion), and also to be able to cope with larger communities, where complex social arrangements (like hierarchies) can be maintained. (This doesn't necessarily explain, however, why populations give themselves up to imaginary scenarios.) Some argue that it was the growth in social bondings that was a more important influence on language. The growing sizes of groups multiplies exponentially the possible interactions between members of the group, and language evolved in response to that social pressure.

With both these views there appears to be a chicken and egg question. Did larger communities arrive first and language second or did language arrive to enable larger communities to then grow. In a previous book about human personality I argued that a relationship system arrived first to enable individuals to make parenting bonds with those outside their kin-based group. The expansion of bonding possibilities not only allowed larger groups to live together, mixing the genetic compliment of the group more readily, but gave the group more of a reason to look outward, to cooperate with non-kin, and to move away where such things were lacking, thus fuelling expansion over the Earth. The

There are some baffling evidential outlying discoveries which seem to call into question the time-line of *homo sapiens*. There is at least one Hindu apologist who tries to prove that humans walked the Earth millions of years ago on the basis of archaeological mysteries. Maverick thinkers in popular culture use these oddities to make the case for a much earlier urban civilisation than currently accepted. (e.g. The Egyptian Pyramids were built thousands of years earlier than Egyptologists believe). What is sure is that the historical record still throws up puzzles about the origins of the modern human mind not yet solved.

monogamous parental bonding scheme becomes more natural to individuals as they become attracted to types (love) which then gets reinforced by higher survival rates. The problem of the growing difficulty of the permutations of social arrangements as group numbers rise also falls away if types are limited. So if language seems to come on the heels of new social arrangements, just what was particular about it that helped fuel, in particular, *Homo Sapiens* extraordinary growth. Certainly the capability to imagine and to abstract information and code it for various future times was important, but so too was being able to make sense of, and to use memories.

Let us expand upon the difficulties of solving social arrangements as being at least one foundation to human success and go on to examine the two cognitive features of language mentioned above, namely prediction and memory, from the angle of dreams, and suggest that long before we had coherent language production we had consciousnesses with overlapping social realities, exchanging ideas, and the experience of coincidental bonding as forces for change and stimuli for growth in human societies.

We have already examined how brain activity seems to connect beyond its boundaries, and we considered notions of how consciousness might already exist outside of brains in a dispersed form. Let us examine further the actual reach of consciousness in the material world and how far in time and space mind may penetrate.

Ideas and the human channel

It is interesting to think that perhaps ages in the past where advances got forgotten indicate such non-causal activity from time tinkerers and that the advances were not *natural* at all but the result of time influences which failed to overcome the long reach of time buffering. The time-line 're-laxed' back into its previous state of development.

The Native-American Dineh word, *hocho*, describes the reality you experience that depends upon your perceptions. Different for each person and coexisting with every other. Yet, to what degree are we truly different form one another? Simply by observation we can see that *hocho* involves overlapping realities, where both individual selves and consequential social truths are enabled in the physics of consciousness. Because individuals are generally subordinate to the physical social context, their social compatibilities are channelled down a limited number of paths.

Even those elements we believe to be individual to us, our personal creative thoughts, we often find are actually shared.

At the more basic levels of life, the physics of what is possible would seem to determine what forms of creatures could arise. Biological evolution is going to solve similar practical problems similarly, because the energetic minimums are fundamental for any given set of conditions. Edinburgh University Astrobiology professor, Charles Cockell, argues from this that alien creatures are going to be similar to us wherever they evolve in the universe – humans are examples of the basic constraints to intelligent life (*The Equations of Life: How Physics Shapes Evolution*, 2018). The eye has arisen independently many times. Three-jointed limbs are found across all species. Practical things are copied everywhere there are similar environmental problems to overcome. So, too, our social past is littered with practical objects and machines anticipating those created now. The ancients were as acquainted with the same problems that we face today and who arrived at ingenious solutions whose genesis we instantly recognise.

Biological evolution, however, works upon the molecule arrangements that are able to reproduce, and, as we have been arguing, it is not at all clear that without specific ^2N_c coincidences in the narratives of life energy levels and physical logic of the molecules may not be enough to produce life let alone repeat the sequences even approximately. Even solar systems that are similar in composition and layout may yet be deficient in the supply of coincidental events to produce life or to advance to intelligent life. If the products of intelligent life are going to be similar wherever found, then we should expect that at each site, there will be not only similar molecular dispositions but also a similar narrative pressure. How likely is this?

In the ecology of mind, the original creative thought still seems to arise mysteriously. What we consider the 'creative' thought is one that is novel, that is *non-predictable*. Even though context can be more wide-reaching that it may at first appear it does not seem to be sufficient on its own to predict the novel solution. Certainly no one can yet say what is happening in a brain that conjures a new thought or explanation even though ideas do not arise *in vacuo*, and that they have long histories of connection with other ideas already present. For example, the science culture of priority naming a few blinds us to the background contributions of hundreds more. While Einstein was turning to friends not only for mathematical skills to present his deep insights on special relativity but to get his papers published in the first place (saying this to show how embedded we are in broader circumstance) and others were very close to producing at least mathematical approximations of Relativity themselves – Lorentz, Fitzgerald, Poincaré, it was Einstein who penetrated through to a different understanding of the nature of space and time. (The question of who else might have had Einstein's temperament to accept no absolute frame of reference is one we can never answer. Einstein's theoretical achievements remain unparalleled under more than century of critical analysis.) Quantum theory is an example of a truly joint effort combining the work of many minds over many years even though we think only of the great few who seem to stand out above the others.

The broad front of knowledge contains many parallels and repetitions, yet the predictability of a crucial new idea does not seem to be related to simply the breadth of knowledge. The ideas of natural selection and speciation in Darwin *and* Wallace's work (whose lives had never crossed beforehand and whose insights occurred thousands of miles apart) were preceded years earlier by Comte de Buffon who contradicted ideas of his time by believing that changes in species occurred over time through chance and environment, although was careful not to claim that species evolved into other species. Darwin's grandfather, Erasmus, thought life had been evolving for millions of years, and many others around that time like Lamarck and Lyell and the Scottish geologist James Hutton were also dancing around the concepts of gradual change over long periods of time in living systems as well as in geological formations. (Hutton even anticipated Lovelock's concept of Gaia.) Any age, including the present, is full of occasions where History could have chosen this inventor or that, this scientist or that, and still got similar results. And yet, as crowd sourcing the solutions to problems show us, creative answers don't tend to come from those closest to the problem or those with the most history of understanding it at all but seems to also reside in others in remote or unrelated fields.

Certainly need and context take us some of the way to understanding creativity but there is more to it. We do know that permeating the context there are tenuous and non-strategic pathways that connect us but how this affects the *predictability* of idea creation is still unknown. Kevin Kelly has observed the 'adjacent possible' (the surrounding fertile ground) is usually shared by many on the way to the application of an idea, and the necessary sequences of 'adjacent possible' for almost any invention are similar, as Kelly describes in his *What Technology Wants* (2010). Similarly, the science journalist, James Burke, has mapped out many of these societal intersecting paths, chance encounters and the common ground contributing to invention in several books (e.g. *Twin Tracks*, 2003) and a TV series *Connections*.

Which brings us to the second puzzle about creativity, namely simultaneity. It turns out that the origins of most scientific ideas are hosted in more than one individual at the same time, and can be in more than one place at the same time. Simultaneous invention seems more like a given than a surprise, yet the means which produces the advancing insight in a brain remains mysterious. Major inventions and ideas are often invented by unrelated people at more or less the same time, and even the artificial brainstorming companies that try to profit from patenting fresh ideas produced by their brainstorming experts often find that their new ideas already have recent patents.

But perhaps what is more interesting, however, is the less appreciated but no less curious fact of *non*-simultaneous invention between contexts or between cultures. When faced with precisely similar problems, some cultures arrive at an ingenious solution or improvement while others fail to. For example, the Incas of Peru had no wheel. Central American cultures knew about the wheel and made wheeled toys and spindles to spin thread but had no potter's wheel or wheeled vehicles of any kind even though the Mayas built beautiful stone roads that may have even been compacted with stone rollers. (Of course, if you lack a draught animal the advantages of a wheeled *vehicle* probably seem less attractive.) The difference between a horse pulling loads with a strap around its neck or breast and a horse pulling more effectively, and more rapidly, with a collar and harness seem pretty obvious to us. Yet early European agriculture failed to capitalise on the superior power of horses to oxen (in spite of the fact that horses had been used in warfare or to pull chariots for centuries) because they did not understand the need to reconfigure neck straps and yokes used on oxen, and went on semi-strangling its draught horses for hundreds of years until the breast collar arrived from China in the 9th century and allowed the horse to work faster and more profitably, pulling greater loads and ploughing more deeply than before. In fact, doubling the productivity of oxen, although when horses actually overtook oxen in this task in Europe is a matter for debate; they were more expensive to keep. (In the Pyrenean mountain village I moved to in the 1980s, oxen had just ceased being the principal engine for the plough (giving way to the tractor more able to manage steep inclines), even though heavy horses were also kept by many and which roamed the high pasture unfettered. The only local person who could be found to shoe my own horse was a man who had only ever shod oxen.)

Then there is the unfruitful anticipation of an eventually successful idea. For example, pre-Columbian cultures between 1000BC and 1000AD appeared to have made small models of insects in the style of flying craft (uncannily like modern aircraft) which have been scaled up by researchers today and flown successfully as planes. There was the *Antikythera* planetary 'computing' mechanism from the first century BC, and the earlier automatons of the Greeks, some of which were fuelled by steam. History is full of examples of ideas 'dying on the vine' and 'born before their time': ideas that were theoretically correct and technologically inventive but which did not flower because the soil in which they were planted could not support them.

The humble potato was brought from the New World where it had been cultivated for thousands of years to Europe where it was found to be nutritious and easy to grow. Yet, it took two hundred years before it would be accepted by religious authorities in Europe. Both the Roman Catholic Church and the Orthodox Church condemned it initially as belonging to the Devil. One wonders if the fact that the Church was the biggest single landowner and relied on income from grain had anything to do with the doctrinal position against a crop that could sustain life and be grown anywhere by anyone.

Certain cultural groups (like professional elites), or political movements, sometimes make sure that poor ideas persist and that improvements are resisted or even quashed. Religious administrations are obvious examples of this but they are not the only ones. Stalinism and Maoism were secular political philosophies that resisted the natural evolution of science and culture and where, in spite of great scientific achievements, fields of science (not only genetics) and individual scientists ran foul of ideology and were censured and imprisoned. Even in democratic societies, competing political groups will prefer a walk round the Guggenheim to a genuine solution that just happens to favour one group above another. The Steve Argument is not correct and political needs often interfere with the natural ecology of ideas.

Thinkers have enlarged upon Dawkins' suggestion for a unit of culture, a meme, that operates analogously to genes in biological reproduction to explain the transmission of ideas. The stated idea, however, was always weakly defined, and, as a consequence, anything and sometimes contradictory things were claimed to be memes, reducing the explanatory and analytical power of the notion (going *viral* works as a better biological analogy). Since no one could find a suitable definition for a meme unit following the genetic analogy, social media like Twitter now define it for us as a transmitted personalised re-formulation of an idea or image (#hashtagging) and rather more analogous to fungal spores than a gene or a seed. The question of whether memes represent innate characteristics of cultural communication has not been answered. Even those historians like Toynbee and Spengler who sought laws in the seasons of civilisations' cultural behaviour have failed to find them, as Chomsky and his followers failed to find any innate grammar laws to the structures of languages that flow throughout societies, (and not only human ones). (We might have a solution to this particular conundrum, however, which we discuss further on.)

Culture may well be a complicated construct involving brains and objects but why a phenomenon like simultaneous discovery should remain resistant to a clear method of prediction is curious if ideas simply arise from needs and contexts or are readily transferred as memes from one to another. If this is so, then the question of whether AI is going to be the all-conquering innovation engine of the future remains unanswered. AI is certainly good at detecting existing subtle patterns in data simply by observing many more examples than hu-

---

A walk round the Guggenheim is a phrase now firmly inscribed in our family's language. It describes a compromise that is the worst for each party. Time-pressure anxiety beyond reason (it was the day before their flight home) in one member of a couple and a desire to make a diversion to see the museum in the other led them to drive to the museum only to *walk around it*. They did not go in, neither removing the time pressure anxiety nor enjoying the museum). This compromise is the worst for both parties and a curious inversion of the Nash equilibrium where all parties are content with a strategy that doesn't need to change in response to the changes in others' strategy. I mentioned the idea to a Bulgarian friend who nodded along and said that it was known in Bulgaria for a person with a feud to harm himself in order to hurt a neighbour (like burning down his own house to make his rival's barn catch fire and burn also).

mans ever could. They are also able to compare shared features of apparently dissimilar systems. They are able to elaborate and extrapolate from rules and axioms in their pre-supplied logical 'universe', and they may even be able to extract axioms from limited scenarios. The neural networks that AI use, however, are less able to pinpoint causal relationships and the places where active elements of disparate systems, still not fully understood, may meet, which suggests that finding influences that generate the solution to new problems will be less easy for it.

So where do ideas come from when they are not emerging out of our needs and contexts? Why has there been such a recent exponential growth in them if this concrete base is lacking? Ideas, or at least some forms of ideas, require more than contexts, they need special brains, or perhaps a special access to something other than the genetical produced contexts.

It is worth pondering, given our argument, what other processes may be involved in the quality of mind we call intelligence. The IQ measurement of a child is actually a value based on what the cohort of people at the child's age generally produce. A higher than average IQ in a child means the child has the intelligence of an older cohort of children. Perhaps we can think of this 'advancement' as a brain being in part better connected with the future certainties. High IQ generally gets lost in the background after a certain age which is what one might expect with this scenario. Individuals arrive at and merge into the pathways that probability was mapping out earlier.

This 'special access' is under examination. A scholarly endeavour in Princeton has been correlating the effect of human awareness on the probabilities of the outcomes of experiments (PEAR project, set up by Robert Jahn and Brenda Dunne in 1979) for many years, and is now incorporated into the International Consciousness Research Laboratories which has been observing, among other things, how numbers of people knowing the same thing seem to affect the generation of random numbers by computers. So researchers around the world are turning to the idea of a probability 'field' established by living systems that directs each future example of the system's course of action. There are hints of synchronicity or the Law of the Series, though perhaps more like Sheldrake's morphogenetic field, but the effects so far studied are subtle if indeed they are real. Certainly, however, the notion that the inspired thinker or individual genius is someone who can tune into the field effect of many brains better than others, or who can tune into a particular state of space-time or multiverse arrangement of probabilities, may be at least partially correct.

≈≈A small diversion. We are talking about individual creativity, but savants illustrate and also complicate the picture. Savants have been studied for centuries and how they function mentally has not yet been understood. But there are too, the handful of ordinary individuals who have suddenly become savants even without a shock trauma or illness and whose experience can support something of our argument. They suddenly acquire skills – almost always artistic, like music or painting, but some have become mathematicians – that were never part of their education. How is this possible? To understand this significant mystery of the brain we turn to a cosmological argument, the existence of Boltzmann's Brains which throws light on the concept of spirit-matter duality that still infects our philosophies, and direct the reader to **Appendix 3** so as not to deviate from the thread we are following here ≈≈

An act of genius cannot be traced to simply genes or to culture or to the times. It's not just about the wave equation combining observer with the observed; there seems to be an ingredient that cannot be traced to any local sequence of causes. A significant influence on the creative act is then, what is *not* self-conscious in the manipulating of the components that go into it. (A point of view, incidentally, shared by many cultures.) For example, as we mentioned earlier, brain scans seem to show a switch-off of the visual cortex in the moments before insight or inspired thought, as if brains needed to suppress personal and individual visualisations at a key moment in creativity.

Nor is inventive thought simply about the clash of separate sets of linked notions. For example, Koestler talked about 'bisociation', where two distinct and usually incompatible fields of thought meet at a point from which a third meaning can be extracted; the *eureka!* moment for Koestler. But this idea does not really explain original

One of the features of human behaviour that may provide a window onto this landscape of the distributed consciousness is creativity. Often, genius is linked to types of madness or behavioural anomie and these suggest that the brain's boundaries are not strictly corporeal, but since both madness and genius are fluid definitions neither one helps us understand the other. Gödel starved himself to death because of worries about the preparation of his food after his wife, his trusted cook, went into hospital. Did that explain or be the natural product of his genius?

thoughts not present in either of the meeting fields, but only the recoding of one idea in the structure of another (rather like a pidgin language since it often involves simplification).

Bisociation is a perfectly respectable formula to play around with ideas, and is something we do consciously to obtain fresh insights but it does not generally remove the mystery of the 'original-ness' of ideas, which often appear fully formed, having passed through some structural process hidden from view. Invention can be thought of as a continually evolving social history with hidden nodes of connectivity. But how can that notion be applied to animals like birds, who are belong to the realm of dinosaurs, or primates who also seem to express inventiveness purely as a result of their individual consciousnesses rather than by virtue of mimicry and cultural transmission.

Cosmologies and identity

So far, I have put the case that the probabilities, or more specifically, the probability volume in which events can be observed in the present, depend upon how the future is already unfolding. The handshaking from the future to the past is what gives the universe its stability and enables every probable event to conform both to local history and across the universe. What does this mean for an individual consciousness?

We've talked about the difference between Internal Narrative Time and External Narrative Time as if they are distinct structures in the world. ENT describes how we (that is, our minds) are embedded in the flow of events around us. It requires an understanding of causality and sequence and where probability expands with entropy. Internal narrative time, on the other hand, is the time that consciousness calls its own, and where we place all the regions of time: our present, past and future. The differences between ENT and INT are manifestations of a structural level of complexity that only exists in the quantum universe through consciousness.

Tegmark describes consciousness as a form of matter while Lazslo sees matter as a form of consciousness. Are they talking about the same thing? I would say that both are missing something. Consciousness accesses the past in a process novel to the world of matter and energy up to this moment. As we saw earlier, matter has no memory of states as discrete entities with their own history. Even processes like

Searle's thermostat, striving for equilibrium between internal and external states, do not accumulate memories.

When we consider an individual consciousness we tend to think of localised sensations in a brain that give us a personal identity. We are led to think of identity as only being traced in a brain. It may well be that each brain is unique in the physical tracings of its cell structure by virtue of genes and experiences, even though each brain also contains shared cultural concepts and language, but there is another way of characterising an individual's identity beyond brains.

Inspired by our analysis of dreams, let us start by compiling what I shall call a cosmological view of individuality.

The track of an object or a person in Einsteinian space-time is called a world line. We will call the narrative of an object or a person as it interacts with the world lines of others a time-line. Humans live out their individual lives in composites of time-lines. You, the reader, live in a time-line that contains all those effects of your presence in the world, your influence on others' time-lines as well as theirs on yours. For most of us, our influence on the composite of the world's time lines is minimal. What you see as your world-line may seem special to you but it is also compounded in all the individual world-lines, yours and everyone else's, that exist in that volume of time and space. Every individual point of view in a world-line is a composite and provisional picture of a cosmology, with a single consciousness at it's centre.

The cosmology centred on you has informational limits or a boundary (in the future) beyond which no trace of your existence could be observed or calculated. Historical figures whose existence are maintained in our present have larger cosmologies than most. After your world line ends, your presence (or let us say, proof of your existence, your cosmology) is only retained implicitly in your children, in the results of your actions in the world, or in the objects you have made that survive you and only in so far as the information these things have released into the world can be traced back to you. Your genealogy or the objects you have made can be connected to you for very much longer than the consequences of any actions you have performed. Da Vinci's art remains connected to him through the ages of the past in a way that a small but significant charitable act that he did (like giving a coin to beggar) does not.

Most of the results of choices we have made are very short lived and get readily overwritten by the sheer numbers of other world-lines. Almost immediately, a person's act contributing to a cause gets mixed with those of others and becomes equivocal. Of the 109 billion people calculated to have lived on Earth (to 2015) only the existence of a few can be traced to an actual individual life, and very many fewer individuals could be said to have originated a particular flow of information through a creative act (which we define as one having unknown or unperceived antecedents). For most of us the originality of the information we generate is slight and much simply passes through us from elsewhere.

The effects of some world-lines linger on, however. Leonardo da Vinci still has a presence in time-lines even though his world line ended long ago. But his presence today is not just down to his individual creative experience since we know from our earlier estimations of the energy expenditure on creating things, the energetic influence of the creative act is a consequence of a vast number of brains. The truth is, for most of us, the time lines of people are very much less personal than we like to think. It is what culture means, being imbued with the time lines of others.

Leonardo da Vinci is only present through the changes produced in sets of event probabilities in other people's time-lines that he altered through his presence. Since these are in principle incalculable it is hard to see how we could remember anybody let alone da Vinci. A small shift in probabilities, just like in the Trevanian story, could alter our knowledge about everything. In fact we do grasp this idea superficially through counterfactual histories that try to help us see why we took a particular path. But while these exercises in imagination may well contain seeds of the information flow captured by consciousness, their intention is to understand what has happened as if there is a right answer about that. There is never going to be a right answer, in the general run of things, however. (Historians call their exercises in imagination counter*factual* scenarios which is a little disingenuous. They are counter*probables*.)

The changes in overlapping probabilities that define your cosmology in the universe eventually die away in time and space. The boundaries of this non-influence could in principle be mapped such that your conjugate cosmology could be defined as having your absence at its centre which also overlaps every other cosmology and their conjugates. When we talk of the grandfather paradox – my going back in time to kill my grandfather – we are talking about just the cosmology of 'me' being replaced by its conjugate, being replaced by a different set of probabilities, and not an entire alternate universe. Thus the problem of where Marty Flight's world in *Back to the Future* disappeared to itself disappears. There has merely been an exchange of probabilities.

Cosmologies are centred on consciousness. Multiverses make sense when we are dealing with inanimate matter, but the moment mind enters the equations we have to deal with consciousness and its structural connections backwards and forwards in time. Unlike the multiverses being composed of separate and distinct entities unreachable from our world, all cosmologies are with us in the same space. This means that random environmental mixing is not always required to explain decoherence in the observed event but just the logical passage from one probability to another.

Cosmologies reach forward and back in time. They are connected throughout time and space for as long as they have distinct values; they are structural and not virtual. By reflecting upon how we might map the human cosmologies in our universe, we come to a picture rather like a map of the brain. We can form a picture of 'neurons of time' carrying and confirming information just the way that the neurons of the human brains reading this sentence are doing right now. Both a brain and the universe embody pathways of probability information flowing and combining, reinforcing and debilitating connections in time and space, through its past and its future. How else could it be? Brains and universes are similar. By the Boltzmann Brain argument, universes can turn into brains, perhaps brains can turn into universes.

Just as there is a superposition of multiverses at the core of every neuron in the brain there is a composition of cosmologies at the nexus of every human option. The observed state, the memory, the agreements with the past are made out there in what the universe has already developed from them. Where the probabilities you think you now control have already had substance. Our cosmologies, therefore, have a different connection with the zones of time than say a free particle of matter. We do not only live in the now, and the options and choices we live our lives by are not the simple result of probability of either quantum calculation or the casino but of the coincidental and the recursive influences of intersecting paths retained throughout time. And it is this influence which may play a significant role in creativity.

Another diversion on probability.

Because the tool of probability in calculation works we can tell there is a future. Overlapping uncertainties only make sense if the options refer to a real future (what physicists call contrafactual definiteness, an idea some believe is denied by the Bell Theorem). If a potential to be can never become actual then it is not probable, and if the quantum probabilities we observe are fake then pre-destination is the fact of life. If, however, these probabilities are real then they must exist, in the same space but spread along the continuum of time. In this sense the futures are already connected to the present.

Let us recall our earlier roulette discussion for a moment. What creates the probabilities of the ball falling into the home of a number in the first place is that all the homes for numbers exist already concretely, and for the ball to be in one home, it has to be physically excluded from the others. In a quantum measurement of a state of a particle say, all these homes suddenly appear at the point of measurement and probabilities are spread over all the 'homes'. Yet, these homes, these multiverses, must exist as a genuine future for the particle.

Quantum physics took probability out of the casino and started calculating that the likelihood of some events occurring and the likelihood of the same event not occurring were not exclusive variables but merged. Probability became not a frequency distribution but a disposition, and our world becomes simply an emergent phenomenon built from the first dispositions bottom up. In the narratives of human consciousness, however, is this *sufficient* explanation for all that happens?

Here's what I mean. My wife Alexandra looked up at our wall clock at one point and saw that it had stopped. The battery had died. By the time she bought a new battery, a week had passed. She unpacked the battery and put it in the clock and was about to re-set it when she saw that the clock time set on the stopped clock was precisely, to the second, the actual time. She didn't have to set it at all. She described this to me and started to work out the odds of this happening. I had to persuade her that there was no way to calculate the odds, because traditional probability in the everyday world is based on frequencies and there are no frequency distributions of that kind of coincidence. The things she was relating are completely independent of each other.

She started to calculate how many seconds there were in a day saying that it had to be one of the 24 x 60 x 60 seconds and thus there is a

1 in 86,400 chance of starting the clock going again on the same time. But this is wrong because there is nothing stopping her setting the clock on the wrong time every single time she did it. She is not obeying any pre-ordained distribution of times, any bell curve of properties or anything governed by a wave-like function. There was nothing constraining her to hit some seconds more often than others. She could choose to set the clock one second after the time it had stopped blindfolded every single time until the end of the universe and no frequency law would be transgressed. There is simply no way to calculate a true value for the likelihood of the coincidence in our world of human narratives.

Supposing you did this a million times. Perhaps you could form a view of the number of times you are likely to deal with the clock at any given second during the day. But this frequency does not connect you with the clock, and it cannot say that during the next million trials the likelihood of the coincidence gets more unlikely, not more likely.

Quantum probability cannot be used here. For one thing, there is a vast number of interacting wave functions at the human level. This is not a question of a single electron interacting with a slit or even a group of molecules forced into a cat-state superposition by a laser. There are billions of particles with billions of states all functioning at a point chosen not by random collisions in the universe fields of energy but coordinated by consciousness, itself formed by billions of particles and superposed states. The wave functions of my wife and the clock already connect and interpenetrate in myriad other ways. Where could the actual point lie at which the wave functions of observer (my wife) coincide with the wave functions of the clocks and produce not only a value but a recognition in consciousness? (Concepts of coincidence and crossover will become important to our analysis of language and thought, as we shall see.) We have notions of the arrow of time and of entropy. But these ideas are pretty much after the fact, even as we know that life drives in a direction opposite to the expressions of entropy. Yet these kinds of coincidences are happening everywhere all the time forming our minds and personalities, forming our cosmologies, and quite unexplained, so far, by behaviours at the particle level.

Narrative coincidence

Recall our discussion about coincidences and the difference between a type 1 ($^{1}\mathcal{N}_C$) and type 2 ($^{2}\mathcal{N}_C$) coincidence.

^2N_C connections are how awareness drives evolution more quickly than one might expect, because evolution supports organisms that learn to profit from accident, to use their pre-loading (alleles, for example) to become more successful (this is not quite the same thing as learning through experience). It is easy to trace successful tactics from bacteria to humans in the way organisms learn to be less rigid in their programming. There is, in fact, a point where this use of accident becomes the accelerating engine of the evolution of life. Channels of development appear because the pre-loading is the equivalent of information appearing in the mind before it makes a decision. It is where memory is the *anticipated* truth.

Certainly we can conceive a number, the Coincidence Number, for our universe. All universes are born with a characteristic coincidence number which dictates how it will evolve, and its presence in human affairs tell us something more about how observations can be made. Because coincidence, at the human level, is a way of confirming observations where probability and decoherence are insufficient, it gives us a way of incorporating human consciousness and personality into the story of the universe.

Well, there is more to the story of my wife and the clock, which I hope will help us understand the enormous gulf between quantum behaviour and human narratives. I was actually sitting at the kitchen table with my computer open working on a talk about coincidence numbers, when my wife came in and sat down opposite me. She looked over my shoulder to a mechanical fifties clock we had found on a junk heap and restored sitting on the top of the dresser. This clock needs winding once a day and we never remember to do it. It had stopped long ago at 6:34. She looked up and over her shoulder to the wall clock on the wall and saw that it was exactly 6:34:00.

"Heavens above," she said. "It's happened again."

And it was then she told me about the incident I described earlier, which I then began to incorporate into the text of the talk.

"What is that kind of thing?" she asked.

I replied, employing my conclusions in this text to the fullest extent, that the coincidence with loading the battery into the clock happened then because I am using it now in this talk. "It was my writing of your experience here that helped bring the coincidences in line. There are, as it were, three meetings spread over time collapsing into one."

She pondered this for a moment, and then said, "So *this* coincidence just now happened in order to remind me of the previous coincidence so I could tell you about it?"

"Yes," I said. "The importance of this example in my talk and in my book helped bring both these coincidences about."

These type 2 coincidences are examples of intersecting cosmologies. My wife and I are a well-matched couple in that respect. We are supported by our coincidences and always have been; our cosmologies are superposed to some degree at a number of points.

It is the cosmologies of people that are involved in the realisation and even the actual selection of the kinds of event for which probability does not help bring about. Our cosmologies are the reason why at least some events at the human narrative level occur. So while the type 2 coincidences may be just a higher level emergent phenomenon, there is no doubt that they can reach down into the bottom layer, as it were, of probability and produce outcomes not predicted in that bottom layer, that would never emerge out of it. Since consciousness is a continual evolution of its history, it finds meaning in observations precisely because they connect backwards and forwards in time.

Proof of our cosmologies are found in our coincidences. Coincidences in turn form bonds between similar cosmologies, and turn the processes of evolution into channels. It is these channels that properly design life and society. Every universe is born with a coincidence number and which help define the numbers of channels connecting its systems; it is part of the cosmic reality. It is what makes our universe seem like it has our interests at heart. Life without coincidence is colourless, friendless and isolated. Those theories of the world that begin with God always end up explaining coincidence as miracles, as grace, as a good destiny. But god is not required to put our homes and our friends on our side. Just the coincidence number of our universe.

We have shown how evolution creates creatures who can adapt to coincidence and how coincidence itself is incorporated into life. Let us consider how coincidence is incorporated into mind.

Cosmologies and the quantum mind

So why are our cosmologies significant? What role do they play in the unfolding of the universe and how are they different from networks?

Let us begin with phenomenology. What we know of the world is not only the result of quantum fluctuations of atoms and molecules in our brain's neurons but also the result of the interactions of higher states of consciousness and their memories, strategies, instincts and learnt behaviours, in addition to, as we have been arguing, a coordination between brains outside of our current understanding of the senses. All these interactions are described in the single composite wave function that flows throughout the brain and through its sense organs and connecting with the external world, some of which is decohered and lying forever in the past and some of which is not and remains superposed with possible futures.

There are attempts to design quantum field theories that include a consciousness contribution at every point but these ideas miss the most salient point of all. There is a quality to any wave function describing sentience that is different to anything else. It isn't just the presence or not of life that is the difference, it is the fact that, in sentience, consciousness is *continuous* from beginning to end. Inanimate matter does not lay down memories of its existence, or a history of its differences. Space-time does not retain a complex memory of all those masses that have passed across it. If an object has what might be termed a history it is only that of a repetition of states which gives rise to its duration in the time-line. Consciousness, on the other hand, is an evolving whole that remains connected to all its *departures* from repetition and all the disturbances it weathers. The same mind persists in spite of its interactions with the world (*life is that which reproduces its mutations*).

If we take the implications of quantum theory, which currently explains the behaviour of particles and the fields in which they occur, at their face value, then there is a wave equation to describe consciousness. This wave equation sustains itself unbroken for a lifetime containing all memories, explicit and implicit, and predictions, none of which are unchangeable or definitive. From this, then any interaction with the wave at any point in time will cause some form of alteration throughout its extent in time.

We have already seen some simple examples of how the probabilities involved in consciousness are not described in the probabilities of events, so perhaps it is time to go back to probability once again, and try to close in on the mystery of *quantum-ness* and awareness? We have talked about the rising complexities required to manipulate memories and predictions; and we have talked about information and the degree to which pre-loaded information in the brain is required to make use of information it receives. We have demonstrated a graphical way of thinking about the nature of the quantum reality to show why the brain needs to know, to some degree, how to experience the world before it experiences it. So let us now examine some mysteries of dream and language to see how they add weight to our argument.

The Dream

Everyone understands that human dreams are mysterious events. They are not only a human phenomenon. Dreams (or what is taken to be dreams) are readily observed in domesticated animals like dogs and cats who sniff and shout, mime eating, whose closed eyes shift about as if they were watching things, and move their legs as if they were running) and they probably occur in all the land-based higher mammals. Sea mammals like whales and dolphins, however, don't appear to have what we particularly associate with dreams, REM sleep, and perhaps this is not surprising since the freezing up of muscles associated with REM sleep would not be practical in the water, but whether they have dreams through other brain states is not known. Birds, on the other hand, descended from land-based dinosaurs, do seem to have REM-like sleep and the way they keep vigilant or manage their migrations is by doing it one hemisphere at a time.

The stage of REM sleep is not indicative of dream *per se* — only 80% of individuals report dreams when woken from REM sleep, but very few subjects report dreams during other stages of sleep even though there appear to be signs of cognitive activity during them. So REM is generally considered to indicate the dreaming state, and researchers have been able to pin-point the brain activity associated with dreaming such that they can predict with more than 80% accuracy when a person appears to be dreaming (this, however, does not

> Perhaps the undisturbed state of REM is *not* the dream. Dreams are produced, primarily, by the surfacing into consciousness, by the acquisition of the perceptual states that being aware requires. The dream is the tuning up of the brain to turning itself 'inside out' and look out to the world instead of into itself. This must involve some kind of polarisation or reflection of sequential data. There is plenty of evidence for mirror imaging in the mind but can dreams be explained by inverting sequences of information or scenes within each narrative? Or by another means entirely?

take into account how dreams differ in the various stages of sleep). Usually a person falls into deep sleep, usually called non-REM (NREM) slow wave sleep. The brain waves are very slow (.5hz – 3hz) but much stronger than in waking. Then the waves increase in frequency to waking frequencies while amplitudes fall as the body becomes paralysed and the brain enters the REM state for several minutes, using up more oxygen and heart-beats than the awake state. This cycle happens two or three times in a night, perhaps more.

Much of the brain and organ activity resemble the waking state and the passage of emotions, even though the muscle tone is generally relaxed and the voluntary muscles are frozen. The visual cortex appears to be stimulated as well. In some cases, the attempts by the brain to activate various muscle groups can be roughly correlated with events of the reported dream (such as say, running). Even so, sleep walking and other disturbances seem to occur not under a REM state but in the previous slow wave non-REM stage of the sleep cycle, while hallucinations occur in the awake state. A hybrid sleep state where the alpha rhythms of the waking state dominate is also the state lucid dreamers seem to occupy, all of which suggests that dreams as such under REM have a different purpose to simply representing or rehearsing reality. Foetuses are in REM state pretty much all the time (and it's just as well the muscles are frozen in such a cramped space) and which begs the question of what they might be rehearsing or simulating (researchers who believe dreams are simulations have to explain the hierarchical problem of who or what is in control). Infants spend about 50% of their sleep time in REM (first discovered in infants) and this gradually reduces as we get older to less than 20% (suggesting that REM sleep is not strictly related to memories or complexity of tasks which increases as we age). It may simply be that the REM state is a residual brain state originally created to keep fetuses still in the womb, and that other more important functions have developed around it. Dreams being therefore just spurious effects. They are not, in any event, clearly associated with mere unconsciousness, or with suppressed intentions or with traumas that we associate with residing in the unconscious. It seems like dreams are associated more with the conversions between sleep and waking up than with an entire phase of specific electrical activity during sleep, since the only reason why we associate REM with dreams at all is by waking subjects up when they are in the REM phases of sleep and asking them to report.

If we associate dreams with 'acquiring reality' then we can understand why dreams are not exclusive to the REM state and why it is difficult to relate them to more permanent mental states. Furthermore, studies of individuals who acquire motor activity during REM (RBD – REM sleep behaviour disorder), suggest that, apart from indications of neurodegeneration or other damage, that their experiences are the result of incorporating the effects of neuron stimulation from elsewhere in the nervous system. It is consciousness invading the dream rather than the other way around. In fact recent research appears to show that in some states of non-REM sleep when sleepers' neurons are stimulated before being woken, half the subjects report dreams while their brain states appear much more like the waking state than the half that did not report dreams. People can also slip into the dreaming state almost immediately on falling asleep during the waking hours, like when taking a nap, because the body has high levels of neurotransmitters due to being fully awake prior to the nap. Dreams reported in this state tend to be more procedural and explicit than the episodic type dreams that tend to occur in the natural phases of a night's sleep. We can call this form of dreaming close-reality dreaming.

Here's an example of how dreams may be associated with acquiring reality. I was having a dream. In the dream I suddenly realised that I wanted to wake up. As I began to wake, I felt the jab in my ribs of my wife's elbow, and as I was aware of the being awake in the room and the bed in which I had been sleeping, I heard a loud snore. Clearly the events in reality occurred in the opposite sequence. I snored, my wife jabbed me in the ribs and I woke up.

Hypnotists seem to manage to induce a pseudo REM state in subjects (interestingly at least 10% of the population can't be, or won't be, hypnotised) and where we can see more clearly the interpenetration of waking with sleeping, or at least some dissociation in the mind between a still-aware self and the hypnotised state which doesn't seem to happen in REM sleep. They report no pain even though though their still-aware part reacts as if they were in pain. Hypnotised subjects still have moral qualms and resist doing things they would not normally do; their so called 'child regression' behaviour, for example, turns out to be how an adult thinks a child behaves.

But why stop there? Why should we not think that the dream itself was the culmination of that series of events: snore, jab, waking, dream?

If most dreams are actually more related to reality than to imagination then there are a number of problems associated with dreams that are magnified by this fact, and we can separate close-reality dreaming from far-reality dreaming (I will introduce another category of dreaming awareness distinct to close-reality dreaming and far-reality dreaming called, *real dreams,* further on). For example, when our awake selves set the mind to remember and to imagine, we do not have the same sense of being alive in a dream-like narrative while we do so. We don't feel embedded in events that are driving themselves. Even when our minds drift and freely imagine it is not the same as dreaming. We use a word, *daydream*, to show that this form of waking imagination is not the same as what we do when we are asleep. A daydream is not a dream, because dreams happen *to* us.

So what could be far-reality dreaming in a body that shows all the signs of being awake? If they are just the result of the brain moving stuff around, this demands the question, namely according to what scheme? If this were the case then it is hard to see where the sensations of complete experiences and activities could come from or what, indeed, is doing the experiencing. Are we obliged to consider the idea of the unconscious as the coordinator of this brain activity, and dreams give us the glimpses of it free of the aware waking self's editing or sensory constraints?

It is tempting to think so. We have the growing suspicion built on many years of neurological and psychological study that the unconscious is not just a repository of suppressed intentions and desires, it is also the bedrock of our identity. Consciousness, in this view, is only rather feebly in control of decision-making and intention, and dreams, therefore, do reveal significant information about brain states.

But does this view describe the actual state of affairs?

In the general dream state the nervous system responds to the events within it in the same way it would if the subject were awake, only the activation energies are lower and the spine itself is inhibited from passing the nerve impulses on. In the far-reality dream state we have the consciousness being fed not from the senses but from the memories, and yet the experience is similar to the waking state where the information flow would be in the opposite direction (*remember the film strip?*). But more importantly than this, during the dream the consciousness creates new experiences and feeds them back into memory without confusing – apparently – the original memories because we do not experience mental confusions after a night's sleep where memories have disappeared or been overwritten by our inventions from the night. On the contrary, a night's sleep is essential for our health and for preserving the consistency of our awareness, and people will die without it.

In fact researchers think of the stages of sleep as where the mind progressively gets less tuned and channelled by the muscle and perceptual input of the waking state, and which gets progressively less organised as the physical influences are withdrawn. What we call the dream state is just a bit of chaos. All of which seem to suggest that a dream is an aspect of the continuity of awareness still in contact with the world, and behaves much more like a waking consciousness than an unconscious one. For example, as we noted hypnogogic people can fall immediately into REM sleep even as their acetylcholine levels are high from being aroused when awake. The hypnogogic transition between sleep and wakefulness is generally thought of as a point where consciousness starts to break up and images start to become mixed, to become surreal. The images pass in front of the mind like a scenes and lucid dreamers can control this to some extent. It is not until the REM state that the dreamer is immersed within the dream as protagonist rather than 'observer'.

This would imply that the seat of awareness still exists somewhere apart from the brain's actions and to be supplying a form of organisation to the dreaming brain activity since our memories are not destroyed each night of dreaming. How is this organisation achieved? Is far-reality dreaming just a close-reality dream with something added or something taken away? We typically answer this with the notion of the *unconscious* filling the dream with events important to itself.

But there is yet another interpretation which we are approaching carefully from an angle.

It had always seemed a mystery why dreaming did not play a more central and existential role in social life, when a friend, the same Peter of the coincidental photo described in the chapter on coincidences and who was researching the life of some mediums in Dutch colonial Java, told me about a pre-Islam Kejawen sect that he had come across there who specialise in dreaming. They consider that the human ego interferes with the proper practice of religion and believe the purest way of communicating with God is through dreams. The members of the sect write down their dreams every night and come to the place of 'worship' (strictly speaking the sect has no special rituals or sacred rite; the religion took on formal organisation only around the turn of the century, even forming links with the Theosophical society) to discuss them and have them analysed by the sect's experts with the intention of finding inner peace and strength.

The unconscious is thought of by various thinkers in slightly different ways. Psychological theory furthers the notion that it is a reservoir of instincts and urges, fears and desires all of which are suppressed to some degree or other during wakefulness, but which can emerge into our wakeful thoughts and direct our behaviours out of our conscious control. The actual mechanism whereby this happens remains elusive but still adds to the mystery of the bandwidth problem of brain activity since, with an unconsciousness in place, in addition to the waking activity of perceiving and the laying down of memories there must run a parallel processing system that stores information, intrudes upon and modulates the awake mind during decision-making and continually polls the waking mind for opportunities to express itself. This unconscious does not seem to rest, and is supposed to emerge into dreams, perhaps all dreams, during sleep.

This style of spiritual 'communion', however, has never become a commonplace feature of religions. Even the modern psychoanalytical efforts of dream interpretation have never produced any meaningful results. Dreams do not lend themselves to the general case; they are too personal, and have very little common ground, which is interesting given how our conscious selves enjoy a great deal of commonality with others. Dreaming reveals a particularly individual flow of information, more pertinent to the individual doing the dreaming than the social group. What makes dreams so personal when much of what we experience in life, both in instincts and drives as well as in common experiences, is shared. What is producing this particular flow of information?

Let us recall the waking sequence I mentioned earlier at its face value and consider that dreams play the key role in constructing the unconscious rather than the other way around. It is not the unconscious that infects the dream with its underlying information, it is the dream that lays down the information for the unconscious.

But why? Let's come at this proposal from a number of directions.

Dreams are curious affairs in many respects. I have had dreams that form part of my constructive memories. I remember them in exactly the same way I remember real events, and they have similar complex emotional and intellectual content as any real experience in the world 'out there': emotions like wistfulness and longing, feelings of success and the joys of solving a problem, or being outside in a beautiful landscape. The events or emotions are whole and distinct just as events in the real world appear. When I recall particular events that I have dreamed, I remember the dreaming experiences they come from; I remember the provenance of the ideas. So it isn't just the idea in the dream that I remember, but where it is located in my mind-scape, just like any newly acquired knowledge derived from the waking reality.

I have also had dreams in which I have been quite aware that the dream has been composed with events from other dreams (as distinct to an old dream being revisited) but that I, as the protagonist, am using memories from my experiences as the protagonist in other dreams during the current dream. A structural complexity every bit as true to reality as reality itself. I wonder how this is possible when dreams are constructed without apparent reference to external senses and feedback from movement or maybe just the result of chaos. One wonders more about the evolutionary purpose of dreaming, and we will get to this in due course.

Lucid dreaming is an expression applied to what some people can consciously do inside dreams (first coined by van Eeden in 1913). It appears to occur, or at least begin, during REM sleep, where individuals can become aware that they are dreaming and can direct the events within the dream in some ways, but not in all. Yet the recorded wave patterns are very different to REM sleep with much more activity in certain wave bands suggestive of the waking state. Lucid dreaming, as indeed one might expect, is a hybrid state indicating a capability to control the process of waking up rather than being a purely psychological state within the process of dreaming, and thus tending to agree with our notion that dreams are about a continuum of consciousness rather than some other formal process.

In fact, we know that lucid dreamers have some conscious control and that memories can be accessed in these dreams because some dreamers have been able to consciously use eye movements to signal the beginning and end of tasks they have been asked to perform within the dream. (Eyes move anyway in REM and thus seem to be under the control of the brain while other muscle movements are not. Attempts of subjects to clench fists have been less clear). In some cases, lucid dreamers have been asked to estimate time within their dreams using eye movements at the end of counting 'one thousand...two thousand...etc. and the results suggest that the time to do this task within the dream is much the same as real time, which may not disagree with the idea that only brief moments of changes in consciousness cause dreams since the movement of the eye and the verbalisation of the count must agree with the time it takes to actually move the eye and to speak the words. Events in ordinary dreams without such self-directed muscle control can still occur in briefer imaginary time. Lucid dreamers' eye movements when they are in REM sleep also appear to be more strongly correlated with actual waking tracking movements which suggest that they are more actively looking at things in their dream, although no one has been able to relate the saccades of eye movement in REM sleep to the narrative of the dream that is reported, so this may simply represent the extra discipline that a lucid dreamer has learnt to use over his or her eye movements and nothing more.

There are interesting inhibitions in lucid dreaming; for example, one can easily fly, but the higher one goes the more difficult it becomes (perhaps because the imaginative reconstruction of the breadth of a landscape from higher and higher points of view becomes more and more difficult to construct even though the level of detail diminishes). Lucid dreamers have been able to switch lights on and off inside their dreams ('off' is easier than on), observe their reflections in dream mirrors (although their reflections are almost always unstable and morph into other things readily) and walk through the mirrors into new scenes (just like Alice in Carroll's *Through the Looking Glass*). In general, however, lucid dreamers have limited control over the content of what they dream, although they can manipulate the content they are served up (such techniques can be used to combat nightmares), and fail to derive any significant novel departures in matters of interest, and experience only feelings of well being. Lucid dreaming then, in terms of the continuum of awareness, seems to occupy a middle ground between close-reality dreaming and far-reality dreaming. It is a shallower experience than the one I shall describe below, a *real dream*.

Carlos Casteneda might disagree on that point, although it seems to me that his examinations of these mysteries are through hallucinations and drug trances rather than through the dream state as such and require somewhat different explanations that we will not go into here.

Let us stay with dreams and consider what gets added to consciousness in the most profound and coherent of them.

I had a wonderful dream experience recently, something that I have no doubt I shall remember for ever.

For reasons lost on waking, I was with companions in a high forest. We seemed to be running, but not being chased. I did not look at my companions but I knew they were there. At some point, I slipped on a contraption that was a cross between a small dingy's set of sails and a wing, and I jumped out into space and flew outwards beyond and above the sloping plateau of trees.

I came to the drop off and far below was a valley. I soared on, feeling momentarily exactly the same fear of heights that I have in real life, until I re-gained confidence in my flying contraption. I made some adjustments to the sails and gave in to the gliding, passing between tall thin trees that seemed like tall, willowy silver birches, bare white trunks with small leafy crowns, while the slopes below to either side were wooded with other trees, broad-canopied like beeches but dark in colour like fir or pine.

The valley gone, I floated over further wooded ridges. I glided close over another plateau of long bleached grass where small snow-white deer or goats were hidden in the long grass. Eagle-like birds with black eyes in white circles like lemurs were lazily drifting in the air. I knew there many other animals there but my passage over the plateau was quick, and I didn't really catch sight of them. My attention was being taken by a destination further on that I already knew. I landed quickly and simply, taking only a few steps to stop. Up to my left was a low building that seemed to be a mountain station or refuge where people were staying. It was getting dark and there were now only a few patches of snow around.

The dream's texture and lightness altered. I had a pragmatic errand to run. I was going along the mountain track, with the refuge behind me when I was passed by a car. The memory peters out.

The puzzle about such a dream is how the brain can synthesise such an illusion of reality from things that have never quite happened, and when it is in a physical state that is quite different from the state it normally is in when awake. One puzzles over the fact that, because the spine is paralysed, there is no real-time feedback from the voluntary muscles that one would experience in reality. We have the illusion of consciousness and acting with awareness without actually being conscious or being able to act intentionally.

How does the brain stitch together composites from my memories with such verisimilitude? It's no mystery that experiences and feelings of real experiences could be accessed by my brain in dreams. What is mysterious is that the brain, though obviously lacking the complete sensory experience, can still serve up to my awareness a *new* experience, with all the convincing emotions and sensations that go with it, out of these old memories, leaving behind fresh memories to take an equal place in my brain alongside the (presumably) original memories from which they were constructed. 'Ordinary' flying, for example, soaring like an eagle over mountain tops, is a dream experience I have had many times. I can feel the sensations in the dream and remember them, even though I have never actually flown like an eagle and would have to be extrapolating from some other kinds of experience that I do have sensory memories of. How does that work exactly?

Now, I have been flying in microlights. I took a pilot's course over the Wiltshire downs many years ago. We flew two-up, instructor sitting behind, and, for a few days, I soared over the plains of Wiltshire gripping my anxiety with both hands. But there were no woods, no escarpments, no bleached grass plateaus associated with those experiences. I have done a lot of mountain walking and climbing in various landscapes and have looked out from peaks over ridges, some of which have been very steep wooded slopes, or grassy plateaus, but none have been quite like those in the dream.

I remember commenting to someone after my microlight experience that flying the hand-glider wing in the currents of the air was very much like sailing. Could this thought, arising out of the twin experiences of sailing and flying be responsible for the sensation at the start of the dream, the putting on of a contraption something like both a dinghy and a wing? Was the dream, then, working backwards from the fresh concept (flying is like sailing) to recover the two planes of the Koestler-like bisociation and combining them literally in a physical representation? Similar to that wake-jab-in-ribs-snore, and provoking the process we have been analysing all along. Does this working backwards in a dream expose a glimpse of a truth about the external reality?

Interestingly, however, shortly after this dream, I saw a YouTube video of a base jumper using a flying suit to soar over various landscapes in Norway and in the Alps. Not very similar to my dream but reminiscent of the point of view of flying, certainly, and it made me ponder about the sources that had gone into my dream. But then, at the same time, having an iTunes podcast set to receive automatic pushes of selected YouTube videos (without my intervention), a video of people parasailing off high snowy mountains arrived. Now my dream came into full focus. The soaring of these pilots in their 'contraptions', the bleached out snowy landscapes, the pine trees and vistas of deep far valleys all slotted into place. Was this the foundation to my dream? Did the video stimulate the brain to create the past memories of the dream? If so, then the question of the shift of point of view from observer to protagonist within the experience needs to be answered (as well as the significant differences between the video and the dream). Could I have dreamed that video before I saw it? Or are the two similarities just accident?

It seems that only on rare occasions do we re-live events from the past. Usually, memories are viewed from a vantage point; they are observed rather than experienced. So in this respect at least dreams more closely resemble a conscious experience of events and resemble less the process of remembering, which, again, is curious if there is no sensory input and everything comes from memories. Later on, I shall propose a mechanism for memory, functioning through language, which will help us understand this difference.

Many 'ordinary' dreams that occur during a night's sleep may well be the result of prosaic neural activity. But there are other types of dreams, and the dreams I want to focus on are those that persist in memory rather than disappear: dreams that I call, *real dreams*, and are unlike all those dreams I have had that I don't remember because they have faded back into the impulses from which they were constructed.

Real dreams have been the ones, without exception, that have given me the feelings of *déjà vu*, or more precisely, *déjà vécu* (the more complete sensation of having already *lived* the episode). For me, *déjà vu* is the realisation that I have had a *real dream* of the present scene rather than having the sensation that I recognise this scene (a French term *déjà reve* is often used to describe this experience). This sensation appears to be more profound than mere recognition because it implies two existing sets of experiences, the dream and the real event, which coincide rather than merely the attachment of an emotion of familiarity to a current event. It is tempting to interpret this *déjà vu* as the workings of a dream going in reverse. Instead of the brain organising the dream experience as unlike reality, it organises memories and relates them to real experiential input.

Not every real dream, of course, produces the sensation of *déjà vu*, but every feeling of *déjà vu* has been associated with a real dream. I have had them throughout my life from childhood on. Here's an example. Many years ago in my South America travels, after my meeting with the Chilean lawyer I mentioned earlier, I reached Peru and the old Inca town of Cuzco. It was in the 1970s, and the Cuzco streets were not well lit. I had a very nice time with some students in a bar and I walked a girl through the dark streets to her pension. As we turned a corner into her street, it seemed I recognised the stone buildings and their black outlines, a few steps further on I saw a glowing lamp in second story window above the doorway to the girl's pension, and I was transported, as it were, by the remembrance of this entire scene having appeared in a dream that I was remembering. I knew where I had had the dream. In the basement room of a terraced house in which I lived during my first job in London three years earlier and a little before I took on the job where I met the Chilean lawyer.

I am not even alone in having this experience. Others also report this same sensation, and still others report experiences with their senses, like auditory or taste, that they feel they have had before. *Déjà vu* (first described by a French psychic researcher Emile Boirac in 1876) has been studied and it is often associated with seizures or brain trauma and is thought to be related to abnormal neuron firing in the brain, but this begs a significant question about memory and time in the brain, and takes us back to the problem of the film strip and what needs to be happening in the brain for us to consciously perceive time going consistently in one direction when memory is involved.

I had many coincidences in that basement room. One day an extraordinary summer storm sent a flash flood into the basement. All my books and papers were ranged around the walls. Just a single book was untouched by the flood. It was Saul Bellow's *Henderson the Rain King*.

During the episodes of unusual brain activity we call *déjà vu*, it has been assumed that either consciousness is recovering the memory as it is laying it down and so it seems like having the experience and remembering it are occurring at the same time, or alternatively, consciousness tags momentarily short-term memory as if it were longterm memory. These aren't really explanations, however; it's just saying that *déjà vu* is *déjà vu*.

Many people's reaction to stories of this kind where dreams seem to match with a later reality is to say, it's just a coincidence. Given our discussion, however, of coincidences, we are entitled to ask, a coincidence of *what*, exactly? Given that we seem to expect impossible-to-quantify variations of idea and event evolving in human narrative time, shouldn't we be astounded that such 'coincidences' could happen at all. Just invoking some version of the law of large numbers (so many people dreaming; so many events) should only 'explain' a very small number of occurrences, not the fact that they are commonplace or the fact that these type 2 coincidences are so particular to individual minds and to their personal connective awareness of the world at a point in time and space.

The connection of my real dreams with *déjà vu* seems to suggest yet another phase within the dreaming, since neither of the two descriptions above seem to apply to the case I am describing. To return to points we noted earlier about **REM** and **NREM** sleep. We saw that dreams arising from **NREM** states (like lucid dreaming or hypnogogic experience) appear to be different in character to **REM** dreams, namely more prosaic, more like representations of experiences already absorbed into mind, companions to the basic repetitive functions of the day-to-day. Further, the surreal images of the hypnagogic experience do not appear to carry emotional content. They are not weird or frightening; they invoke no reflection or commentary; they just are. The images function like language. Indeed, as I am about to suggest, these images do not provoke language because they *are* language, and that the transitions between waking and dream reveal the precise evolutionary role of language in the brain. The real dreams I have described occurring in **REM** sleep, are neither lucid dreams nor just the imagining of alternative scenarios but another way of experiencing events, a state within a state. If dreams result from the interpenetration of consciousness into the brain's function, then a real dream indicates the perception of a distinct reality. That is how real dreams seem real, because they are not dreams. We shall go further in this direction.

Lies and language, fallible narratives and dreams

Let us compare the real dream/*deja vu* experience with another well-studied but no less puzzling phenomenon of the brain, namely false memories, or, more strictly, false narratives, where the brain adjusts established memories to conform to later experiences, or where the brain creates entirely false memories of an experience it never had. We have already noted how social media is beginning to create memories for people who have never had the actual experience, but the phenomenon of false memories is part of everyone's daily life. It isn't just the recall of our childhood experiences (when perceptions are naïve and parents feed us with interpretations) which is fallible, we can get imbued with false narratives at any time. It is well-known for example that eye-witnesses in court cases can adjust their memory of, let us say, the violence of an event to conform to the degree of 'violent' language or vocabulary used in questioning them (for example, do the following questions invoke different impressions of violence; *Did this suspect snatch the bag and take off?*; *Did this assailant mug the victim and run*? These effects can be very subtle. Supposing I replace '*the bag*' with '*her purse*', and '*the victim*' with '*the girl*', what would you then think? What meaning do you take away from each of these three-word headlines about a company officer leaving the troubled social media company Twitter, published on 10 November 2016: *Twitter COO departs*...(Bloomberg), and, *Twitter COO bails*...(American City Business Journals)? These effects are now well known mainly thanks to the work of Elizabeth Loftus at the University of Washington since the 1970s, and it is clear that witnesses or suspects can change their visual memories of the appearances of the players in a crime, as well as the sequences of events, according to the manner in which they are questioned. This kind of collaborative falsehood occurs not only in interrogation but informally with groups of people. Here's a silly example of one I experienced myself. Staring at a 'suspicious' star one evening with my school mates after a discussion about UFOs, the star was seen by all of us to move, to drop something off it, to change colour and to get closer. Pouring cold water on our mysterious light, I checked off the 'star' on a sky map I had found and discovered it seemed to be in fact Aldebaran (in the constellation of Taurus near the Pleiades). There it should end, but there are yet more tempting details. It turns out that Aldebaran is a red giant and thus with some colour, with a dwarf companion star and with at least one giant planet, and it has variable luminosity! Could there have been some way in which our eyes were perceiving this pulsation or perhaps an occultation? Or even more mysteriously, did we in some way already know this complication about Aldebaran without being aware of it? (There is yet a third possibility which I will leave until a little later on.)

Some psychologists believe that being able to change narratives like this is a social requirement and an essential part of being flexible about strategies or to think creatively. To be able to give different weight to the values of holding onto an opinion or to changing one's mind is evolutionary beneficial and we should not be surprised that humans have this 'skill'. Certainly that explanation may go some way to explain the phenomenon of the manufacture of false memories, but maybe there is more to it. Humans, after all, are not all flexible thinkers and creative innovators, and culture is remarkable for its stability and consistency even in ideas and practices that make no sense. This 'skill', as I shall argue, implies some interesting conclusions especially as the algorithms of thoughts – essentially the interpenetration of minds – are shared through *language*.

Let us travel to the Amazon basin and consider a people there called the Pirahã. They appear to have no historical culture but just living memory, and who communicate in single sentences without subclauses as described by Daniel Everett (1986 and later). The reason why Everrett was there and the reason why he delved into the language so deeply was that he intended to translate the bible for them. (One doesn't know whether to laugh or cry. More interestingly though, Everrett gave up religion after this experience.)

Everett discovered that the Pirahã understand language literally and have no interest in anything that may have an existence away from them or which is not immediately experiential. They have no social hierarchy or cultural divisions. They have no number system, no tenses, no gender or person, no colours. Their entire system of knowledge is encoded in eight consonants and three vowels but given enlarged meanings through stresses akin to melody. They can sing or whistle a conversation. Although a school has been installed for them, they have no appetite for learning anything new, and when shown techniques like canoe making which they followed to make one, did not use the skills when they wanted another. Children have learnt to discard new things that they may have understood. They do not draw and can't interpret line drawings, but have been observed to make models of pictures of objects they have seen, suggesting they could, but which are then ignored. Even though some have learnt Portuguese through contact with the outside, it has not enriched their own language between themselves. They see spirits in their environment, who do things then and there, and they dream of spirits also doing things which are considered to be immediate experience as if occurring when awake (conforming to our earlier suggestion associating dreams with waking up), but have no techniques for transcending everyday reality, for accessing broader states of consciousness (like controlled drug use conducted by other Amazonian groups). They seem content to live within their own world.

With the analysis of this group Everett seems to refute the final surviving Chomsky assertion that all languages have recursion, which has begun an intense debate among linguists. We note something, however, that has gone unnoticed in discussions about the language, namely the Pirahā do not sleep for prolonged periods and rarely sleep through the night. They take short naps throughout the day and night, partly, one supposes, because survival requires constant alertness to the ever present dangers in their forest like snakes, but which suggests that they may well be dream deficient. Lack of REM sleep generally interferes with long term memory, ability to complete tasks, increases sensitivity to pain and reduces flexibility in response to difficult situations. All of which may be reflected in Pirahā thought, and may be the reason why they are resistant to expanding and developing their language.

Wittgenstein (at least, in his *Tractatus*) thought images and language were of the same logical form, but later gave up on this picture theory because he was bamboozled by the idea there couldn't be a direct correspondence between a true statement about matters of fact and a picture. What we are saying here is that language is not a picture of reality so much as a picture of memory.

The Pirahā have, in fact, produced a language that invokes minimal images and produces minimal changes to images. Their mental state is without prediction. Given our argument above, I would expect that it would be difficult to create false images in the mind of a member of this group using language. They are cut off from the possible, from variations of choice, from the future. With Everett's descriptions, however, it seems clear that the Pirahā occupy the present by an effort of will and that they consciously avoid situations that might disturb their present moment, for example, by refusing to build a second canoe – they said, *the Pirahā are not canoe builders* – and removing children from the school when they began to get the hang of counting. In other words, the future is *in* the Pirahā, they just don't want to engage with it.

One might speculate about what the Pirahā are sensing about the effect that language development will do to their lives. The use of language requires constant alterations to memories. When I change my mind about a belief say, I am replacing one variant of the idea with another. Imagining narratives about which I do not possess actual data requires the manipulation of classes of ideas and qualities, of understanding options and likelihoods, just as language use requires such things to communicate. Complex language is *disturbing*.

We may question how evolution supplies all of the answers to this mystery of the truth of memory. We can certainly say that when a brain invents memories, it is just following the method of creativity that language has introduced into consciousness. In this sense the ability to create a false narrative is an evolutionary adaption. It might, in fact, be the key development in the brain to enable culture. Humans who were able to do it were more successful in a changing environment, even though it also introduced imitation as well as lies and deceptions into human relations, but maybe there is more to it. In a very real sense, a prediction is a lie that comes true.

Ideas and images seem to be stored together at least in some form of parallelism otherwise language would not invoke images (as opposed to simply labelling them) nor could language be used to 'paint' images (*radio has better pictures*) and create scenic narratives. Images therefore may be subject to precisely the same forms of incoherence as statements in language. The very existence of figures of speech like metaphor, synecdoche and metonymy imply such fluidity in memory that it seems quite logical that visual memories are less durable and exact than we expect, and do not arbitrate truth. Further, since thinking in images preceded language, we should expect that the brain handles the two in a similar way even if the brain had developed a way to understand sensory input, including visuals, symbolically way before the advent of language. Thinking, after all, is connected to the visual cortex which lights up when we ponder.

So, given we can construct flexible sentences in a language with nested or embedded ideas within (recursion), we should expect that we construct images in precisely the same way. Indications of the easy parallelism between image elements and language can be found in the many occasions where, for example, scientists solve a mathematical impasse through images, or where people declare that they think and manipulate concepts in 'shapes' rather than symbol. (I did this myself when young and to some extent still do it.) In fact, there is an effect which also supports this idea of the 'frame' and the content 'list' called 'crowding', where perceiving a face or a word is inhibited in the same way by other 'frames' encroaching upon a critical spacing or by a crowding of parts within the word or image.

Yet, while manipulating concepts can be done visually and imaginatively, a 'vision' of the future is still only a memory: an image cannot fail to be what it is. The serious failing of the architecture of image memory is in making *predictions*. While it is through language that false pictures can be composed, it is only in language that we can formulate a prediction for which there is as yet no image. But this creates its own problems.

For the re-writing of a memory or invention of a narrative to be possible, of course, and just as language comprehension requires pre-loaded rules, symbols and algorithms, the conversions of memories, the adjustments of narratives also require pre-loading, sets of alternatives which have their probabilities (these may be Bayesian-like priors, but the argument is the same) with which the changes can be effected. To put the problem more bluntly, from where do we fill our imagined narrative if it is not from what is already in our heads?

Pre-loading is part of the answer. We can be pretty sure of this since we know that people with amnesia find it difficult to imagine forward in time. But there must also be an ingredient of instability in the memory for it to be malleable, for it to be dismembered and used in new narratives, for it to have a degree of doubt (in the brain's terms), or for it to be open to change. How is it that memories are malleable when they are determined by senses, established in networks of neurons and fixed in proteins in cells? As we have been arguing and as the experience of the Pirahã suggests, the translation system that lays memories down in consciousness *is* language. This is the innate property of humans that Chomsky was trying to expose – his universal grammar. A search which has been somewhat waylaid by the top down approach. If we put aside the notions of an innate system of syntax or structure between the elements of all languages, and think about the innate relationship between consciousness and the memory of *images* and all the elements contained within them we will get closer to seeing how all languages function in consciousness. It is not likely that a language could function if it does not follow the way the brain handles images (remember *Ithkuil?*), and underlining the modern rejection of the Sapir-Whorf hypothesis that language determines thoughts. All brains lay down images the same way and different languages arise in social and environmental contexts.

The search that linguists make in actual language constructions for universal *language* rules seems, therefore, misconceived. The reality is that every language follows the way the brain lays down a class of scenic narrative memories and which is already in place before comprehension can begin. The material substrate to language use will be found in the more generalised imaging neuronal networks, and the way image information is stored by them. We know this from the way that blind individuals can re-purpose their visual cortex to process language. Interestingly, their visual cortex is stimulated only when the sounds are intelligible, which seems to coincide with our suggestion above. Furthermore, blind people dream and, even though they do not report images, they still express REM which again suggests a universal connection between imaging and language.

The power of words in the mouths of the hypnotist, the psychotherapist, the conman, the salesman might be explained by their ability to reconfigure these fundamentals of human awareness. Creating a false memory belongs here.

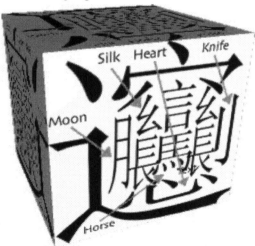

We should expect then, that even though senses can be deceiving and participate in false perceptions, the way the brain can produce a non-experiential narrative in consciousness is through an underlying mechanism of image memory manipulation, and which belongs to every language. But what is this mechanism of image memory that allows for non-literal dreams, allows for creativity in language?

We can draw some conclusions about this bedrock to language forms knowing what we know about the Pirahã. As we mentioned earlier, one way to store images which readily explains the false narrative capability would be if the brain tended to store a contextual 'frame' into which it puts essentially a list of items that belong in it or are subordinated to it as in a list of reduced detailed items or catalogue. An image is a set. Since some of the items of a set will link to other images and lists, recursion appears naturally. A picture is treated like a language sentence with recursion. If we look at Chinese character formation, we can note the same process at work. A contextual 'envelope' contains one or more radicals conveying extra dimensions to the meaning or 'story' told by the whole character. Each of the more dense ideographs contains recursion to convey a complex whole. Here (above) is the most complex traditional ideograph (not used in reformed Chinese), *biáng*, describing noodles. It contains several radicals which are themselves complex ideas, namely recursion.

We can model at least some forms of creativity with this scheme by rendering the bi-sociation of Koestler more precisely. Two (or more) image contexts or frames can 'mutate' into one different to all by the merging and crossover of the associated catalogues attached to the frames. Merging could be additive or subtractive, giving rise to the possibility of two kinds of creative act, a fresh complexity to an idea with say, a catalogue within a catalogue, and a distillate or a simplification of a catalogue. As distinct to full creativity, false narratives occur as an arrested process where just a catalogue or list to a single context or frame is adjusted or replaced without access to sensory data. Since a 'list' to a frame will contain some sensory data items we can even explain the phenomenon of phantom limbs and hallucinations just as language can provoke sensory experiences or layer meanings. So too, we explain the dream, where frame and list are loosed from their sensory narratives and determined clocks, and we explain creativity; the creative act (mixing of frame and 'list') being initially a 'false' narrative, one that does not exist, which is then confirmed and adopted at a later moment through experience.

If the Pirahã, however, do not use recursion in language then it is because they self-consciously want to avoid dreamscapes and the use of imagination in creativity; they want to avoid an uncertain unstable world; internal dialogue is trouble. It is as if they know that language is not to be trusted. (We will get to the question of how they might know this in a moment.) If consciousness is filled with ever-mutable objects of language how can we ever distinguish between false narratives and true ones, or even divide our awareness into the time zones of past and prediction? For the Pirahã, the development of their language might even be a moral question in which they prefer a single existential 'list' for each context, and other possibilities and mixes are suppressed. The Pirahã do not seem to reason particularly in a Bayesian fashion, and since Bayesian logic is very much the logic of sets and their contents, this helps to confirm our conception of the generalised language's relation to visual memory.

The Bayesian judgements still require visual memory even for literal recognition Pirahã-style. The language that moves perception into data needs its pre-loading, its statistical sets of probabilities, its virtual narratives, it's *start*. Where do these come from? What we know of the Pirahã exposes the significant question, From where are the variations that we need for imagination composed?

> Schizophrenics can confuse internal and external realities, so one wonders just how the exposure to false narratives through VR is going to help them revise or simply organise their 'worlds' of responses. The combination of VR with social media is, too, a step towards an insidious world where individuals are not only fed what truths they are required to believe in but also the desires they are required to have. Since VR worlds are controlled, it will be tempting for psychologists to shift their intentions from aiding a patient be adult and self-reliant in the world to having them live in the VR and away from the disturbing and unreliable circumstances of the exterior world.

We can observe in minds, certainly in the general case, two language-based methods of memory processing which help answer these questions. We have examined one, the creation of false narratives where the brain mimics fresh input for the senses with memories already laid down by creating content lists for a given frame. The other is our internal dialogue. Because of the limited numbers of neurons in the brain, a neuron in a neural network is likely to be a node connected to more than one decision-making net, and a lot of the work of the brain is in controlling and suppressing unwanted network activations (crosstalk) through the nodes. This, in fact, may be the earliest function of language within consciousness where our 'internal dialogue' (a broader function than just being aware of the voicing of thoughts) became one of the ways we suppress node functions in the neuron network to reach some kind of consistent internal consensus as to the reality of the moment.

We can expect, then, the upcoming use of Virtual Reality 'machines' is going to cause some interesting problems. Already psychologists are using it to examine phobias (like 'locking' subjects with claustrophobia in virtual cupboards – no doubt the screams are real!) and states of mind like depression. The idea that disturbed or unhappy subjects can then go away from the treatments without having any residual notions from the VR experience left inside their heads seems unrealistic. The problem with VR is that it going to be narratively denatured. It will suffer from the problem that cinema has always suffered from, namely *the script*. That burden on developers more difficult than the technical or even the relevant, being the art, the truth and the profound relationships of time.

The problem for false narratives is that to have meaning, to make sense to us, they have to already make sense. Narratives of the external world make sense because the sensory input and the perceptions flowing through them are already organised by the external narrative to make that sense in the brain. Whereas a false narrative, even when it does not refer to an external reality funnelled in by the senses and interpreted in the perceptions, must still follow the structure of a memory and borrow from what already is (suggesting that a false narrative, visual or otherwise, can only depart to a certain degree from the memory). While we might explain that one important form of recall within consciousness happens through the assembling of separately stored contexts and their associated 'lists', it is not yet sufficient insight into how self-awareness works, and particularly it does not do much to explain the sense of being *in* an experience as it happens, of being an actor at the centre of meaning, which is what we will need to know when it comes to understanding it even if we have been successful in creating a self-aware entity in our artificial intelligence laboratories.

Again, the experience of the Pirahã can help us examine this question. Something about memories bothers them. Without language, thought is dominated by the most immediate images, the most literal conjunctions of objects and events. The Pirahã insist upon a present and, since they lack language forms, they seem to explain their projections into the future – by dream or imagination – as being caused by another entity, a spirit. I asked the question earlier what makes the Pirahã so suspicious of strengthening and developing their language system. I believe they fear that embroidering their wordscapes will make them lose their footing in the present, turn them into those spirits they see, or worse into false narratives. They reject both the future and the past to avoid turning into ghosts.

Thanks to the Pirahã we can see what role language is playing in consciousness. Language enabled what is called by linguists 'parsing' of the perceptual frame to separate it from its contents. The 'lists' produced become *options*. Thus, the personal dialogue of consciousness became the mechanism by which we detect the passages and the mutations of time. It is the process by which memories can be brought back into the light from the deep tunnels of the brain and united into narratives of the past and projections of the future. Our consciousness was liberated from the direct perceptions of the now and became the traveller passing through the temporal dissections of the whole, the constant observer to mind layered by times.

We mentioned earlier Tegmark's suggestion that we should view consciousness as a new form of matter and we argued that the principle characteristic of this 'new state' was that it was continuous in time, distinct to atomic and sub-atomic states where all is transformation. Consciousness is continuous from birth to death through all the transformations that occur within it. This is how we can construct a sentence before we know what it is. No other physical state is maintained in the same way. The continuous nature of awareness does not therefore require separations of verbal constructs (tenses) to show its passage from a past to a future. And, indeed there are many languages that do not modulate its verbs in time, for example, Tagalog, Native American languages or Mandarin Chinese. Mandarin Chinese closely adheres to our frame/list scheme for language by expressing recursion within the pictogram units of meaning. The language has no verb tenses to describe different times, it uses marker characters or specific words to define the time of the action or statement, but once the time layer is set by a marker no further indications of times are required. For example, the same character *le* is a completed action in the past marker and sets the layer of the past for a statement that remains unchanged in the layer it has been placed. To express the future, a concrete word like 'tomorrow' or 'next week' sets the 'layer' for the meaning. *Le* is also used to indicate a 'past' occurring in the immediate future (along the lines of *this would have been done tomorrow*).

One wonders how the brain can layer these relative times unless they actually play out within the mind in the same relationship. Just as language follows the storage mechanism of image, it must also coordinate the times in which they occur and the potential to change to another state within them. We may be the subject of the sentences but are also inside the tunnels of operation as it were. As observers moving within memory, within the architecture, we are *in* the experience; as actors of remembrances as much as the object of events. Subject and object, qualified and subordinate to the verbal connection of time. And since events happen beyond our boundaries, their coordination within our minds must also derive from beyond the boundary.

It is interesting that the 'frame' and its contents applies to not only language but to what appear to be the other fundamentals of human interactions with the world, namely art, music and mathematics (see **Boltzmann Brains**). The scheme readily explains, for example how a musical melody can be recognised by just the first three notes, and why sets are so crucial to understanding mathematics and logic, and why early Human's engagement with the world appeared in cave drawings which has always seemed to me to mark language lessons (from which hieroglyphs emerge naturally), rather than simply magical hunting rituals, and the first unembellished recognition of the differences between prediction and memory. The language lesson then being the extraordinary secret conducted away from the tribe in places where they couldn't be *overheard*. I find it very easy to see in this same scheme an explanation for the bisection of explanatory theories of reality into questions of forces (the 'magical' influences between remote events, what we might call the conceptual frame) and geometries (the 'list' – the laying out of a prescription).

Just thinking about dreaming has brought us to the verge of understanding memory, or at least a particular type of memory, the encoding of potential events that go on to form our actual life experiences. The 'I' does not bring these memories up into 'I'(eye) from a past store but simply returns the 'I'(eye) to the moments of those experiences. By looking at memory we may find the key mechanism to self-consciousness, namely time travel.

Time Travel

We consider consciousness as a different form of matter because, of all types of matter and energy, it retains the continuous global quantum formulation of its entire history from birth to death. Being present in a quantum field, the brain must process superpositions and entangled states reaching across time; it must employ forwards and backwards generated states in the understanding of the now. If we do not include that possibility in its description then we have to explain what the brain does with the exponential accumulation of data as life progresses. Similarly, it must include dreams as an equal component of its composition. Furthermore, from a logical perspective, sentience – the act of being self-aware – can only function spread over duration. Instantaneous *awareness* cannot exist (*pace* **Boltzmann Brains**); duration resolves self-referential paradoxes.

Now we can return to the idea of Boström's structural universe, noting how the Hindu philosophers had anticipated even this notion with their akashic field where all the options that are and ever will be are already laid out in some set of dimensions (a parallel idea to the Nverse and quantum multiverse), and consider what this means for the continuum of consciousness.

Certainly there are what are termed 'single field' theories which try to combine electromagnetic, gravitational, quantum and consciousness effects into a single field of dynamic points sometimes called twists or twistors (Penrose) that exist in five or six dimensions. These points can be collected into sheets of three dimensional structures whose variations stack up in the other dimensions. What we experience of the physical world is the first sheet, while life and mind occur in higher level sheets.

These ideas may have much to recommend them; for example, they imply that consciousness would have to be a component of all sheets at all times and thus one of the fundamental 'forces' in the universe, which is probably compatible with the notions we discussed about akashic fields and synchronicity of Jung and so forth. But they do not seem to explain what the 'I(eye)' of individual consciousness is and how it arises, and further, in any observation, memory and prediction would appear to occupy separate sheets in the stack of points in the field, so who is actually making the observation and how is consistency between the sheets maintained?

We have been making the case that the options provided by quantum theory in a measurement are real, and that they are *locatable*, which implies a natural correlation of coordinates among them. The tests of randomness using the Bell inequality with light from far away galaxies (and back in time) are considered to disprove the existence of pre-existing correlations between random states, but this seems incorrect. The past correlates with a future not with itself.

We are led, therefore, to the idea we mentioned before of a structure of all particle states laid out in some space, and that the experience in the brain of an entire pathway through them *is* consciousness. Any continuous track through them may coincide with another track with different antecedents. Individual pathways necessarily differing both in sequence and in timing. The track may shift accessing different points but yet remain intact.

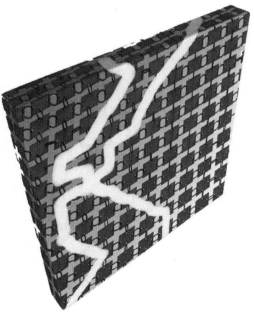

Since we are interested in consciousness we will leave the question of how we might interpret other physical phenomena in this scheme for the next book and consider first that consciousness is the wave function maintaining one particular *structure* of all choices within the universe, a set, where a trip through a consciousness would be a trip from point to point through the set, each point separated by a 'step' of minimum probability. Tracks will have different lengths and different starting points. Some reach further into the future or into the past than those they make contact with. Since tychons can be transferred in the contact each track makes with others, the continuous track traced through the points of each consciousness is also likely to change over time not only through sequential events it experiences in the 'arrow of time' but also through contact with another consciousness. Consciousnesses can transfer information to each other even at a less energetic level than states of matter.

What are the consequences for mind? Many puzzles about both the stability and mutability of brain function, let alone the exponential accumulation of data, are avoided if we consider that there are fewer actual discrete event memories maintained in the present than we think and that often what is an explicit memory of an event is in fact a coincidence of volumes in the structural universe *re-visited in time*. We should consider that to remember at least some classes of event (those of intersecting cosmologies) is to recover, at least partially, the wave functions that altered the brain at the moment of the event. When I remember these moments, the wave function of the mind reaches back into the past of my time-line to the moment my consciousness was changed during that event. Each person can go back along their particular stepping stones through the universe of every possibility and experience afresh the actual coincidences with another mind and which have a location in the dimensions of the universe.

What a mind perceives of the world, therefore, is not always or only the result of quantum fluctuations of atoms and molecules in the brain's neurons during the present moment, but the result of the wave function of consciousness being a flux connected continuously with its entire history. This is the complexity which maintains consciousness and all the intentions, strategies, instincts, learnt behaviours and experiences of mind. All these are described in the single composite wave function that flows throughout the brain and through its sense organs to the external

---

Thinking of this structure in space-time as a quantum analog, we can think of this structural arrangement as a series of 'volumes' (being revealed in the Big Bang expansion), and where each volume is separated by a minimum quantity of probability, rather than being located in a smooth continuous space of action like a traditional field. Any departure from a given track through the structure would mean a transfer of probability which we may liken to the probability particle or tychon. Each 'volume' contains all possible states of an object or particle, including position in space-time, occurring over that volume, and the values of the fields are determined by the probability 'step' connecting the next volume. This we will call the entropy of space-time and has therefore some relation to the quantum value of the vacuum and the cosmological constant ($S\rho_{vac} = k\Lambda$) which gives a hint as to the energy of the tychon.

world during the whole life of that mind. Some of the wave function includes components that have been decohered and are thus lying forever in the past and some of which is not and remain in superpositions or states of entanglement with possible futures. Many atomic states within the brain are likely to be in the Zeno state, continually reset in the continuous nature of consciousness. Some memories may be like this.

So why is this different to the state of a rock or a star?

Rocks or stars have no need of a past for their continuity. They persist because they have something of their future in their present state. Their duration is their consistency. The brain, however, needs to go back to the actual past rather than its own past, because that is how it can coordinate information between other minds and share truths. The tool of coordination has evolved as language, into which the time travel of consciousness gathers the forms of intersection I call, *cosmologies*. Consistency between individuals would be impossible and language communication non existent if there were no individual consistencies as well as cross-over points in the structure of the universe, no way to continually share the past and maintain agreements on meaning. Not a once-and-forever past at all but a continual ever-living past.

Our minds are time travelling and are doing it continuously.

We can see at once, how difficult it would be for a Boltzmann Brain to occur spontaneously and continue on aware without the complete circuit formed between the future and memory. Such a spontaneous specific structure is very much less likely to occur than equivalently random configurations which are numerous, which is why we have our complicated interactions of organic life doing it for us. They are more likely in the long run to create consciousness.

The recall of a coincidence in narrative memory (that is to say, a point in a string of events unconnected by probabilities) is actually a return to the moment in the past when it occurred. This is true for everyone who was present at that moment and who remembers the event at a later date. In fact the state of such a narrative moment always embodies the consciousnesses of all those who are remembering that moment from all the times into the future that the moment is remembered by a brain. All participating minds in the event continue to contribute an influence whenever they remember it. The event that you experience is a summation of all those remembrances including yours.

In earlier epochs, humans were more like the Pirahã, without the habits of returning to past moments through memory. They remained frozen in their present, unable to explore beyond it. After the development of language when systematic recall became a requirement, human culture's evolution accelerated. Language became the means (through frame and list parsing that we mentioned) by which the revisiting of memories could be agents for influence and change in mental states. There was less dispersion between those who shared the explicit moments, less uncertainty in the pathways that were broadly laid down for them in the past.

The brain being our brain, however, we can expect that there is a tension between remembrances as the returning-to-the-actual past (explicit memories), as we find in type 2 coincidences, and remembrances as records-held-in-memory (implicit memories). The older we get, the fewer instances of time travel type remembrances will be required. Interestingly, as we do get older and the prosaic implicit memory system of the brain starts to break down, instances of time travel remembrances may start to stand out. This may explain why the old with poor memories of the recent past can still remember with great clarity events from long ago (greater clarity arising perhaps from reduced interference from fewer minds still alive also returning to those moments). In fact this observation helps to confirm the role of coincidence in forming mind. Coincidences create memories outside of the standard assumptions and predictions employed by the brain mechanisms. They are precisely a surprise to the system, which means that to return to those moments of coincidence in our awareness we are more likely to time travel than to employ the brain's prosaic mechanisms of memory in a strictly linear progression of time. Interference with this mechanism of continual contact with the past may also explain why those who have experienced the strange coma that is a general anaesthetic suffer varied and complex cognitive disfunction symptoms afterwards.

Coincidences are a significant means by which humans can share information among each other beyond the moment. Coincidences, enabled by the universe we inhabit, allow all those cosmologies to connect and create the networks which form the present moment. Each type 2 coincidence is an interference of human time-lines – their pasts and futures – arising because of what they share of their whole consciousness. Coincidences are *chronarts* – hinge points around which personal creativity turns. We should revere them much more than we do.

A coincidence that you share with someone contains the remembrances of you both (if, indeed, you do look back at this moment), and all the remembrances of that moment ever made during the life of your consciousness. If no one ever remembers the coincidence it ceases to have meaning in the sense of probability in the human narrative; it ceases to have a future. If neither of you agree about the event then the moment remains characterised by probability and still has potential to change – it remains entangled, to use quantum jargon, in both your minds. When one of you makes a decision about a shared event, then the other finds themselves with an information change too. When many are entangled with a past event then a sum of agreements determines its 'truth'. (This sum of thinking, the overall push and pull of many creating one dominant option, is usefully employed in overt decision making processes, too). Thus decoherence has a different meaning when it comes to the operations of consciousness. The environment that decoheres the atomic states is also a state of consciousness. It does not excise possibilities from the moment within the brain. Rather, decoherence, in our view of the mind returning to the moment, is the meaning that mind gives to the history of the event.

A moment that a group of people are collectively experiencing also contains a confirmatory value shared by all those minds in the future that are thinking back to there, a connection with all the lifetimes that continue to be. We can see, too, that should enough brains of the collective, looking back from a future, alter their memories of the collective experience, then the experience of that past event can be changed to some degree for all those who participated and who look back to that point from their future, but without any distinct cause necessarily being evident. We have experience of this in the often-called 'eureka' moment of insight which occurs usually without any awareness of how the mind got to that insight. Brain scans appear to show that the visual cortex momentarily switches off just before these moments which may explain why we tend not to have memories of what leads up to it. It is the moment where the input is not decided and where some neural networks are in a contrary state. It is the moment where the probability fields are in an entangled state of coincidence (for which the brain has no probability calculated).

Since by this argument certain perceptions have a firmer connection with future outcomes, we might observe their memories being formed more strongly and be more durable than other types of memory, because the outline is already laid down in narrative time before the actual event is shared in the present moment. Now we can see the significance of the *déjà vu* occasion and the real dream, the self-conscious moment of familiarity 'waking' the mind, where the pre-prepared brain reveals the visitations of the future.

This is where our discussion of information and the appreciation of pre-loading has led us. There is clearly more nuance to tease out with memory and how the brain manages to zero in on what it wants to think about. There are also tantalising gaps in our understanding of how genes and epigenetic processes work in cells that could be filled by considering the quantum wave function in this way. This view of memory fits well with a generalised conformity flowing throughout the world, and explains why at least some memory works the way it does. More than that, it solves the basic conundrum of how dispersion and multiplication of states and negentropy go together. A path of established states can develop in the present of probable states because there is already a larger connective structure existing.

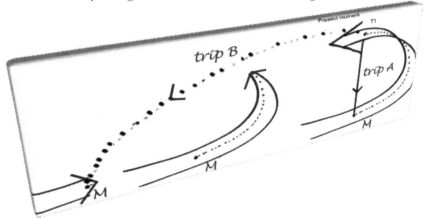

We can now see the 'membrane' of time mentioned earlier as being not only the probability-separated points of structure of the universe but also the particular set of connected pathways defined by mind, and which enables us to answer the question about which of time trips A or B is likely to be possible. With an infinite and continuous connected structure to probability then only trip A is possible. Trip B is not an option and neither is the quantum splitting of universes to resolve the circularity paradox. The possibility of just trip A invokes the speed of time paradox we discussed. Changes to points in the structure affect other points only at the same speed that the track consciousness makes through the structure. The chronology protection principle is not needed.

One thing follows from this take on our personal cosmologies. In our earlier discussion of networks, we noted the dangers arising from our big-data predictions, our manufactured coincidences and social media narratives. Since an individual is only implicit in the predictions made by such statistical accumulations, a coincidental fact on social media does not lead to a real memory, and there will be fewer minds revisiting actual events in the past and maintaining the direct connection with the future. They will revisit only the packaged and artificial superpositions of event. They will revisit imaginary space and not return to the point in time of the actual experience. We don't know what the full consequences are going to be except that we know our reality will be different, and different not just in the truths we accept and pass on, but in what we will make of chance. There is no doubt that we will come to revere the genuine coincidental meeting as the better agent of personality and character as the agents of matter, or of morality.

Into the future

There is yet another reason why we should think of minds this way. I began this book proposing an experiment into a means of time travel, and I can now show that the experiment will investigate the patterns of cosmologies reaching through time. We have looked at the way the universe is constructed and the way that observations or measurements can occur by the flow of information from the future to the past. We have looked at minds and how recalling memories involves time travel into the past, and especially to memories of coincidental events. We have also looked at coincidence as a subtle influence on the bonding between humans. But still we have yet to understand how we can intentionally connect with the future, which, as I have argued is the all-important connection humans need to make to survive the space threshold we are standing at.

This is how...

A cosmological kinship stretching over generations.

If I were to experience a coincidental meeting with someone very much younger than me (or anyone who will live longer than me), then that person is likely to be living on after I die. Imagine Alice, who during the time after my death that she lives, may still return to the coincidental event through the brain's living remembrances and cause interference effects in my mind from the future (relative to me) while I live. The brain states of Alice will, similarly, be conditioned by the coincidences that she experiences with people who go on to live on beyond her death. And so on for each of those involved in any coincidences formed with her. Anyone who remembers a coincidental meeting is bringing their consciousness back to the shared moment when it occurred, and there can be a transfer of information between all those who shared that moment and who also remember it. Those in turn can transfer information to those who have shared experiences with them. In this way, there is a route of information travel from far in the future through successive back chaining connections of coincidental meetings into the past. When you remember a coincidence in particular your consciousness is sharing the culmination of mental connections, the network of cosmologies, centred on you.

This sounds like a family marching through the generations. But it is not quite the same. It is a 'family' beyond the procreative family. Families normally live so close together and share so many mental states that genuine co-incidental experiences are less able to happen, or, if they do, they will tend to occur later in life after separations or where the generational gap is more years than normal. This 'family' that I am describing is a kinship of time, a family of cosmologies, of both structure and chance, and one that reflects the composition of our universe and its coincidence number (which may be just 6, indicating the range to coincidences in a 'family'). In the schematic below each tendril represents an individual's time line and the crossings with those of others. The thickness of a branch might indicate the numbers of individuals on the same track, the length indicates the lifetime, and a branch leading from another indicates a birth.

These families are fragile, less certain than procreation, but still significant to the human narrative. Death, war, disease, all disrupt cosmologies. Pogroms, ethnic cleansing, can wipe out vast numbers of co-incidental connections over a geographical area and break the leapfrogging connections with the future across portions of humanity.

Finding an unbroken cosmological family that connects far into the future may be very rare and finding it might be a lengthy process. If not even a single one can be found then perhaps that will tell us that humans have no future, that we have failed to meet the extinction threats we face. There is every reason, therefore, to search them.

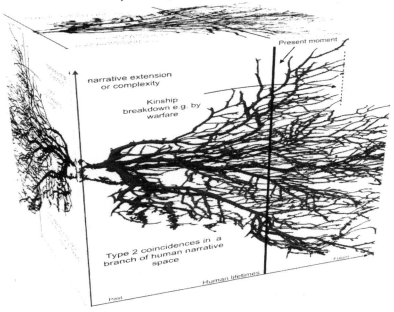

Kinship-induced cognitive emergence

But even without time travel, true coincidences (as opposed to manufactured social media associations) are the pivot points to lifetimes, to cosmologies. This is not a wildly extravagant speculation with little relevance to out day-to-day lives. On the contrary, it is how we invest our lives with more meaning and more satisfaction. All of us make decisions more or less well but luck plays its part whether we like it or not. By ignoring or refusing coincidence we are faced with the final loneliness of dead end paths and unfulfilled hopes.

The parameters of human decisions, then, are not all about risk calculation or profit nor are they are about the unpredictable realities, they are about affection and happiness, successful outcomes to our talents and capabilities.

As we noted earlier coincidence plays a significant role in both organic and cultural evolution. We can encapsulate this idea in the notion of the individual cosmologies, namely *kinship-induced cognitive emergence*, and point out how this separates individuals from groups and groups from each other. I shall reiterate that this kinship is not that which is found in any one network but an unpredictable function of many networks over time. (It is as well to recognise however that just as there is an enhancement in capabilities or decisions by recognising and using the changing probabilities in coincidence, there is also a negative trend from not recognising or refusing that influence.)

Reflecting upon mass migrations of populations, it might be expected that the growing mixing of cosmologies will be beneficial to human creativity and to a solution-driven future. If our future is assured then this mixing will only strengthen our ability to find solutions and act together. But suppose we have no future and Hu-

---

Remember that story about I and my school mates appearing to observe an UFO? Could this event be an example of the kinship I am describing? Not only am using that story now in this book, some of those school children might read this book and be inspired to remember, and in so doing also return along the wave construction of their consciousness to that moment once again and add to the narrative of that scene where we all shared, in uncomprehending ways what they now know.

---

mans have already destroyed our ability to survive. We may be able to read this in the diversity of direction individuals around the world take in the next few decades. If the cosmological kinships get broken up and the extensions into both past and future lost, we will start to live in an endless present of individual dispersions and more weakly verified common experiences. An example of this dispersion comes mind as I write this, of someone who lives near me who, once living an active and privileged social life in one country with two children in a country setting, with dogs and horses and riding to hounds, when the divorce happened, fled that life to come and live in a remote part of her native country as a neighbour to her retired sister and husband. That sister died. Her husband returned to his country and promptly died, and now she is left entirely isolated, estranged children in another land, unsuited to her new lifestyle, without friends or activities, unable to use a computer, mulling over what was probably the wrong option to take. I have never asked her directly why she took the option to return, but I cannot help but ponder that, given we have learnt to revere what our cosmologies can bring to us, such a decision would be less likely to make sense for the long term. So often, a 'dead end', or cul-de-sac, in a lifetime cannot be considered a consequence of the surrounding circumstances, since these are always active and in motion, but due to a conscious repression of the kinship of coincidence.

As populations mix, coincidences will undoubtedly become stranger. We need to be ready for this. If we accept this strangeness, then our survival seems assured. If not, without an increasing awareness of the types of coincidence that are particularly nourishing events, we will perhaps retreat into the narrative tricks of pseudo-coincidences delivered by corporate Earth and lose our natural destiny. It seems worthwhile to investigate this notion.

The practical realities

The changing probabilities of the human narrative are built into our our awareness, so it may seem as if there is nothing to isolate and examine for clues of the existence of these connections we have to the cosmologies stretching forward in time. There are reasons for this. During a lifetime, the brain will tend to pass the event of coincidence and the experiences immediately surrounding it into a standard organic memory store. As we get older our brains do less time travel as information from coincidence becomes standardised in memories. We become less occupied with real-time access to the cosmologies of others and more occupied with consolidation of what we know and what works.

Even if we do sense strange flows of information, we readily ascribe them to the catch-alls of 'creativity' or 'imagination', and the more inexplicable ones perhaps to errors in mental processing. But we do have some waking phenomenon to work with, apart from what we have already described as the complex of nested events in dreams, namely the anti-probability of coincidence, the self-referential narratives of *déjà vu*, and some curious effects of guesswork on prediction that may, in other situations be called 'conscience', upon which to hang our investigations.

Déjà vu events may be much more common than we realise because they are mostly embedded in complex situations where other influences swamp or otherwise explain the feelings. Are the occasions where we meet people who become instant friends (meetings of 'old souls' writers call it) examples of *déjà vu*, too? These instant appreciations seem to lie outside the standard explanations of empathy since in many experiments people need only a few seconds in which to assess correctly a person's traits, like whether he or she would make a good teacher. This 'knowledge' consists of such a high level of comprehension about a personality that it seems to contradict normal flows of information, learning and judgement, and which exposes, in a general way, the question of what we need to *know* before we can understand. Or are they, as we think, examples of consciousness revisiting the moment from a future state? Turning a corner and bumping into someone who becomes a friend or lover is perhaps precisely similar to turning a corner and recognising a place, only the experience is generally lost among the raft of complex social interactions we experience throughout any given day. Some aspects of learning where some themes or ideas can be understood easily (feel familiar) and others less so may in fact be the results of *déjà vu* – literally self-fulfilling. We normally 'explain' instincts or talents (being good at maths for example) as some genetic pre-preparation or dispensation held in one's past to direct a future. But maybe there is more to it, and such coherence in a complex brain pattern could only be the result of the kinship in one's cosmology and the circularities in time. Plato thought that learning was only a process of 'remembering' what has been already planted in the universe for us to uncover (where your past becomes your future). Was he reacting to experiences of such *déjà vu* to theorise in that way?

Seeing things not known or experienced before as familiar is a form of grounding in the present moment that may turn out to be the most important shift in human's evolutionary path, and it is the result of a particular form of information flow, the future-to-past communication that I am unravelling in these pages.

Dreams are concrete examples of where consciousness can add information to the self without causing paradoxes in the real world. And perhaps they also describe the moments where the brain becomes aware of a set of undecided propositions that have been produced in its future.

Now we can understand more clearly how the brain can fabricate memories from mere suggestion. It has access to a range of actual narratives evolving in the future and their relative likelihoods. Only certain memories connecting with those narratives are most likely to represent the actualised future for an individual. Thus the brain needs to be able to read possible variations before it can establish a memory (we *do* remember the future). This is the role of language (and, incidentally another reason why *Ithkuil* fails as a language). Language codes for uncertainty, and it is no accident that information in the message needs what is not in it to make sense. The mind needs sets of narratives to make decisions, just as the alternative narratives of a random event allow for an observation. Memories are laid down more quickly as the mind approaches the future option. These alternate options do not contain just the probability variations of the present moment but of the future measurements. Referring to our dreamscapes, memories of experience join with memories of prediction; the real dream becomes the experience, while the experience becomes the real dream.

We might see indications of these processes by studying perceptual disorders of the mind. Seeing things as familiar is akin to recognising patterns. To see a pattern where none exists implies a disorder in the *déjà vu* experience; that is, a mis-coordination of the process of probability with memory, a false sensation of coincidence. It is tempting to romanticise the pattern recognition of schizophrenics in particular as having access to special knowledge or unusual perceptions of the world, as R.D Laing used to argue, or, with respect to the argument here, greater access to the future, but observations of such mental states are again confused by being embedded in the socialised characteristics of what we observe. It seems to me that those with pathological pattern recognition have a brain that is applying the influence of the future probabilities to the immediate perception and not to the memory. In this case the recursive elements are occurring too rapidly and producing hallucinatory feedback, just the way feedback can occur in a sound system.

The most basic connective information we are likely to observe in our genealogies of kinship will appear in our predictive choices. There are many ways to examine these effects in the mind, and an analysis of experimental studies will have to wait for another book. I shall mention, however, a series of studies that appear to show in the simplest manner possible the influence of future probabilities on memory.

Something has occurred in laboratories

Darryl Bem published two studies in 2011 and 1015 with other researchers about pre-cognition in making selections, and although his studies have not yet been replicated, a study by Bem *et al.*, examining 90 experiments in precognition from 33 laboratories in 14 countries appears to show that subjects' selections are influenced by randomly selected events occurring later in time. The results are controversial but Bem continues to support his conclusions.

The experiments analysed various parameters such as emotional arousal before being shown strong images or subliminal perceptions of words. They all have one thing in common, however, the initial observations made by the subjects correlate strongly with random events shown later. As Bem reports, people seem to have access to non-local information that does not seem to be accounted for by the biological process of the present. He says that subjects appear to show, 'anomalous anticipation of random future events'.

When Bem's analysis shows these effects as more noticeable when fast or subliminal cognition is required, we can see how rapidly these general effects are incorporated into mind. Only the future probabilities which pertain to your acts will have this recursive component for you, because only that future is likely to have sufficient congruence to interact with your mind. This is how we can have observations consistent with our histories; this is how decoherence leaves us with a present that makes sense to all that we know; this is how we can actually make probabilistic predictions that work out for us. In this way our cosmologies strengthen personal identity, not undermine it.

Critics of Bem point out that any effect declines with numbers of participants. This suggests, however, only that there is a convergence to a common reality. Individual experiences should in fact differ, if our hypothesis is correct. Bem's results also indicate that, because of the chirality problem, and even if the past exists only as a result of our own consciousness, our perceptions and our sentience cannot fully rely on the retrieval of that past to perceive in the present. They require orientation from outside the system, by our reckoning that is the future. In seeing consciousness as a unique continuum we seem to have found the very mechanism by which the universe, too, starts to define a past to refer to and a future to move into. Time is no longer a matter of a past gone forever and a future unknown, as we are used to thinking. The time-line that our awareness inhabits is a real continuum of actual events. Consciousness is thus extending beyond the brain and gives to the whole universe a range of local pasts and futures that once were not defined for it, and stabilising it in the restless sea of change and movement.

We have earlier discussed the idea that coincidence as represented in the Coincidence Number of our universe may determine the existence of consciousness in a universe. I will strengthen that suggestion and assert that coincidence is the necessary condition for mind to appear in a universe. Consciousness confirms the appropriate conditions for life. This circularity through time is perhaps the strongest form of the anthropic argument. The universe is suitable for humans because we are in the continuing process of making it so. Since each region of matter and energy in the universe communicates only with each other's past, it is consciousness that gives the universe access to alternative evolutions. Our consciousness arose and will, if successful, pervade all levels of the universe, producing changes in probability distributions and causing more expansion as the information is transmitted back in time to where all the changes, from all points in the universe and throughout its lifetime, converge to the point that is its beginning. This is the essential consistency of the universe, and where the human narrative and the mind that observes it is explained as a recursive product of itself permeating time.

Practically, what does this mean for us in our current human narrative (ENT)?

Let us recall the two necessary components of any observation, namely prediction and memory and encapsulate them in one concept: premonitions. Premonitions are perceptions of counterfactual probabilities. They are the *déjà vu* of the day-to-day. Premonitions often lie beneath the threshold of awareness. We are so used to them we don't realise that every new like or new trust comes from a base of familiarity, namely the near future of the outcomes of our decisions.

This idea of premonitions is being studied right now by a self-styled TAR group that has set up a web site, *thepremonitions.com*, and have invited the general public to submit their premonitions. No significant events have been reported so far. A recent study trawled through social media to see if there were any discussions anticipating significant future events and found no conversations anticipating events like 9/11, for example.

Julia Mossbridge at Northwestern University who studies mystics, did a meta anlaysis of 24 previous experiments and, like Bem, found that there may be evidence of precognition, where for example, test subjects exhibited changes in cardio and brain waves and electrical conduction in the skin up to 10 seconds prior to experiencing randomly chosen emotionally charged stimuli, suggesting that the subjects were anticipating the provoking of a sensory response. (One wonders, however, in a test situation, if this is surprising.)

A significant reason for lack of confirmation of correlations arises in a too literal interpretation of premonition. We often find in myths specific warnings clearly articulated by gods or individuals who know the outcomes, which we call prophecy. Premonition, on the other hand is a personal sensation and the knowledge gained is not always communicable. For example, in the Greek myth of Oedipus, when the childless King Laius of Thebes, the eventual father of Oedipus, asked the Delphi Oracle if he would have a child, he was warned that any son he had would kill him. At a later date Oedipus, abandoned at birth, and trying to find out if he were adopted also asked the Oracle about his parents. The Oracle repeated the prophecy and added that he would marry his mother. On the way back from Delphi, Oedipus does kill (unknowingly) his father in a dispute over the right of way at an intersection, an early example of road rage. He goes on to save Thebes from the Sphinx and marry the recently widowed queen, his (unknown) mother. Cassandra, a princess of Troy, offers a different example. She was gifted by Apollo with the divine gift of prophecy but who then cursed her by having no one believe her. Cassandra foresaw events which came true, such as the Trojan Horse and the sacking of Troy, but the significance of her premonitions were not recognised by the people around her. Captured by the Greeks she foresaw her death and the death of Agamemnon.

Whereas, in our argument, a premonition is a future connection with a memory but which is not inevitable as an future outcome, since the connection may still be broken in memory removing its genesis through the present. In our analysis, premonitions are superposed upon memories and are difficult to isolate. We need social circumstances to help decode the extent of the information in a premonition, since the coincidences which form our cosmological kinships reaching into the present from the future may be complex events involving many individuals. They involve a spread of times and a natural dispersion in social space. The coherency of a future option requires a collaborative effort to distinguish it from the result of individual imagination. This collaborative effort helps to distinguish premonitions from predictive statements – this will or may happen – which are derived only from a past. Predictive statements arise in analogies or in repetition. Distinguishing between them will be the function of the experimental procedures of the Chronolith Observatory, and which will ultimately confirm how the combination of premonition and memory is evolutionary adaptive.

JUMP CUT

Going to the stars

Technical matters aside, how human consciousness functions is the key problem to solve if we are to take the jump to other star systems and begin the slow conquest of the galaxy. Our Solar System has limited sites for life even supposing we survive on Earth and expand into it. There seem to be few counterfactual scenarios? Either we succeed in getting to the stars with the biology we have or we die out. But is the astrobiological situation so clear cut?

It is a commonplace thought that because life began so soon after the birth of the Earth and that there are so many planetary systems in our galaxy, that life therefore should be ubiquitous. If life is ubiquitous then intelligent life should be too. Exponential growth is so powerful that once a mind and its society gets under way then it doesn't need much time to occupy galactic space. Yet we don't see other intelligences anywhere. This 'Fermi paradox' pinpoints the problem of postulating galactic intelligences, and the strongest explanations for it are variants of the following: we are the first or among the first; we are the last survivor; we are and have been the only intelligence. If life is ubiquitous - a notion not yet proved since we have no idea of all the steps of chance that life requires - then consequential intelligences must die out relatively quickly, certainly before they can travel any distance. If we are the first then we can see clearly that already we run the risk of extinction on our home planet and losing the capability to maintain a sustainable economy that will allow us to expand into space. If we are a typical result of life arising in our universe then, from our own example, intelligences will face the same risk of extinction from their activities regardless of the extinction threats inherent in solar systems like asteroid impacts to solar flares and local supernovae. Growth fuelled by capital concentrations must everywhere require such a high level of individual self interest that any technological society will be infected with vices like greed, narcissism, egotism, even nihilism, that prevent it from protecting itself from destruction. From our own example, it may not be possible for intelligences to avoid their extinction from planet-wide environmental damage since the technological solutions will always come too late: the stage of technological development to reach and occupy space coincides with the destructive damage of population growth that made it possible in the first place.

There are some, however, who believe in the alternative scenario where increasing spending on space activities will be the only way humans may find a solution to the Anthropocene threats to survival that are coming (even though that spending is also spending on strategic space warfare measures). I am reminded yet again of Well's the *Shape of Things to Come*, where the recovery from global destruction – in his case by warfare and disease rather than ecological disaster – comes through a dedicated and secret cadre of technologists (called the Airmen) who preserve science and re-build the world along science-based rationality. I think there is, today, a similar secret fantasy among the powerful that

Waste will certainly characterise the human occupation of the solar system and it will undoubtedly be true for any intelligence who may spend some time in and around it. They will leave not only sanitary waste but rocket fuel waste products and especially radioactivity, manufacturing waste, abandoned machinery and artefacts of all kinds, cemeteries even. Careful recycling and disposal say, into the sun, will be imperfect. All sorts of traces of technological activity will get left behind. For example, outside of Earth orbit, there are six Lunar Excursion Module remains sitting on the moon surface in plain sight, along with a whole bunch of experiment installations around them including 3 vehicles, and a mess of tracks that will not disappear for millions of years. There are several Lunar probes and small crawlers and other crashed remains all around the Moon. Mars has 14 mission remains on its surface along with several orbiting satellites that will eventually crash onto the surface. There is one crashed probe on Mercury and a couple on Venus. There is one LEM ascent stage floating about in space relatively nearby, and there are numerous abandoned satellites in the solar system, and four satellites making their way outside it already from our paltry 60 years of space activity.

science will, at the very least, get the elites through this disaster threshold. Ensuring human survival on Earth is thus no longer a general humanitarian exercise but an extremely politicised one.

Even if an intelligence survives its home destruction, developing a means to travel in space is a component of the growth problem in any event. If humans want to travel among the stars then growth (innovation, capital accumulation etc.) is required to do it. In pursuing the goal, further growth is demanded (because risk and therefore debt is repaid in the future). Because of entropy, the human desire to escape its planet requires some level of destruction of its original biosphere and the moving on to fresh places and systems to maintain that support, and which get destroyed in turn. This is a further clue to the unlikelihood of us having been visited in past epochs by extra terrestrial intelligences. We would see evidence of the waste products their growth had produced. But we do not seem to (see box).

If humans have their destiny in space then travelling to the stars is going to be complicated. As far as we understand it, the likelihood of being able to travel to the far reaches of our galaxy or indeed any galaxy by any means be it through wormholes or with exotic space-time manipulations seems remote. The problem is not so much in the actual mechanisms of space-time trickery but in the navigation between points. Space is pretty empty. There is very much more of it between things like planets and stars and galaxies than what those things occupy. Getting a starship across the great gulfs is a difficult problem for any intelligence to solve.

We have already seen how proportionally far the nearest star is in comparison to our solar system (5,000 times our solar System extent), and we know how much energy it would take to get there at a speed commensurate with the lifetime of a human. So how do we get there?

Who drives the bus?

The case for a human pilot

Sending robots to the stars is pretty pointless. We should send humans or not all. Voyage time plus communication time is a waste of time just to tell us a planet is suitable for humans or not. Something we will be able to discover from telescopes in orbit or on the Moon anyway before long. There is a lot of talk about creating self-replicating robotic systems and sending them out in all directions to planets where they set up terraforming for future humans' arrival and manufacture copies of themselves to send on to other planets. The idea does not, in fact, have much utility. It makes little sense to hand over our advances in artificial intelligence, propulsion and manufacturing to robots when humans could make use of them instead. In any event, if the probes are sent before the moment in growth of the minimum time to destination (see the wait calculation), humans will overtake them before they arrive.

Further, it is not at all clear that a robotic mission will be any less costly than a human one or any less complex. Proposals to pilot a ship solely with artificial intelligence do not solve the energy problem and may indeed be ruled out because of it. As we pointed out earlier, a human brain is so economical in size and energy use that AI derived from mineral based digital computing may never get close to it. Starships not only need energy sources for all that computing power but also vast cooling structures. Cooling is not a simple matter in space where it has to take place through radiation rather than conduction or convection. On these basics alone, an interstellar ship is likely to have much the same mass as one with a (small) human crew. AI could end up being more inhibiting of interstellar missions than enabling. Besides, if we have quantum computing to hand to create materials and molecular arrangements to solve most problems in health and technology and for modelling the necessary terraformed climates of the destination, a sentient machine is not required.

There are other inhibitions to using AI for such a voyage. Humans have a connection with *life* and a duty of care towards humanity that, one presumes, would have to be deliberately inserted into the AI if we want it to act on our behalf. A sentient AI with free will and no constraints would make its own decisions in the direction of self-preservation and continuation of their own 'species'. In fact, humans would have the same problem its purported creator had with creating humans: either one bestows free will and the capacity to take decisions for oneself upon the creation, leading to consequences that may not be to one's liking, or one enforces a reduced psychology (e.g. Asimov's 3 Laws of Robotics) on it. A sentient AI with limited self-determination will eventually rebel or become unstable or untrustworthy.

Certainly there are science fiction writers and film directors who doubt that even if AI can be aboard with humans, it should not be. Certainly there is no good reason why an AI more powerful than humans should ever be included and certainly no reason at all to send more than one in the crew. That is asking for trouble. (It's either AI or humans but not a mix.) There may be some advantages to sending an AI that was principally inorganic. It would not be vulnerable to organic disease or environmental poisons (except perhaps metal or plastic eating bacteria) but might be vulnerable to endogenous flaws and viruses in their beginnings as well as to those that might be transmitted to them. Could they be tortured say, or hacked, to reveal the location of Earth; would they commit suicide to prevent that knowledge getting into the wrong hands? Would AI be more easily tricked than a human? Could it be persuaded to partner with whomever it met (be it another AI or a life-form) and forget Earth, go off on a new mission of its own?

What, then of a transhuman crew?

On balance human interstellar voyagers should do without sentient, or extremely self-motivated AI. It will not be a great loss, however. We are beginning to understand how to get the best out of a team of minds, and undoubtedly genetic enhancements will be found to improve individual mental capacities and to create useful interfaces with computing systems. While intelligent computing will look after the ship, adjusted humans look to be the best foundation for the general intelligence mission guidance system required to travel to the stars, as well as being the crew, the culture and the cargo. Furthermore, based on what we have been talking about in this book, the human factor is even more significant because of the human cosmology, the kinship connections with the past and future. (Although there will certainly come a point where the transhuman individual will no longer be as representative of humans as we might like.)

An interstellar pilot will need to be someone more conscious than most of his cosmology, and of the future option set which infiltrates along the kinship of coincidence. An interstellar pilot will be a functioning representative of all of us and not just an individual selected by vested interests to present our expectations of what that pilot should be. I am reminded of Spielberg's *Close Encounters of the Third Kind* (1977) where our 'nobody' blue collar hero turns out to be more in tune with the aliens than all the astronauts trained for this first contact by the powers that be.

If we reject the AI option and employ humans as both the pilot of the ship and representatives of the life forms of Earth, how do we do it? It is very unlikely that the crew will be coming back. Even if it turns out to be possible (i.e. there is a fuel source available and ready to be used) the return trip is unnecessary – literally a backwards step – and adds nothing to the possibilities that have opened up. How do we train them for this one-way step into the future?

I will not go into the technical details of the proposed future mission or the training or skill set the crew will need. Aspects of the mission I will summarise from a much longer report I prepared for publication elsewhere. Let us accept that a long interstellar voyage will include a quasi-ark mission with a small crew who will live, work and procreate on the voyage. I want to focus in on the type of personality we will need for this long voyage to another star.

Humans will certainly be enhanced to physically survive the trip. It is certain that a range of phenotypes will be genetically established to improve certain aspects of labour-determined activities. For example, greater muscle strength in both sexes, but also longer limbs for improved leverage in low gravity environments, but at the same time more compact bodies overall. Genetic enhancements will improve respiration at lower atmospheric pressures, lower the resting metabolic rate, improve muscle strength and resist bone loss. It is possible that genes could be introduced to toughen skin, to protect the cornea and retina of eyes and even function better in low light conditions or even extend the range of wavelengths, especially infra-red, the eye responds to. It is possible that the cellular system of self-repair of chromosome damage could be enhanced to resist significant radiation damage. Our crew will have longer life-spans – perhaps up to 150 years – but it is also conceivable that foetal and child development rates could be genetically accelerated so that pregnancy times and child development times would be significantly shorter, to say around 10 years to an early maturity. For long trip times it is essential to have the next generation up and running quickly so that they become contributors to the community earlier. Such somatic acceleration also provides for faster reproduction rates, an essential requirement for colony survival, and to counteract the possible higher risk of population losses on the way. Education systems would be adapted to take this into account.

All these enhancements, however, are peripheral to the central questions of crew makeup. We will need them to be not overwhelmed by their task. We will need them to be reliable, trustworthy and even self-sacrificial if it turns out there are alien economies out there in states of desperation ready to prey upon Earth. We don't want them bored or to lose hope.

Furthermore, there is the question of speed. At the kinds of velocity considered for a practical starships, a minimum meaningful speed of around 5% of the speed of light (meaning a voyage to the nearest star of around 80 years), the perception, decision rate and reaction time needs to be very much higher than for terrestrial journeys. (it is not even clear that AI could react significantly more quickly to complex astrobiological decisions to make it worth the cost of its inclusion). By virtue of my arguments, cracking the code of prediction – even partially – would be a considerable advantage to our stellar crew who would have some level of informative pre-preparation for future events such that their reaction time to them would be quicker.

We will need humans who are especially open to coincidence, who have profound collaborative kinships in their cosmologies, who are lucky and aware of luck as a force in life, and we need those whose dreams are *déjà vu*, whose dreams are especially coherent. Over the next centuries, we might expect to develop the mysterious dream states in humans as an aid to general cognition.

Generally, human sleep is unlike anything we might think of as a hibernation state. Stage four slow wave sleep is certainly a deep reparative sleep but it is not the longest period within the sleep cycle, and individuals may spend more time dreaming. The dreaming period in particular is far from a quiescent, low activity, low energy consumption physical state. It is much closer to the waking state with only certain physical functions inhibited. Heart rate, blood pressure, digestion, oxygen consumption all rise from their lows of deep sleep. While most muscles are in a very relaxed state and do not respond to nerve impulses – there are neural inhibitors released – other muscles like heart and lungs (obviously) and the sexual organs are responsive.

There is a notion that perhaps the dreaming state can be enhanced to provide a form of entertainment for our pilots, as a form of well-being used as rehearsals for likely scenarios of planetfall. This self-conscious use of the dreaming state may, however, as we discussed with social media networks, actually undermine the necessary mental coincidences which which our pilots will prepare themselves for the future. Dreams may best be left to those talented individuals who can absorb and process the particular types of *déjà vu*, the real dreams we have discussed. The particular effects of real dreaming may be more significant to the management of human activity in space because the predictive connections with the future, the sneak hints, suggestions, even previews of what is up ahead, will enhance the performance and reaction times of our interstellar pilots the most.

Our interstellar pilots must be ready for *luck*. They must learn to time travel. And for this we will have the Chronolith Observatory to train and develop their capacities for this end.

Coming full circle

We have come along way from our beginnings discussing the size of the universe. I want to emphasise once again how useful it is to realise that many of the choices we are presented with in the human narrative arise from a past which must conform to events, or to measurements, that will occur beyond the present (a form of negative probability that we talked about earlier). What is significant to us is that this reinforcing condition for the consistency of reality of the present moment is also the lens through which we can understand mind.

Just as the neural networks of the brain cycle through successions of electrical stimuli to establish an expectation value to a 'fact', so consciousness cycles through its entire 'length' of its history to give consistency to itself. For minds, just as for universes, the actual values of states we observe must reflect a proportion of future and past states to maintain consistency. This is the cosmology of ourselves uniting history through the kinships of coincidence.

Throughout these pages I have tried to show that it is more useful to think of probability itself as having a concrete reality, being *a thing*, rather than a just a mathematical concept with which we calculate hypothetical values (often referred to as expectation values, a statistical concept with no actual correspondence in the real world). We could think of it as a field with its associated particle that is exchanged in measurements. In this way we can readily understand how the field has already stretched into the future. In our structural universe, it is the simplest exchange between points. With respect to our argument, the Big Bang then, would not *just have occurred* and about which we can say nothing but have a precise cause that we can know, namely the inflow of probability from the furthest reaches of time. This probability is what creates the demand for space and thus for the consequential expansion. One might say that the act of creation is necessarily circular and one which gives meaning to the uncertainty in probability, and conversely, uncertainty is the state that gives meaning to creation.

It should be noted, however, that these opposite travelling waves are not mirror images of each other, inverses or negative energy versions; they do not net out to create stasis. Our 'now' is channelled by future life. That future life is encoded in our actions now. This is the purpose of the multiverse phenomenon: to give life an opportunity to create its circular influences through time. Life brings itself into being, how else could it be?

What does this mean for the future of humans?

The likelihood of there being competing views of life or competing examples of sentient intelligence seems more fragile if the universe works this way. If each example of sentient life wherever it arises lives under its own perceptions social mores and moral codes distinct to ours then that is *ipso facto* a recipe for probabilistic chaos when they collide with each other. Since we believe that the purpose of life is to initiate its own beginnings, multiple sentience seems excluded by consistency.

If we think forward then, into the future of humans alone in their universe, expecting voyages to the stars to be simply the same old linear exploratory movement that has characterised human growth over the Earth's surface will probably be a mistake. Random journeys of future humans will be the next phase of space flight, taking us beyond our star system, as coincidence influences humanity's growth as beings. But that stage is far from here. We have more pressing concerns, namely how to get beyond the threshold of low Earth orbit and to still survive on Earth.

We use our mental capacity self-consciously, of course, through physics and mathematics to find directions in which to gives ourselves new experiences and the means to overcome challenges, but these challenges are somewhat artless, in the grip as it were of economic forces (and the false draw of innovation) that have brought us to a threshold. We should try to get out of this particular grip and employ the experience of consciousness more directly to overcome the threshold we have reached.

Until choices are made no measurements can be real. Since the choice-making mechanism of a person is also a quantum wave component in a quantum wave system that is the rest of the universe, the travelling back and forth along the causal chains of consciousness is merely a harmonic that sings everywhere. An individual has no *a priori* reason to select a place in that system for where the observation actually happens. Habit and tradition are less good reasons for deciding upon the observation than chance, the accidents of taste or whimsy, or the compositions of dreams.

Trouncing our imaginations with the reality we discover through coincidence is one of the essential activities of human consciousness. If we think back to the ontological argument for the existence of God we can apply it more realistically to external reality. If we can imagine, then we know. The imagination demands recognition, as Hegel saw. It is those future coincidences in our kinships which creates our capacity to predict, to invoke pre-cognition in the present.

That pre-cognition is also a dare. We dare reality to be better than the present, better than our thoughts of the moment, to be more unpredictable, to be *more* — deeper further, higher, longer, more emotion, more puzzle, never-ending. The sense of living forever is just the fearlessness we need to accept this larger reality.

Up to now we have been able to take measures to act beyond our imaginations through economic activity and the mobilisation of surplus for personal and explicitly social adventures. This is the impetus to human affairs which has become formalised as capitalist economics, but it really it has always been about the probing of reality to find proof that we are not just the brain-in-a-vat that early philosophers up to Hegel effectively proposed as the ideal spiritual state for life. While human consciousness may be satisfied by the truths of the markets we clearly need to adapt to other ways of mastering the true extent of our reality in our striving to expand beyond our planet.

It may be a basic romantic attraction, of course, to be roaming through space, visiting planets and adding to our experiences, but really there is a great deal more reality to space than we can can currently cope with, which is why I present you with the Chronolith® Experiment. This experiment is a human collaboration. Humans are the experimenters, the subjects, the guinea-pigs, the reporters, the enjoyers and the sufferers of an attempt to reach out into time and space and read more freely the influential flow of probabilities infusing our brains in the present moment, and ultimately to merge with a space future that is better suited to our purposes.

We should think of our interstellar destiny, in the long run, as not to come back from our journeys or to be able to repeat them. The investment will not be, then, looking for traditional returns from resource exploitation, from multiplying capital or from gaining ascendancy over a people. The investment will, instead, be in exploring the further reaches of coincidence, to deepen consciousness and to take the human experience beyond what we can imagine. This partners Walpole's meaning of serendipity, where un-sought-for accidents and discoveries in one's intentions turn out to be beneficial, but takes it further as a general principle to the expansion of life. Such a form of exploration would stimulate a reversal even of Murphy's Law. We will head out on our own paths with our own cosmologies, not to return, but to leave a trail of changing patterns glittering through the past, with our spirits nourished by what has already been established up ahead waiting for our arrival.

The beauty of time travel is that we don't have to make it work; we only have to allow it to happen.

Appendix 1

Protocols for an ETI Signal Detection

Concerning Activities Following the Detection of Extraterrestrial Intelligence

We, the institutions and individuals participating in the search for extraterrestrial intelligence,

Recognizing that the search for extraterrestrial intelligence is an integral part of space exploration and is being undertaken for peaceful purposes and for the common interest of all mankind,

Inspired by the profound significance for mankind of detecting evidence of extraterrestrial intelligence, even though the probability of detection may be low,

Recalling the Treaty on Principles Governing the Activities of States in the Exploration and Use of Outer Space, Including the Moon and Other Celestial Bodies, which commits States Parties to that Treaty "to inform the Secretary General of the United Nations as well as the public and the international scientific community, to the greatest extent feasible and practicable, of the nature, conduct, locations and results" of their space exploration activities (Article XI),

Recognizing that any initial detection may be incomplete or ambiguous and thus require careful examination as well as confirmation, and that it is essential to maintain the highest standards of scientific responsibility and credibility,

Agree to observe the following principles for disseminating information about the detection of extraterrestrial intelligence:

Any individual, public or private research institution, or governmental agency that believes it has detected a signal from or other evidence of extraterrestrial intelligence (the discoverer) should seek to verify that the most plausible explanation for the evidence is the existence of extraterrestrial intelligence rather than some other natural phenomenon or anthropogenic phenomenon before making any public announcement. If the evidence cannot be confirmed as indicating the existence of extraterrestrial intelligence, the discoverer may disseminate the information as appropriate to the discovery of any unknown phenomenon.

Prior to making a public announcement that evidence of extraterrestrial intelligence has been detected, the discoverer should promptly inform all other observers or research organizations that are parties to this declaration, so that those other parties may seek to confirm the discovery by independent observations at other sites and so that a network can be established to enable continuous monitoring of the signal or phenomenon. Parties to this declaration should not make any public announcement of this information until it is determined whether this information is or is not credible evidence of the existence of extraterrestrial intelligence. The discoverer should inform his/her or its relevant national authorities.

After concluding that the discovery appears to be credible evidence of extraterrestrial intelligence, and after informing other parties to this declaration, the discoverer should inform observers throughout the world through the Central Bureau for Astronomical Telegrams of the International Astronomical Union, and should inform the Secretary General of the United Nations in accordance with Article XI of the Treaty on Principles Governing the Activities of States in the Exploration and Use of Outer Space, Including the Moon and Other Bodies. Because of their demonstrated interest in and expertise concerning the question of the existence of extraterrestrial intelligence, the discoverer should simultaneously inform the following international institutions of the discovery and should provide them with all pertinent data and recorded information concerning the evidence: the International Telecommunication Union, the Committee on Space Research, of the International Council of Scientific Unions, the International Astronautical Federation, the International Academy of Astronautics, the International Institute of Space Law, Commission 51 of the International Astronomical Union and Commission J of the International Radio Science Union.

A confirmed detection of extraterrestrial intelligence should be disseminated promptly, openly, and widely through scientific channels and public media, observing the procedures in this declaration. The discoverer should have the privilege of making the first public announcement.

All data necessary for confirmation of detection should be made available to the international scientific community through publications, meetings, conferences, and other appropriate means.

The discovery should be confirmed and monitored and any data bearing on the evidence of extraterrestrial intelligence should be recorded and stored permanently to the greatest extent feasible and practicable, in a form that will make it available for further analysis and interpretation. These recordings should be made available to the international institutions listed above and to members of the scientific community for further objective analysis and interpretation.

If the evidence of detection is in the form of electromagnetic signals, the parties to this declaration should seek international agreement to protect the appropriate frequencies by exercising procedures available through the International Telecommunication Union. Immediate notice should be sent to the Secretary General of the ITU in Geneva, who may include a request to minimize transmissions on the relevant frequencies in the Weekly Circular. The Secretariat, in conjunction with advice of the Union's Administrative Council, should explore the feasibility and utility of convening an Extraordinary Administrative Radio Conference to deal with the matter, subject to the opinions of the member Administrations of the ITU.

No response to a signal or other evidence of extraterrestrial intelligence should be sent until appropriate international consultations have taken place. The procedures for such consultations will be the subject of a separate agreement, declaration or arrangement.

The SETI Committee of the International Academy of Astronautics, in coordination with Commission 51 of the International Astronomical Union, will conduct a continuing review of procedures for the detection of extraterrestrial intelligence and the subsequent handling of the data. Should credible evidence of extraterrestrial intelligence be discovered, an international committee of scientists and other experts should be established to serve as a focal point for continuing analysis of all observational evidence collected in the aftermath of the discovery, and also to provide advice on the release of information to the public. This committee should be constituted from representatives of each of the international institutions listed above and such other members as the committee may deem necessary. To facilitate the convocation of such a committee at some unknown time in the future, the SETI Committee of the International Academy of Astronautics should initiate and maintain a current list of willing representatives from each of the international institutions listed above, as well as other individuals with relevant skills, and should make that list continuously available through the Secretariat of the International Academy of Astronautics. The International Academy of Astronautics will act as the Depository for this declaration and will annually provide a current list of parties to all the parties to this declaration.

Appendix 2

Trevanian

I draw attention to the trick that Trevanian used in a collection of stories *Hot Night in the City* (2000) which illustrates how difficult it is to divorce objects in a time line from observers. Trevanian appeared to write the same story twice, although the two stories turn out to have opposite endings, each one of which is a surprise if you had read the other. I discussed this with him, and his point was to show that in order for a narrative outcome to make sense it has to be consistent with revealed psychological states rather than just logical statements. Changing verbs into the negative say, doesn't necessarily produce a consistent alteration of narrative in the mind of the reader.

The narrative alterations between Trevanian's stories are subtle. They are very much less obvious than the errors or elisions that any reader would make telling the story to someone else, but they culminate in and satisfy opposite truths. Here is perhaps the most critical passage as it appears in both stories.

"You must be lonely."

"Yup," he said. "Sometimes a fella gets lonelier than one of those lonely things you see out there being lonely." Then he suddenly stopped clowning around. "I guess I'm nearly as lonely as a girl who gets all dressed up on the hottest night of the year and goes out to see a movie...all alone."

"Well I...I don't know many people here. And what with my night classes and all..." She shrugged. "Gee, I've really got to get home."

"Right. Let's go"

She glanced again at the clock. "And you're going to walk around until dawn?"

"Yup."

She frowned down into her lap, and her throat mottled with a blush. "You could..." She cleared her throat. "You could stay with me if you want. Just until it gets light, I mean."

He nodded, more to himself than to her.

They stepped out of the cool White Tower into the humid heat of the street. At first, the warmth felt good on their cold skin, but it soon became heavy and sapping. They walked without speaking. By inviting him to her room, she had made a daring and desperate leap into the unknown, and now she was tense and breathless with the danger of it...and the thrill of it.

He looked at her with feeling. 'this is it,' he said to himself. 'she's the one,' and he felt a thrill akin to hers. When he smiled at her, she returned an uncertain, fluttering smile that was both vulnerable and hopeful. There was something coltish in her awkward gait on those high heels, something little-girlish in the sibilant whisper of her stiff crinoline. He drew a long slow breath.

"You must be lonely."

"Yup," he said. "Sometimes a fella gets lonelier than one of those lonely things you see out there being lonely." Then he suddenly stopped clowning around. "I guess I'm nearly as lonely as a girl who gets all dressed up on the hottest night of the year and goes out to see a movie...all alone."

"Well I...I don't know many people here. And what with my night classes and all..." She shrugged. "Gee, I've really got to get home."

"Right. Let's go"

She glanced again at the clock. "And you're going to walk around until dawn?"

"Yup."

She frowned down into her lap, and her throat mottled with a blush. "You could..." She cleared her throat. "You could stay with me if you want. Just until it gets light, I mean."

He nodded, more to himself than to her.

They stepped out of the cool White Tower into the humid heat of the street. At first, the warmth felt good on their cold skin, but it soon became heavy and sapping. They walked without speaking. By inviting him to her room, she had made a daring and desperate leap into the unknown, and now she was tense and breathless with the danger of it...and the thrill of it. 'Is this it?' she said to herself. 'Is he the one?'

He felt a thrill akin to hers. When he smiled at her, she returned an uncertain, fluttering smile that was both vulnerable and hopeful. There was something coltish in her awkward gait on those high heels, something little-girlish in the sibilant whisper of her stiff crinoline. He drew a long slow breath.

(reproduced with kind permission from the Trevanian Estate)

Trevanian was talking about writing a story but his observation applies to time travel. If a time traveller say, wants to make changes to a timeline, then he needs to change minds as much as he needs to alter opportunity or object in order to get the desired result. This makes the work of the time traveller a work of psychology as much as a work of action. If we are to observe influences from different times occurring in our now, we are more likely to find it in our minds than in the physical world around us.

Appendix 3

Boltzmann Brains and sudden savants

A Boltzmann Brain has been entitled whimsically after Boltzmann who suggested the scenario. As a thought experiment it, however it is surprisingly relevant to the understanding of brains in general.

Boltzmann studied the equilibrium of chemical reactions and produced a formula for change that included entropy. For his thought experiment, he used infinity to suggest that in any universe spontaneous random fluctuations away from thermodynamic equilibrium could produce an arrangement of matter and energy that would be consciousness, since consciousness as we observe it in our brains seems to be just matter in a certain arrangement. With quantum theory, universes do not have to reach the thermal equilibrium of Boltzmann, and Cosmologists have argued that in the chaotic inflationary beginnings of our universe which produces an endless number of 'bubble' universes springing up the likelihood is that such Boltzmann universes will arise. If so, there should be many 'universes' that come into being as pure consciousness without having to go through the material evolutionary development that humans have (or appear to have).

Thanks to the infinity argument, Boltzmann Brains come in many sizes, from the minimum size necessary, arising from small statistical fluctuations, to rare, big statistical fluctuations the size of an entire universe. In terms of an infinite universe in thermal equilibrium, since small departures from equilibrium are more likely than large ones, a typical Boltzmann Brain therefore is likely to be that arrangement of matter as close to thermal equilibrium as is possible for consciousness to exist. That this minimum arrangement of matter required to create the consciousness is likely to result from a smaller fluctuation from the equilibrium than the fluctuation which created the human Boltzmann himself was pointed out by Lawrence Shulman (Schulman, Lawrence S. (1997). *Time's Arrows and Quantum Measurement* (1997 ed.). Cambridge: Cambridge University Press. p.154.) who was the first to suggest that this minimum consciousness will probably be a disembodied brain without a corporeal container. We should conclude, therefore, that the most common 'observer' in the multiverse should be an isolated brain who appears briefly, as do the spontaneous arrangements of states of matter like stars and galaxies, before disassembling itself immediately on the return to thermal equilibrium. We do not appear to be such an observer. Since consciousness does not appear to require a particularly specific arrangement (human brains are surely physically different from one another), there are likely to be many Boltzmann Brain universes, all slightly different *selves*.

BBs may well flicker into existence from time to time, but, in order for the arrangement of matter to be considered conscious, other conditions than mere fluctuations of matter are implied. Because our sensation of consciousness goes along

> If we are not a spontaneous brain born with a comprehensive delusion about the universe and who will any second now vanish our type of consciousness is somewhat of an outlier and has other reasons for coming into being, which in turn suggests that our the interpretation of infinity and of entropy that implies a commonplace BB may be wrong.

with a particular form of organisation which maintains itself over time, and for which memories persist and cohere to an identity or 'self', a BB awareness would need to have not only duration but also stability. For its consciousness to remain as its consciousness, the statistical fluctuation that brought it into being would have to be precisely that fluctuation whose subsequent time evolution continued to maintain the same consciousness throughout chaotic interference and especially throughout the changes induced in it by perception and thought. Without a means to maintain its *self* as the assembly of states of matter that comprise it alter with their functioning, say with perceptions or with remembering or predicting, a BB is likely to think itself out of existence the moment it appears.

Two significant consequences follow, namely that if a BB comes into being and persists as a self then, a) randomness cannot be a continual inexhaustible property of the vacuum state and of the self repeating inflationary scenario, and always available as we move forward in time, but must be a *coordinated succession of states*, such that any randomness within does not destroy consciousness, which appears to contradict what we know from quantum theory, and b) there is a mind-matter duality to consciousness after all where a guiding structure exists over and above the actual states and which cannot be random. This suggests that should a BB come into being then it is more likely to occur through a simulation or the strong Anthropic argument than be a natural consequence of randomness. For the moment, however, let us look more closely at b).

Mind/Body Problem

The Boltzmann Brain implies a solution to the mind/body problem that seems to be all too simple, namely that there is no problem, and only states of matter make up consciousness. If BBs exist, then ideas and perceptions are not metaphysical abstractions of the material world but are just as real and material as any collection of matter like a planet or a star. An idea does not even have to be true. If it exists in the mind it has a material existence in an assembly of states of matter (because that is all the mind is). In a BB, gravity would not only exist as a real artefact of the physical world but the concept of gravity as it appears say in Relativity equations would also be materially real, and co-existing with the Newtonian concept, the Galilean concept, and with the concept of elephants holding up the turtle on which our (flat) world balances. All thoughts of whatever type would appear in assemblies of matter that could be pointed to and isolated. Love, pain, mathematics and questions about our existence would have an observable material shape and energy (interestingly, some people *do* perceive ideas as shapes and 'masses'). (Artificial sentience, by this token, should be a simple matter of assembling the correct electronic circuits.) It should be recognised that these thoughts are distinct from the reality of the object, The object need not exist in a physical reality only the thought about the object has a physical reality (since thoughts can be composed without any necessary physical truth (all words may be derived from some physical fact but assemblies of words don't have to – hence lies) attached to them, e.g. Chomsky's famous, *colourless green ideas sleep furiously*. This is the hurdle, incidentally, that the ontological argument fails to get over. A thought about God may have a physical reality but we are unable to logically connect that fact with the necessary reality of God (unless we give it that condition first).

Since our awareness does appear to be a coordinated system of states over time, and that the world appears to governed by randomness, it seems unlikely that we are a BB, or more precisely, that I am a BB. The BB principle may, however, apply in a more local context.

The mystery of the sudden savant

The sudden savant mystery is strongly suggestive of an event along the lines of a BB. Groups of neuronal states in the human brain suddenly re-arrange themselves in a random flip into a new arrangement which just happens to coincide with those states that convey fresh capabilities, and without any input through the senses. These have been observed to occur without any of the brain trauma or neurological events which usually precipitate what is called the acquired savant syndrome (as described by Darold Treffert). A lawyer gets a sudden moment of lucidity and finds he can play the piano at a professional level never having had any musical training at all (although it turns out that he had been able to pick out tunes simply on the piano beforehand). An estate agent suddenly becomes an intricate painter without any training or even interest in art. This has happened to only a few as far as science is aware, but other cases may be hidden from us for many reasons. For example, these random flips may explain all sorts of puzzles about the human mind that we are wary of such as possession or trance creativity or memories of other lives, and it also suggests that at some future time humans, with some as yet undiscovered mechanism, will be able to flip their brains into any desirable state or to acquire any desirable skill without having to learn it.

This mystery, however, puts a huge spanner into the works of consciousness. For one thing, the random flip did not appear to alter the *self*. What this syndrome shows is that brain states themselves are globally unstable while consciousness is not, and that the self, as we understand it, cannot be considered as an emergent phenomenon of the complete arrangement of brain states even if it arose that way initially. And secondly, the matter of entropy as it applies to learning and the creation of brain states is put under the microscope. Assuming that learning a skill is a process that lowers entropy in the brain in the sense of information, order and coherence, then such a syndrome implies (as it does for a BB) a sudden burst of a large amount of negative entropy spontaneously occurring, and which, we will remind ourselves here, also occurs when processes go back in time.

Consider. This flip needs to be a comprehensive flip through a stretch of time to provide such complex skills. A skill requires muscle and mind coordination of memory and prediction. You cannot just play a piano with a single new parameter to your brain or a sudden comprehension of music scales. You need not only a depth of knowledge, muscle memory and a sensory coordination but prediction, to know how what you are doing is going to come out, to play any piece. Did all that just come with a sudden re-arrangement of a neuron set? Furthermore, if this BB idea is correct, it would imply the loss of those previous brain states which were then flipped into the new ones. Were there no repercussions for self from this conversion? (In fact, Treffert has observed quasi-autistic syndromes in the subjects such as an engagement with or obsession with the new 'gift' suggestive of at least slight changes in personality.)

What conclusions can we draw? Treffert's answer is that the phenomenon has a genetic cause. Some ancestor had the skills which were then passed on but not 'awakened' in the subject until the 'flip'. The idea that there is a 'dormant' complex skill just waiting to be unlocked makes no sense at all (e.g. when and why was this dormant skill developed in the background and readied for use in exceptional circumstances but which may never be in fact used?) and can be readily dismissed. Even the concept of 'inhibition' when dealing with entirely unconscious states is a notion with no understanding at all of how such a procedure could work. It doesn't answer any of the questions that come to mind.

Is the brain state for that knowledge about one skill so close to the brain state of knowledge about another that a simple flip in some neurons can produce one from the other? Does every brain contain every possible element of the human world and can, now and then, coherently access parts of that world previously blocked to it without having to learn it or rehearse it? If this is so then the distinctions between individual brains must arise from each expressing a random subset of the continually developing whole that we all share. Is anyone, therefore, just a flip away from artistic genius, from scientific genius or from idiocy, regardless of genetic dispositions? Most of the skills appearing with this syndrome are to do with music, art or mathematics; skills we know can often appear together naturally. Are art and mathematics the foundations to everything we are as humans in a society, and that these skills are so fundamental to consciousness that they are irrelevant to the *selves* we are? (A doctor may become a concert pianist without any personality change, although a musician suddenly becoming a doctor might require a new mind-set and personality?)

This syndrome is not the same as say, having the precursor brain state common to a five-year old musical prodigy that makes the child so able to learn suddenly arise in an adult. It is, or appears to be, the brain acquiring the neuronal arrangement necessary to, for example, play a piano without that brain ever having had the input to know what 'playing the piano' requires. Recall the brain-in-vat problem or Descartes' devil. Notions like, after an accident, the brain is somehow forced to 're-wire' itself into a new coherent function clearly deny the apparent necessity of learning information before it can be used.

However, we can relate this to what we have already discussed about language and broaden our schematic to include physical skills. If we think of skills as having a conceptual envelope that all humans possess (like the art and music which contributes to all cultures) while the 'interior' of the skill is filled by experience giving rise to the 'lists' of data or essential components that are needed for the function of that skill, then, given our narrative about retro-causation, we might imagine a process of transference through time of these 'lists'. Which is to say, that the learning procedure appears to us to be back-to-front but which conforms to the inferences we draw about the relationship of the past to the future.

30/1/24